# Mechatronics Innovation & Intelligent Application Technology

# 机电创新智能应用技术

Li Rui
李 蕊 主 编
Xie Jintao Yue Zhenli
谢金涛 岳振力 副主编

Beijing
Metallurgical Industry Press
2020

## 内容提要

本书分为工程实践创新和工业机器人两部分，共分 6 个项目 17 个任务。本书主要介绍了工程实践创新技术的认知、工程实践创新项目应用、工业机器人的认识、工业机器人的基本操作、ABB 工业机器人的 I/O 通信、ABB 工业机器人的程序数据、ABB 工业机器人的编程实战等内容。

本书可作为职业院校机电相关专业的国际化教学用书，也可作为有关企业员工的培训教材和有关专业人员的参考书。

**图书在版编目（CIP）数据**

机电创新智能应用技术：Mechatronics Innovation & Intelligent Application Technology：汉、英/李蕊主编. —北京：冶金工业出版社，2020.8

国际化职业教育双语系列教材

ISBN 978-7-5024-8583-2

Ⅰ.①机… Ⅱ.①李… Ⅲ.①工业机器人—高等职业教育—双语教学—教材—汉、英 Ⅳ.①TP242.2

中国版本图书馆 CIP 数据核字（2020）第 153354 号

出 版 人 苏长永
地　　址 北京市东城区嵩祝院北巷 39 号 邮编 100009 电话 （010）64027926
网　　址 www.cnmip.com.cn 电子信箱 yjcbs@cnmip.com.cn
责任编辑 俞跃春 刘林烨 美术编辑 郑小利 版式设计 孙跃红 禹 蕊
责任校对 李 娜 责任印制 李玉山
ISBN 978-7-5024-8583-2
冶金工业出版社出版发行；各地新华书店经销；三河市双峰印刷装订有限公司印刷
2020 年 8 月第 1 版，2020 年 8 月第 1 次印刷
787mm×1092mm 1/16；19.75 印张；475 千字；293 页
**59.00** 元

**冶金工业出版社 投稿电话 （010）64027932 投稿信箱 tougao@cnmip.com.cn**
**冶金工业出版社营销中心 电话 （010）64044283 传真 （010）64027893**
**冶金工业出版社天猫旗舰店 yjgycbs.tmall.com**

# 国际化职业教育双语系列教材

## 编 委 会

# Foreword

With the proposal of the 'Belt and Road Initiative', the Ministry of Education of China issued *Promoting Education Action for Building the Belt and Road Initiative* in 2016, proposing cooperation in education, including 'cooperation in human resources training'. At the Forum on China-Africa Cooperation (FOCAC) in 2018, President Xi proposed to focus on the implementation of the 'Eight Actions', which put forward the plan to establish 10 Luban Workshops to provide skills training to African youth. Draw lessons from foreign advanced experience of vocational education mode, China's vocational education has continuously explored and formed the new mode of vocational education with Chinese characteristics. Tianjin, as a demonstration zone for reform and innovation of modern vocational education in China, has started the construction of 'Luban Workshop' along the 'Belt and Road Initiative', to export high-quality vocational education achievements.

The compilation of these series of textbooks is in response to the times and it's also the beginning of Tianjin Polytechnic College to explore the internationalization of higher vocational education. It's a new model of vocational education internationalization by Tianjin, response to the 'Belt and Road Initiative' and the 'Going Out' of Chinese enterprises. Tianjin Polytechnic College and Uganda Technical College-Elgon reached a cooperation intention to establish the Luban Workshop to carry out vocational education cooperation on mechatronics technology and ferrous metallurgy technology major in 2019. The establishment of Luban Workshop is conducive to strengthen the cooperation between China and Uganda in vocational education, promote the export of high-quality higher vocational education resources, and serve Chinese enterprises in Uganda and Ugandan local enterprises. Exploring and standardizing the overseas operation of Chinese colleges, the expansion of international influences of China's higher vocational education is also one of the purposes.

The construction of 'Luban Workshop' in Uganda is mainly based on the EPIP (Engineering, Practice, Innovation, Project) project, and is committed to cultivating high-quality talents with innovative spirit, creative ability and entrepreneurial spirit. To meet the learning needs of local teachers and students accurately, the compilation of these international vocational skills bilingual textbooks is based on the talent demand of Uganda and the specialty and characteristics of Tianjin Polytechnic.

These textbooks are supporting teaching material, referring to Chinese national professional standards and developing international professional teaching standards. The internationalization of the curriculums takes into account the technical skills and cognitive characteristics of local students, to promote students' communication and learning ability. At the same time, these textbooks focus on the enhancement of vocational ability, rely on professional standards, and integrate the teaching concept of equal emphasis on skills and quality. These textbooks also adopted project-based, modular, task-driven teaching model and followed the requirements of enterprise posts for employees.

In the process of writing the series of textbooks, Wang Xiaoxia, Li Rui, Wang Zhixue, Ge Huijie, Yu Wansong, Wang Lei, Li Xiujuan, Gong Na, Zhang Zhichao, Jia Yanlu, Chen Baoling and other chief teachers, professional teams, English teaching and research office have made great efforts, receiving strong support from leaders of Tianjin Polytechnic College. During the compilation, the series of textbooks referred to a large number of research findings of scholars in the field, and we would like to thank them for their contributions.

Finally, we sincerely hope that the series of textbooks can contribute to the internationalization of China's higher vocational education, especially to the development of higher vocational education in Africa.

Principal of Tianjin Polytechnic College Kong Weijun
May, 2020

# 序

随着“一带一路”倡议的提出，2016年中华人民共和国教育部发布了《推进共建“一带一路”教育行动》，提出了包括“开展人才培养培训合作”在内的教育合作。2018年习近平主席在中非合作论坛上提出，要重点实施“八大行动”，明确要求在非洲设立10个鲁班工坊，向非洲青年提供技能培训。中国职业教育在吸收和借鉴发达国家先进职教发展模式的基础上，不断探索和形成了中国特色职业教育办学模式。天津市作为中国现代职业教育改革创新示范区，开启了“鲁班工坊”建设工作，在“一带一路”沿线国家搭建“鲁班工坊”平台，致力于把优秀职业教育成果输出国门与世界分享。

本系列教材的编写，契合时代大背景，是天津工业职业学院探索高职教育国际化的开端。“鲁班工坊”是由天津率先探索和构建的一种职业教育国际化发展新模式，是响应国家“一带一路”倡议和中国企业“走出去”，创建职业教育国际合作交流的新窗口。2019年天津工业职业学院与乌干达埃尔贡技术学院达成合作意向，共同建立“鲁班工坊”，就机电一体化技术专业、黑色冶金技术专业开展职业教育合作。此举旨在加强中乌职业教育交流与合作，推动中国优质高等职业教育资源“走出去”，服务在乌中资企业和乌干达当地企业，探索和规范我国职业院校“鲁班工坊”建设和境外办学，扩大中国高等职业教育的国际影响力。

中乌“鲁班工坊”的建设主要以工程实践创新项目（EPIP：Engineering，Practice，Innovation，Project）为载体，致力于培养具有创新精神、创造能力和创业精神的“三创”复合型高素质技能人才。国际化职业教育双语系列教材的编写，立足于乌干达人才需求和天津工业职业学院专业特色，是为了更好满足当地师生学习需求。

本系列教材采用中英双语相结合的方式，主要参照中国专业标准，开发国际化专业教学标准，课程内容国际化是在专业课程设置上，结合本地学生的技术能力水平与认知特点，合理设置双语教学环节，加强学生的学习与交流能力。同时，教材以提升职业能力为核心，以职业标准为依托，体现技能与质量并重的教学理念，主要采用项目化、模块化、任务驱动的教学模式，并结合企业岗位对员工的要求来撰写。

本系列教材在撰写过程中，王晓霞、李蕊、王治学、葛慧杰、于万松、王磊、李秀娟、宫娜、张志超、贾燕璐、陈宝玲等主编老师、专业团队、英语教研室付出了辛勤劳动，并得到了学院各级领导的大力支持，同时本系列教材借鉴和参考了业界有关学者的研究成果，在此一并致谢！

最后，衷心希望本系列教材能为我国高等职业教育国际化，尤其是高等职业教育走进非洲、支援非洲高等职业教育发展尽绵薄之力。

天津工业职业学院书记、院长　孔维军

2020 年 5 月

# Preface

Tianjin Polytechnic College and Uganda Technical College-Elgon reached a cooperation intention to establish the Luban Workshop to carry out vocational education cooperation on mechatronics technology and ferrous metallurgy technology major in 2019. In order to strengthen the cooperation between China and Uganda in vocational education, the two colleges plan to compile a series of international vocational skills bilingual textbooks.

This book is one of the international vocational skills bilingual textbooks. The book is structured with 'Introduction' 'Relevant Knowledge' 'Implementation' 'Evaluation' and 'Development'. It adopts the writing ideas of project units that are now introduced from engineering related knowledge to practice to innovation, from shallow to deep and sensibility to rationality. It also lets teachers and learners understand and experience the teaching and learning methods of electromechanical innovation and intelligent application to enrich learners' practical knowledge, enhance the application ability of technology, strengthen the innovative ability of practice, and expand the profession of learners vision, and develop good professional qualities.

This book is written by Tianjin Institute of Technology Mechanical and Electrical Innovation Intelligent Application Teaching Team, the specific division of labor is as follows: Li Rui is responsible for Task 1.1 of Project 1 and Task 2.3 of Project 2; Xie Jintao is responsible for Project 4 and Project 5; Li Lianliang is responsible for Task 1.2 of Project 1, and Task 2.1 and Task 2.2 of Project 2; Yue Zhenli is responsible for Task 2.4 and Task 2.5 of Project 2 and Project 6; Hou Yapin is responsible for Project 3. This book is written by Li Rui, Chief judge Kong Weijun and Li Meihong put forward many valuable opinions on this book. Here express my gratitude to the authors concerned.

Due to the limited level of the editor, there is something wrong in the book. I hope readers to criticize and correct.

The editor
May, 2020

# 前 言

2019年天津工业职业学院与乌干达埃尔贡技术学院达成合作意向，共同建立“鲁班工坊”，就机电一体化技术专业、黑色冶金技术专业开展职业教育合作，双方计划编撰国际化职业教育双语系列教材。

本书是国际化职业教育双语系列教材之一。本书以“任务引入”“相关知识”“任务实施”“任务评价”和“任务拓展”的结构编排，采用先从工程相关知识介绍到实践再到创新的项目单元的编写思路，由浅入深、由感性到理性，让教学者和学习者了解、体验机电创新、智能应用的教学和学习方式，丰富学习者的实践知识、提升技术的应用能力、强化实践的创新能力、拓展学习者的专业视野并养成良好的职业素养。

本书由天津工业学院机电创新智能应用教学团队编写完成，具体分工如下：李蕊负责项目1任务1.1和项目2任务2.3；谢金涛负责项目4和项目5；李连亮负责项目1任务1.2、项目2任务2.1和任务2.2；岳振力负责项目2任务2.4、任务2.5和项目6；侯娅品负责项目3。本书由李蕊统稿，孔维军主审。李梅红对本书提出很多宝贵意见，在此表示诚挚的感谢。

由于编者水平所限，书中不妥之处，希望读者批评指正。

编 者

2020年5月

# Contents

# 目　录

# Project 1　Knowledge of Engineering Practice and Innovation Technology

Innovation is the soul of a nation's progress and the inexhaustible power of a country's prosperity. Innovation covers a wide range of areas, including political, military, economic, social, cultural, scientific and technological innovation. Using the capability source kit to complete the engineering practice innovation task, this project introduced more than 1000 components according to the structure parts, connectors, transmission parts and electrical components. The VJC graphical interactive development system is recognized and the program is designed using the VJC graphical interactive development system to realize the logical control of the complex electro-mechanical system.

## Task 1.1　Engineering Practice Innovation and Capability Source Kit

### 1.1.1　Introduction

More than a hundred years ago, an inventor went on a trip. He saw an old lady carrying a bag whose mouth had been crushed and his things scattered all over the floor. Back home, the remembered what happened to the old lady that day, and he finally invented the first zipper in human history to solve the pocket zipper problem. Most people just think that it is a small matter of life, and it is not worth making a fuss, but the inventor has a relatively strong curiosity and exploration spirit. Now, almost everyone uses his invention. The story tells us to pay attention to many trivial things in life, diligent in thinking, dare to innovate, and then there will be gains.

Before embarking on a practical journey of energy source innovation, we need to understand the tool for innovation——the competency source innovation curriculum suite. The components and features of the capability source innovation curriculum suite can be felt through a number of typical cases of the power source innovation curriculum suite powerful function. The kit components are presented in four categories of structural parts, connectors, transmission parts and electrical components. At the same time, some common building methods and techniques are introduced to lay the foundation for the next stage of project practice.

### 1.1.2　Relevant Knowledge

#### 1.1.2.1　Engineering Practice Innovation

Apply the theory of natural science to the general names of various disciplines formed in specific industrial and agricultural production sectors, such as water conservancy engineering, chemical engineering and civil construction engineering. More manpower and material resources are used to carry out larger and complex work requires a longer period of time to complete, such as urban reconstruction project and high-speed railway project. A research, on engineering is called 'Engineering', a project, about engineering is called 'Engineering Project', and a comprehensive, large and complex project with each sub-project is called 'System Engineering'.

**Works**

Engineering is a general term for applying the principles of natural science to the industrial and agricultural production departments to form various disciplines. The narrow concept of engineering is that the process of transforming an (or some) existing entity (natural or man-made) into an artificial product of intended use value through the organized activities of a group of persons, using relevant scientific knowledge and technical means, based on a set of envisaged objectives, and the generalized engineering concept is that the process of performing collaborative activities over a long period of time for a group of people to achieve a certain purpose.

**Practice**

Practice is all the material activities of human beings to transform the objective world, with objective materiality, subjective initiative and social history; Practice is the unique activity of human beings, which is different from the instinctive activities of animals to passively adapt to nature, and is objective material activity; Practice is the activity of changing objective things, which inevitably leads to the change of objective objects, not pure subjective thinking activities; Practice has objective, material, active, conscious, social and historical activities.

**Innovation**

Innovation refers to the behavior of improving or creating new things, methods, elements, paths, environments, and achieving certain beneficial effects in a specific environment. Innovation refers to the use of existing thinking models to provide insights that are different from conventional or ordinary people's ideas, using existing knowledge and materials in a specific environment, to improve or create in accordance with idealized needs or to meet social needs new things, methods, elements, paths, environments, and behaviors that can achieve certain beneficial effects.

The types of innovation are: (1) breakthrough innovation, which is characterized by breaking stereotypes, changing traditions and striding forward; (2) gradual innovation, which is characterized by taking the next logical step to make things better and better; and (3) re-application innovation, which is characterized by the use of horizontal thinking and the application of the original things in a new way.

### 1. 1. 2. 2 Energy Source Identification Kits

**Capacity Source Innovative Curriculum Kit (POWERON)**

The Capability Source Innovation Course Kit (POWERON, shown in Figure 1-1) covers all aspects of mechanical, electronic, sensor, computer hardware and software, control and other professional knowledge and technical skills. The POWERON involves professional knowledge and

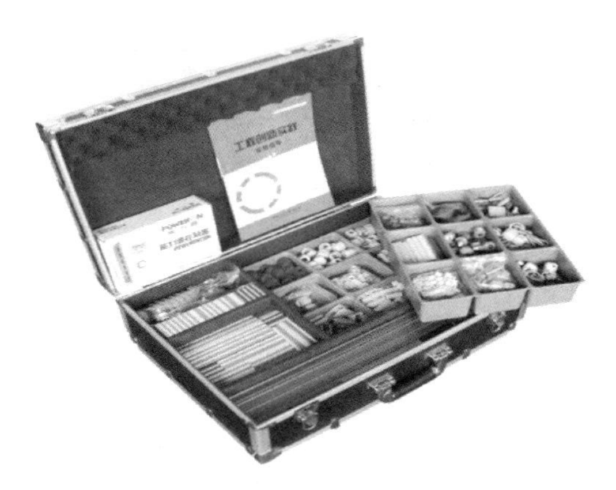

Figure 1-1 Capability source innovation course kit

technical skills in various aspects of machinery, electronics, sensors, computer hardware and software, and control. It adopts project-based teaching in the form of small groups and implements well-designed engineering practices from shallow to deep. Innovative projects, which allow students to understand actual engineering, on the basis of understanding and learning real engineering projects, extract the core technologies of electromechanics in real engineering projects, and use the POWERON as a carrier to stimulate students' awareness of innovation and strengthen students teamwork ability. Students' various professional knowledge are used to solve practical engineering problems through hands-on collaboration, to reproduce the knowledge points learned, and to help students build a wide area of engineering knowledge system.

The POWERON can help simulate engineering projects, understand these engineering projects through books, and experience in the process of hands-on practice. The engineering composition structure and working principle are also deeply understood by the POWERON. According to your own ideas, innovation and expansion of various engineering projects are unlimited.

**Description of Typical Components of POWERON**

According to the different functions, the components of the innovative curriculum kit can be divided into four categories, which are as follows:

(1) Structural elements —— Basic components of three-dimensional structures: Structures, such as bricks for building a house, are the most basic components for building an engineering project. They can be divided into three types, such as point, line and surface (shown in Figure 1-2). They can be connected directly to each other or with the help of connectors and three-dimensional extensions. They can also be used flexibly according to the requirements of the project.

(2) Connectors ——3D extension facilitators: Connectors are similar to cement or lime used to build a house, providing a suitable way to connect the structures to each other.

Some structural parts can be connected to each other without additional connecting components, such as cubes and beams. Most structural parts are connected by means of connectors. The corresponding connection mode of structural parts is divided into: point and point, point and line,

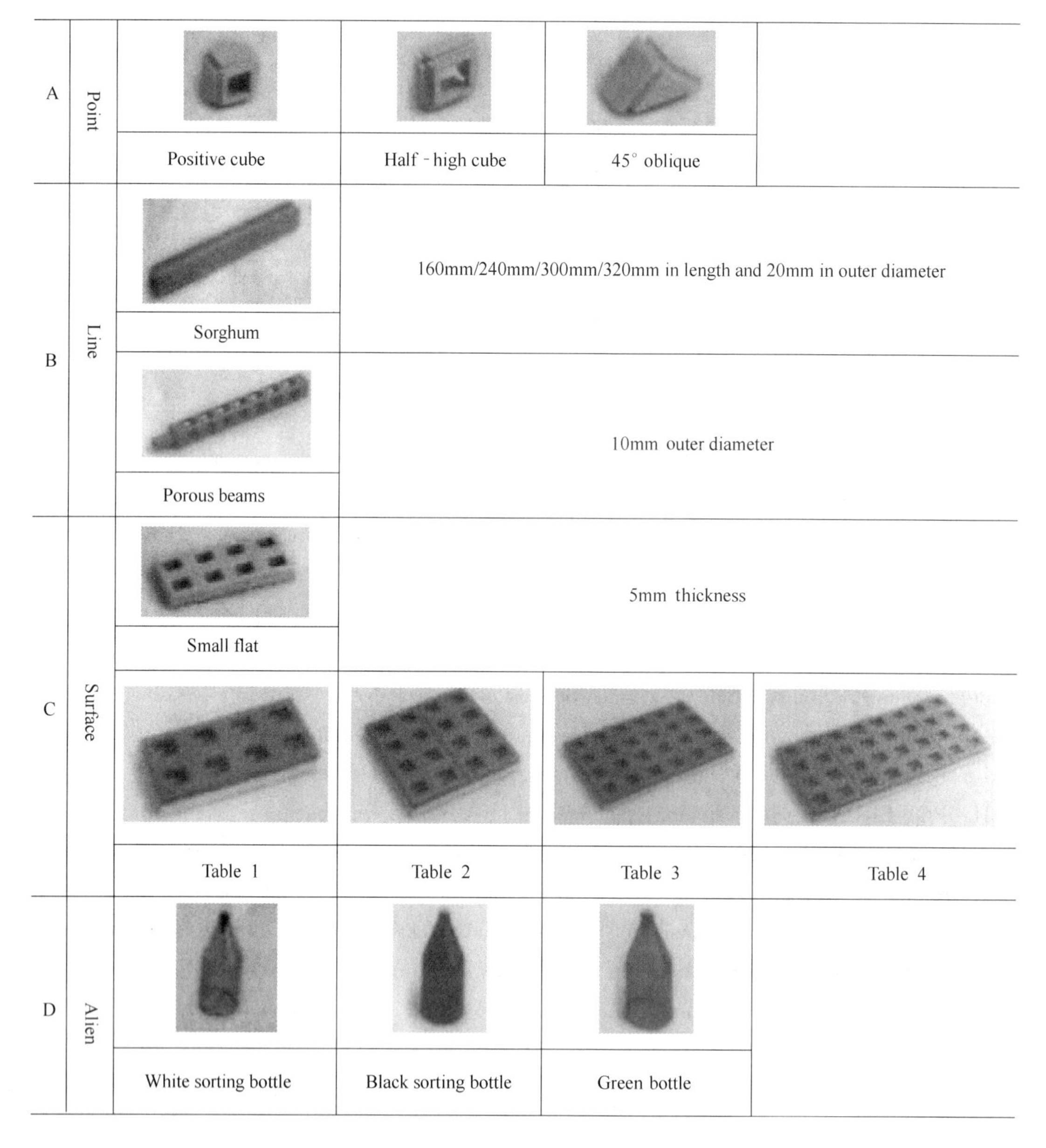


Figure 1-2 Types of point, line and surface

point and surface, line and line, line and surface, and connection between surface and surface, as shown in Figure 1-3.

(3) Transmission parts——Power transfer hand for transmission parts: The transmission parts are mainly components that transfer power or change the direction and form of motion. These components are designed with flexibility and ease of use. Common transmission modes cover gear drive, gear rack drive, worm gear drive, belt drive, screw drive, etc., as shown in Figure 1-4.

| | | | | | |
|---|---|---|---|---|---|
| A | Point-to-point connection | Cube connector | | | |
| B | Point to line, point to surface | Short bolt | Long pin | | |
| C | Wire and wire connections | 40 round tubes | 80 round tubes | Beam support | |
| D | Surface connection | Central L connector | Central H connector | | |
| E | Other connections | Central A connector | A connector | A connector | Five-hole ladder |

Figure 1-3 Relationship between points, lines and surfaces

(4) Electrical components——The basis of automation: The electrical components in the POWERON refer to a variety of sensors, actuators and motor wires. All the sensors and actuators in the kit are integrated in the cube to facilitate 3D construction and expansion, and the connection with other components can be completed without using any firmware.

As the main component of collecting information, the sensor passes the input information to the controller. For example, the AGV car uses the gray level sensor to detect the function of ambient light to realize the line patrol, the automatic door is opened and closed by touching the switch operation door, and the limit position of the door fan is controlled by the magnetic sensitive switch and magnet matching. The CNC machine tool detects the exact rotation number of the lead screw by rotating counter, thus controlling the motion range of each degree of freedom.

Common actuators include LED lights of different colors (digital output type actuators). Electromagnets as well as motors (adjustable output type actuators), actuators are also widely used in engineering projects. For example, whether industrial manipulators are electri-

| | | | | | | |
|---|---|---|---|---|---|---|
| A | Moderators | 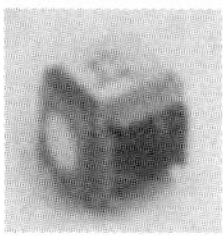 | 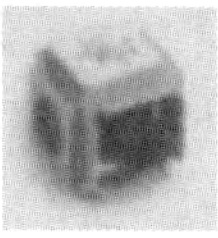 | 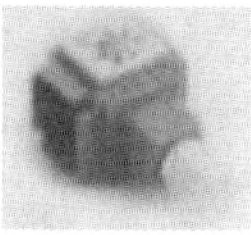 | 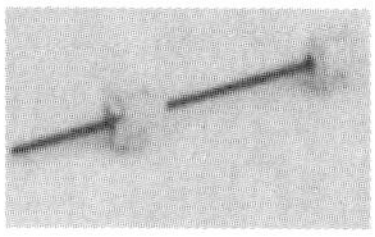 | |
| | | 5:1 slow down | 1:1 turn | 1:1 with axis | Lead screw assembly | |
| B | Gear | 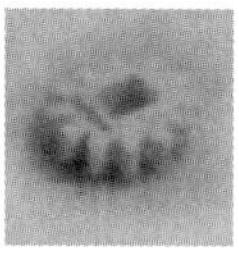 | 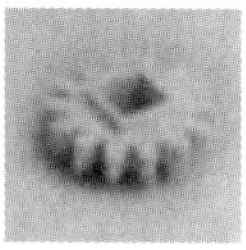 | 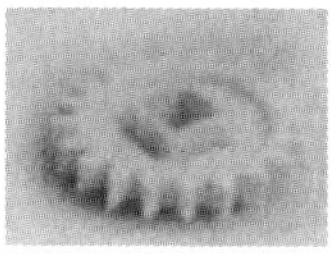 | 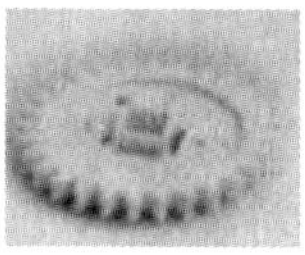 | 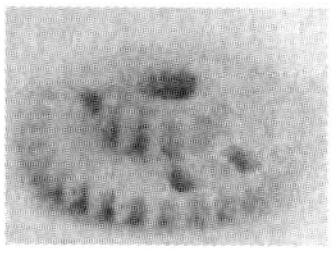 |
| | | 12 teeth | 14 teeth | 20 teeth | 28 teeth | 12/28 teeth |
| C | Wormgear | 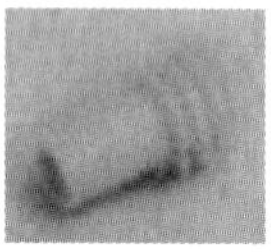 | 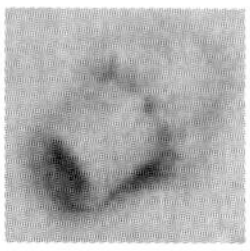 | | | |
| | | Worm | 12 gear worm gear | | | |
| D | Rackrack | 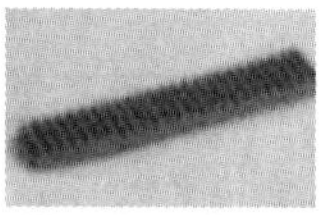 | | | | |
| | | Rack rack | | | | |
| E | Bearingandshaft | 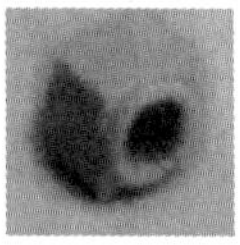 | 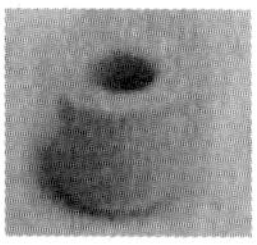 | 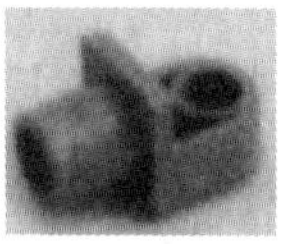 | 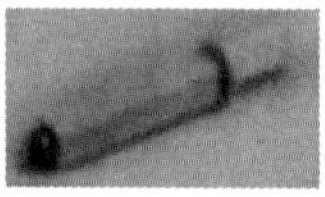 | 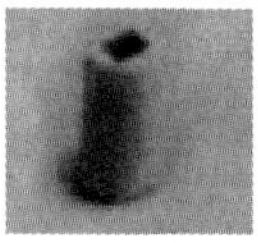 |
| | | Bearing | Sliding bearings | Joint component | With step shaft | Inner outer circle tube |
| F | Hubs | 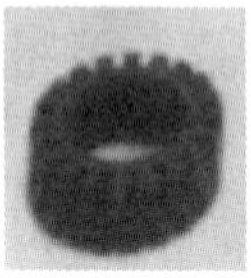 |  | 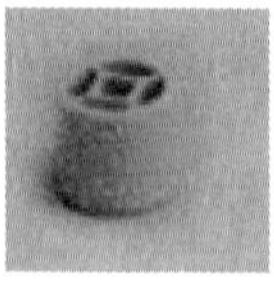 | 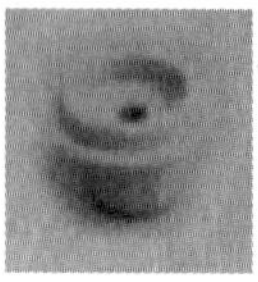 | 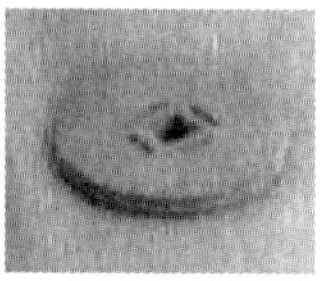 |
| | | Driving tyres | Drive hub | Belt pulley | Steering wheel | Pulley |

Figure 1-4 Transmissions

fied to achieve 'Grip' to take, put sorting bottles, feedback different working conditions with different colors of LED lights, etc. The motor wire is a special wire connecting the motor with the controller.

In summary, electrical components (shown in Figure 1-5) are the top-down components to achieve automated control of engineering projects.

| | | | | | |
|---|---|---|---|---|---|
| A | Analog sensor |  | 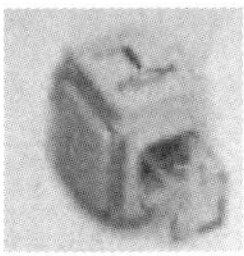 | | |
| | | Photosensitive sensor | Temperature sensor | | |
| B | Digital sensor | 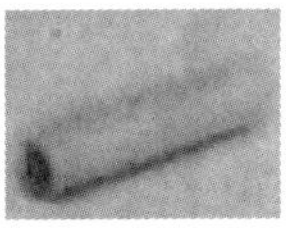 | 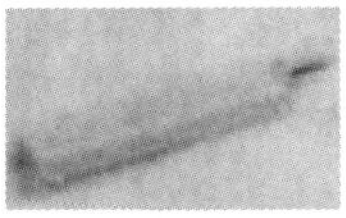 |  | 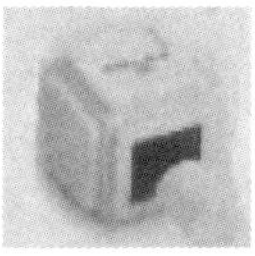 |
| | | Magnet | Magnetic sensitive switch | Rotating counter | Touch switch |
| C | Adjustable output actuator | 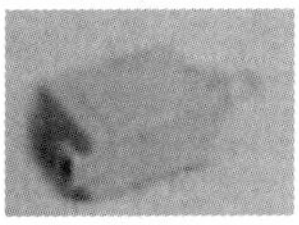 | 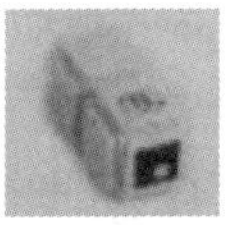 | | |
| | | Motor motors | Electromagnet | | |
| D | Digital output actuator | 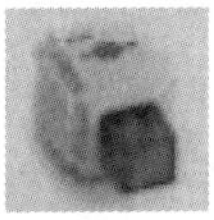 | 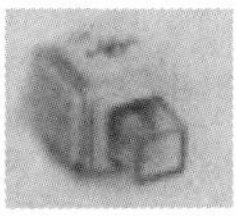 | 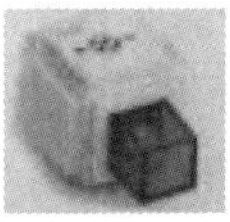 | 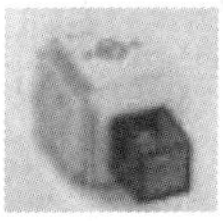 |
| | | Red light | Yellow light | Green light | Blue light |
| E | Connecting wires | 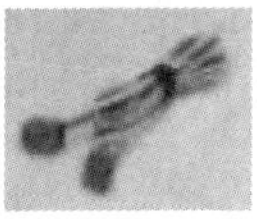 | | | |
| | | Motor line | | | |

Figure 1-5　Electrical components

**Controller**

As shown in Figure 1-6, the controller is the control part of the capability source innovation curriculum suite, allowing more than 1000 energy source components to work together into a whole project, equivalent to the human 'brain'. The internal processor has a 32-bit ARM processor, and master frequency is: 72MHz, 64K SRAM, user memory 3.96Mb. The POWERON contains 12 I/O interfaces and 4 DC motor interfaces. The 12 I/O interfaces support analog input, digital input, digital output, counter, 485 communication, etc.

The detailed parameters are as follows:

(1) USB download port, communication between the controller and the host computer through the USB download port: The program written in the VJC system of the host computer is downloaded to the controller through the USB download port through the download line, and the controller power needs to be turned on during the download process.

(2) Run key (〈Enter〉), used for program execution: The program can only be run after downloaded to the capability source controller and pressed 〈Enter〉.

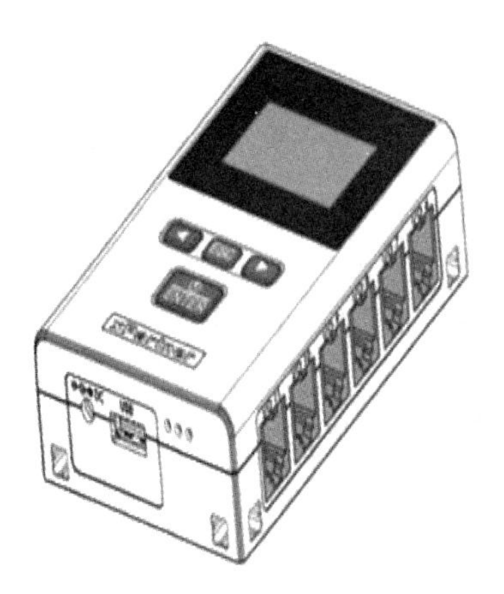

| No. | Definition | No. | Definition |
|---|---|---|---|
| A | Power+, Battery voltage | D | AI/DI |
| B | +5V Power | E | RS485 D- |
| C | Power -/Land/Do output | F | RS485 D- |

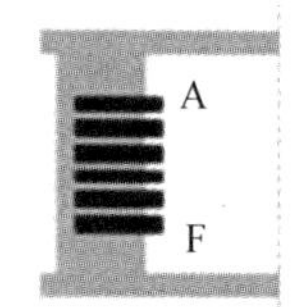


Figure 1-6  Controller

(3) Motor port, a total of 4 channels: It can control the speed and direction of the motor, and support ordinary DC motors and closed-loop motors. The output voltage is the battery voltage, and the maximum current for a single channel is 1.5A.

(4) Power supply, which comes with 8.4V 1A adapter: It can be directly connected to the power port to supply power to the controller, and can also be powered by a dedicated lithium battery. The dedicated lithium battery is 8.4V 1500MAH, and the maximum discharge current is 6.5A, coming with a protection circuit.

(5) LCD screen, 128×64 dot matrix LCD screen, with backlight: It can display graphics and characters. In addition, through the setting interface, you can read and adjust the EEPROM value, and set the switch status of the controller screen backlight and buzzer.

(6) Exit key (〈Esc〉): It can be used to reset the source controller.

(7) Others: Ccon102 controller is downloaded by U disk, providing up to 3.96M of user program storage space. You can select different programs to run by using the left and right buttons on the program interface.

Controller operation includes:

(1) Main interface: The interface entered after the controller is turned on, and the content of the interface contains the power display and the main menu. Electricity display icon can roughly display the current battery power, with the main menu containing 4 items, and it can also be switched between the items by the left and right keys. The icon from the black box is the currently selected item, and press 〈Enter〉 to enter the next level interface of the project, as shown in Figure 1-7.

(2) Program interface: The controller uses U disk way to download, and provides up to 3.96M of user program storage space. When there is no program in the controller, you cannot enter the program interface. There are 2 programs in the controller as shown in Figure 1-8: PROGRAMA and PROGRAMB. The serial numbers of 1 and 2 in front of the program are the program numbers, which are sorted by the download time, in order to facilitate the user to find.

At this interface, users can select different user programs through the left and right keys. The selected user programs will be displayed in reverse color, and the selected user programs will be read into memory and start running after pressing 〈Enter〉. During the running of the user program, the Esc key is pressed to terminate the operation of the user program, and the interface returns to the main interface as shown in Figure 1-8.

(3) Port interface: After entering the port interface, the left and right keys can be successively switched between AI、DO、DI、steering gear and counting. AI、DO、DI、steering gear and counting will be displayed in the lower left corner of the screen. AI analog input port, as shown in the detection interface. 0~11 represents the I/O port number. The following number '0' represents the AI input value collected in real time by the corresponding port, and the picture shows '0' indicating no sensor access. The return value range of AI is 0~4095. The port function of AI is sensor input test function, and just read the value, no next operation (shown in Figure 1-9).

In DO digital output port test interface (shown in Figure 1-10), 0~11 represents the I/O port number, and the following number represents the current switching state of the DO. '0' indicates disconnection, and '1' indicates connection. The DO state of a single port can be controlled separately by 〈Enter〉 and left and right key. To ensure the security of the project model, when exiting the DO control, the DO port returns to the disconnected state.

Figure 1-7 Main interface

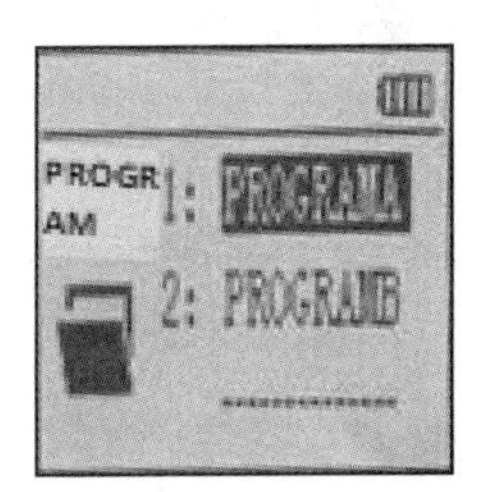

Figure 1-8 Program interface

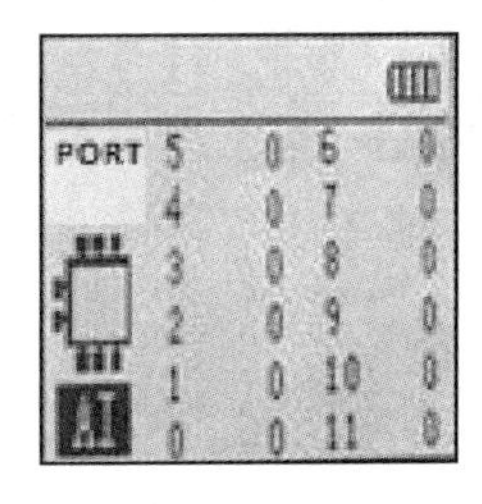

Figure 1-9 AI interface

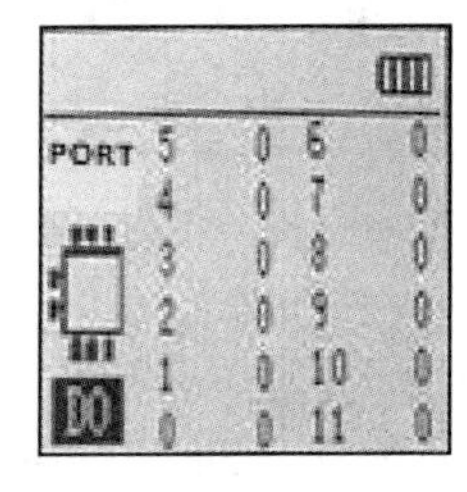

Figure 1-10 DO interface

In DI digital input port, as the detection interface shown in Figure 1-11, 0~11 represents the I/O port number, and the following number '0' represents the DI input value collected in real time by the corresponding port. The return value of digital quantity sensor is only 'on' and 'off', so the corresponding return value is only 0 and 1. There are two cases when displayed at 0: No sensor is connected or the digital sensor is disconnected. DI function of the port is digital quantity sensor input test function, and only read the value, no next operation.

The servo, as shown in Figure 1-12, is the digital servo detection interface. This function can only control one servo, and the lower right side is the control parameters of the digital servo. IO: I/O port connected to the digital servo. The purpose is to turn on the 485 power of this port. ID: It indicates the ID number of the digital servo to be controlled. A: It read and set the digital servo angle.

Counting, as shown in Figure 1-13, is the rotation counter detection interface. This function can only read the return value of one rotation counter. Because the rotation counter needs to be driven by the motor, and it is also necessary to control the movement of the motor under the interface. On the right are the control parameters of the rotation counter. IO: I/O port number accepted by the rotation counter. DC: The DC port number accepted by the motor driving the rotation counter. S: The controller's motor turns. When S is reversed, press 〈Enter〉 to enter the motor control function. The default is on the stop symbol in the center. Switch between the symbols (right arrow), and the current speed of the motor will be displayed after S. CNT: The return value of the counter is displayed. When the value is reversed by key operation, press the 〈Enter〉 key to clear the value.

(4) Motor interface, as shown in Figure 1-14: In DC testing interface, DC M0 ~ M3 correspond in a list of motor mouth on DC0 ~ DC3. The speed value in a column represents the corresponding to the mouth of the DC motor speed value (-100 ~ 100). The encode value in a column represents the behalf of closed loop motor encoder return values (only available when turn on the closed-loop motor. According to the motor and reversing, the numerical range from 0 to 65535 is increased or decreased).

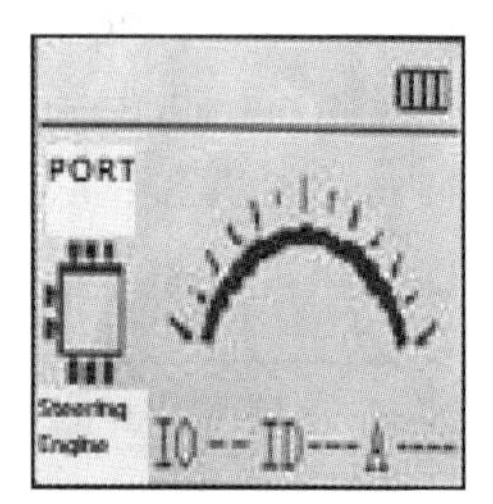

Figure 1-11 DI interface

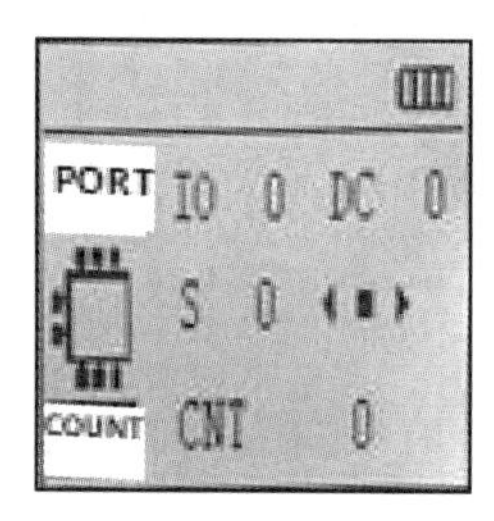

Figure 1-12 Servo interface

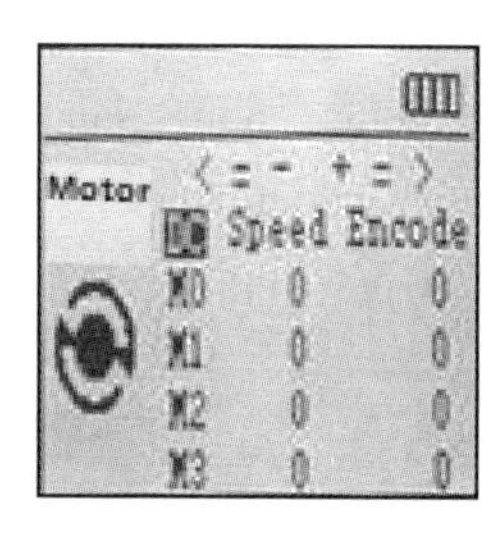

Figure 1-13 Count interface

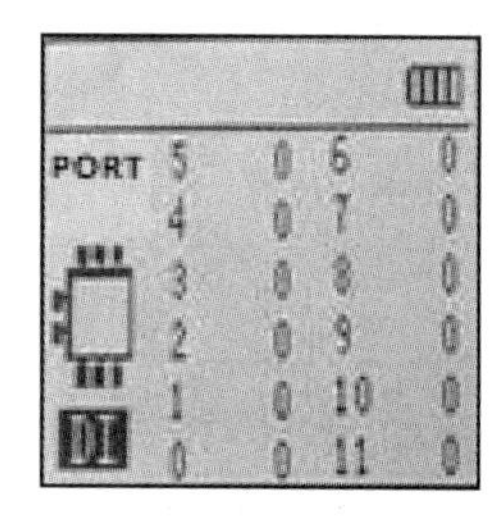

Figure 1-14 Motor interface

## 1.1.3 Implementation

Once you are familiar with the capability source innovation kit, you should be able to use the components to build a simple structure and master the common building methods and techniques of the kit.

### 1.1.3.1 Construction Method of Structure and Connectors

Next you can use the kit for some common kit connections as follows. Firstly, you can illustrate how to connect directly between the point, line and surface, and how to use the connector connection.

For example, cubes (points) and cubes (points) are directly connected to each other (shown in Figure 1-15), and cubes (points) are connected by means of cube connectors (shown in Figure 1-16). With the help of medium, L-shaped connections connect the cube (or half-height cube) (point) to the flat plate (face) (shown in Figure 1-17), the cube (point) and the flat plate (face) by means of rectangular bolts (shown in Figure 1-18), the cube (point) and beam (line) by means of long plug (shown in Figure 1-19), two beams (line) by means of beam support frame (shown in Figure 1-20), and the beam by means of long bolt (line) and flat plate (surface) (shown in Figure 1-21), directly connecting two flat plates (surface) with the help of medium L connector (shown in Figure 1-22).

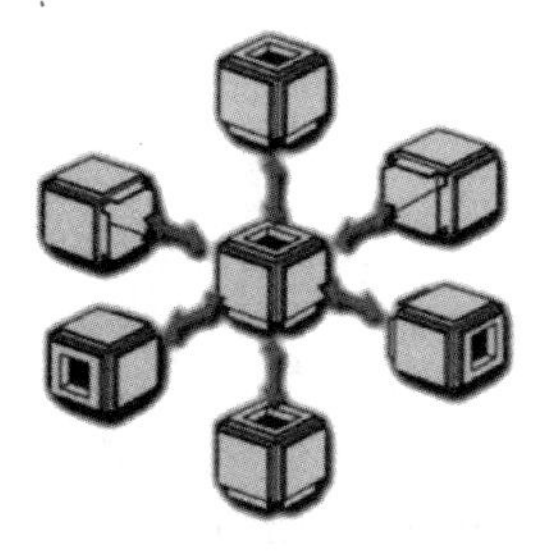

Figure 1-15 Points-to-points (direct connection)

Figure 1-16 Points-to-points (with cube connectors)

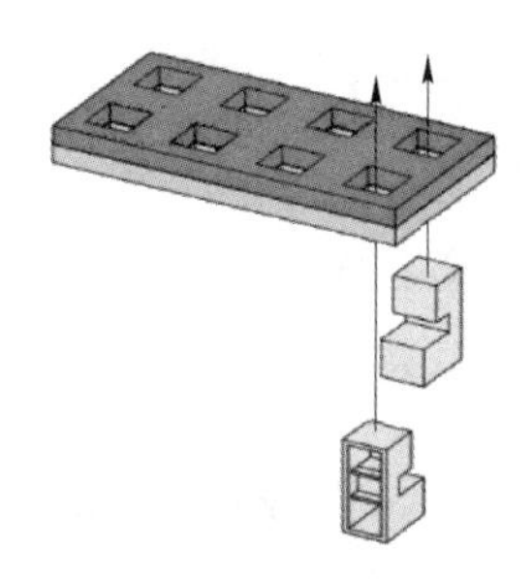

Figure 1-17 Points and face (with L connector)

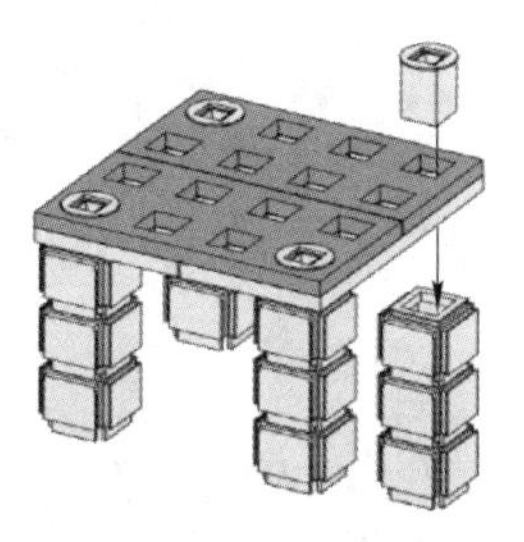

Figure 1-18 Points and face (with short pin)

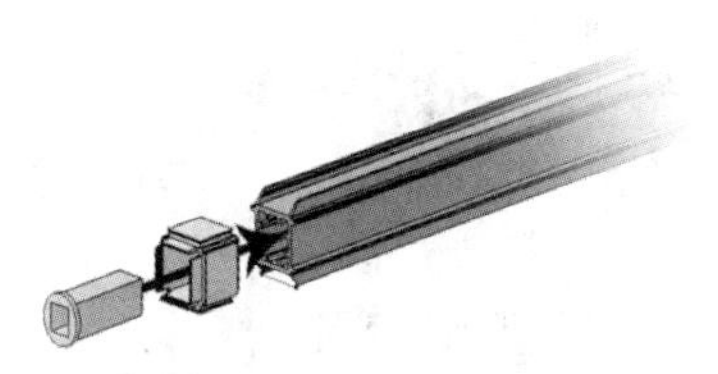

Figure 1-19 Points and line

Figure 1-20 Lines and lines

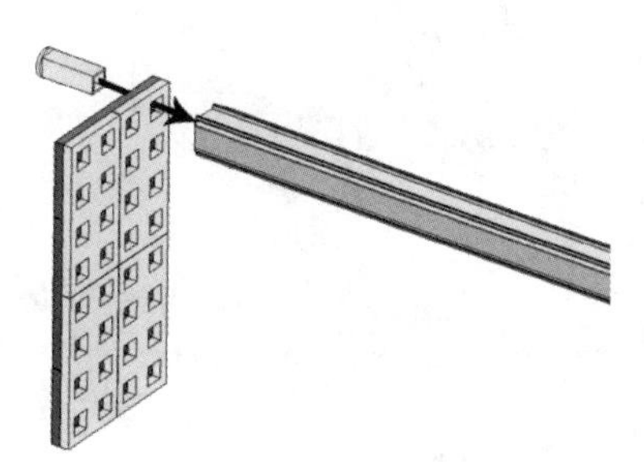

Figure 1-21 Line and surface diagram

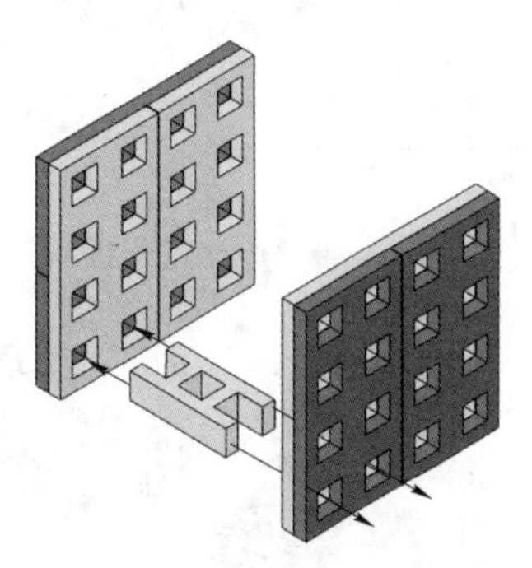

Figure 1-22 Face to face

Only some common construction cases are listed above. The construction of mechanical structure is very ingenious. The use of connecting parts of each part is very flexible and the connection mode and the use of structural parts of the same functional model are various.

### 1. 1. 3. 2 Construction methods of Electrical Components

The construction methods of electrical components are as follows:

(1) The motor is the source of power in the innovative assembly of the power source, and all movements must be supported by the motor. The motor is usually connected to other parts with the help of a cube or plate. Figure 1-23 shows the most commonly used connection mode of the motor.

Figure 1-23 Motor

The distance of the motor refers to the torque output by the engine from the crankshaft end. Under the condition of fixed power, it is inversely proportional to the engine speed, and the faster the speed, the smaller the torque, and vice versa. In order to increase the torque of the motor, we often need to add the deceleration mechanism to the motor to increase its torque.

The deceleration box and gear set are often used in the capability source innovation course kit to achieve the effect of the deceleration box, as shown in Figures 1-24 and 1-25.

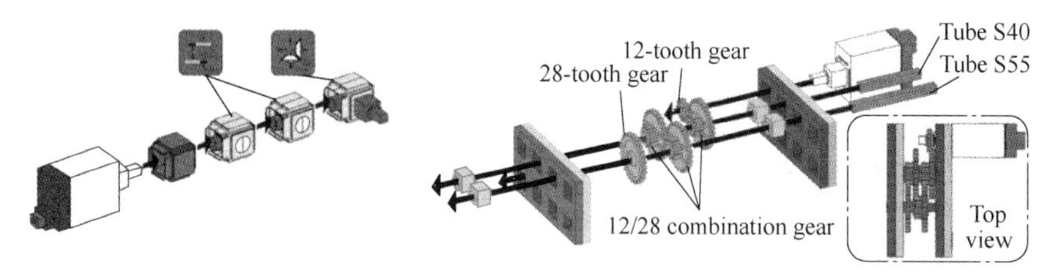


Figure 1-24 Moto connection with 5 : 1 reduction gear case

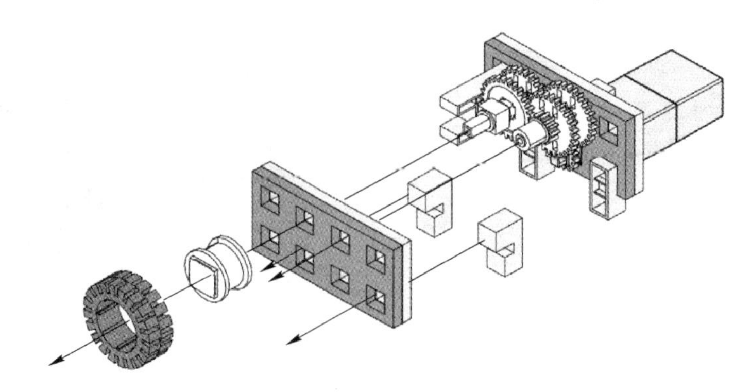

Figure 1-25 Moto connection with gear trains

(2) The sensors of the POWERON (except the magneto-sensitive switch) are all integrated in the cube, the usage is complete with the cube, the magneto-sensitive switch is integrated in the small square tube, and the usage is the same as the small square tube. The magnetosensitive switch can be connected to the cube by a short pin (shown in Figure 1-26), or with a five-hole ladder by two small connectors (shown in Figure 1-27). The touch switch can be connected directly to each other with a LED lamp or temperature sensor, a photosensitive sensor, etc., or with a cube connector (shown in Figure 1-28). The magnet is connected to the cube by a short pin (shown in Figure 1-29).

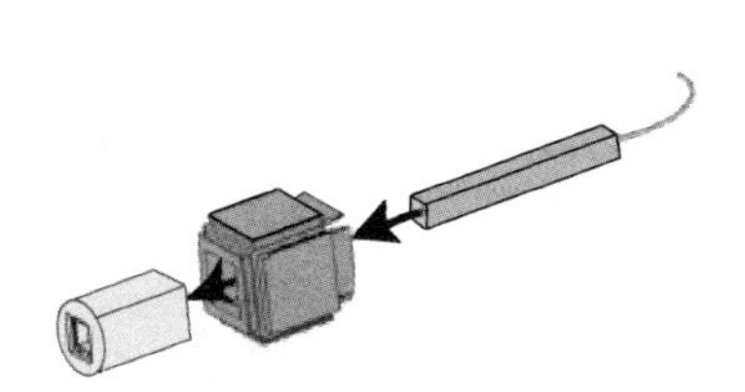

Figure 1-26 Connection of magnetic sensor (1)

Figure 1-27 Connection of magnetic sensor (2)

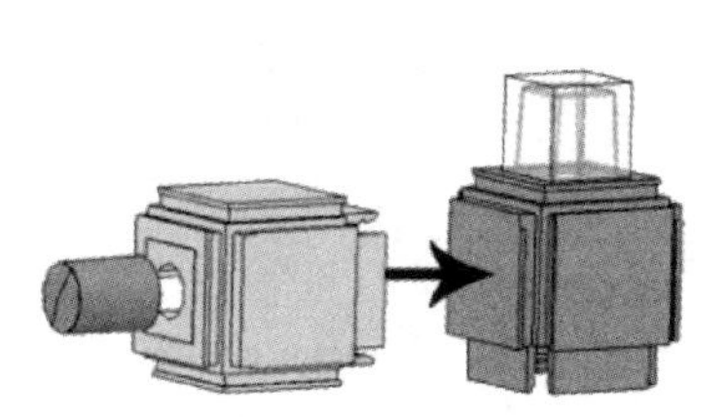

Figure 1-28 Connection of lamp to touch switch

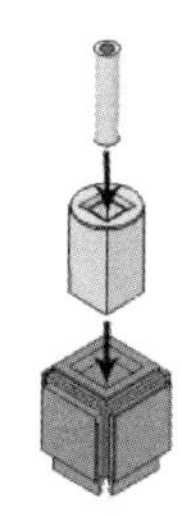

Figure 1-29 Magnet connection

(3) The construction methods of transmission parts are as follows:

1) Gear transmission: Gear transmission is the most widely used in modern machinery. It can also be used to change motion and speed by meshing teeth to transfer motion and power between any two axes in space. The gears in the transmission group are divided into main gears and slave gears.

In the gear transmission structure, there are two forms of transmission between straight gears: deceleration and growth. The number of teeth of the main gear are more than that of the slave gear, and the number of teeth from the gear are more than that of the trunk gear. It is different for the torque output from the two transmission groups.

2) Gear and rack drive: The rack belongs to the gear with infinite radius in machinery, and the gear rack transmission strictly belongs to a special form of gear transmission, which is the mutual conversion of the two motion forms of linear motion and circular motion. Typical application is the structure of the walking wheel system in the automatic door project (shown in Figure 1-30).

3) Worm drive: The worm drive consists of worm gear and worm gear, which is used to trans-

Figure 1-30　Steering wheel system of the automatic door

fer the motion and power between the staggered shafts in space, and the staggered angle between the two axes is 90″, in which the worm is active and the worm wheel is driven from each other, as shown in Figure 1-31.

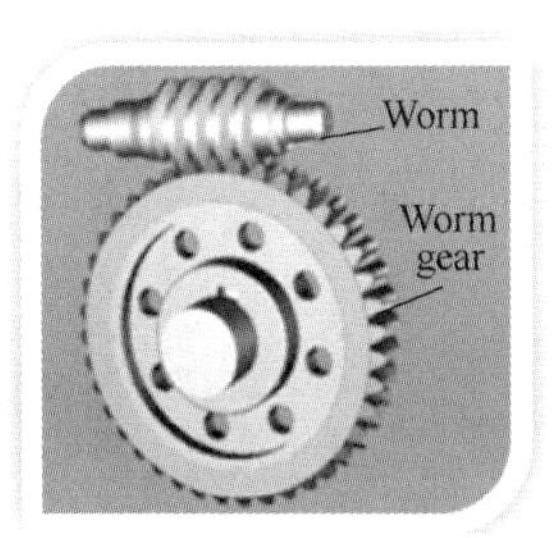

Figure 1-31　Worm gear drive

4) Belt drive: The belt drive is a commonly used, with low-cost power transmission device (shown in Figure 1-32), and the characteristics of smooth movement, clean (no lubrication), and low noise. At the same time, the belt drive has the function of buffer, vibration absorption, overload protection and easy maintenance.

The belt drive is composed of the active belt wheel, the driven belt wheel and the drive belt, and it is the use of the annular belt to tighten the two belt wheels. Friction is produced between the moving belt and the belt wheel, and the motion and power of the active belt wheel are transferred to the slave wheel.

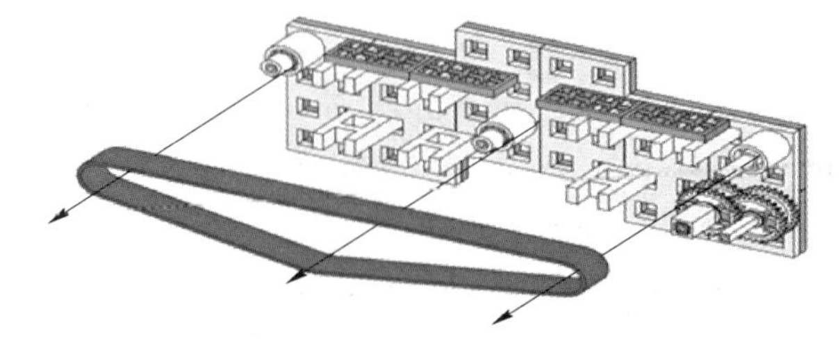

Figure 1-32　Belt drive

5) Screw drive: The screw drive is a mechanical drive composed of a screw and a screw nut. It

is mainly used to convert rotary motion into linear motion and torque into thrust.

The lead screw is the typical representative of the screw drive in the POWERON (shown in Figure 1-33). The screw drive is used in industrial manipulator projects.

It is not suitable to use belt drive and gear drive, while the two axes are parallel, the distance is long, and the power is large. Flat drive more accurate occasions multi-use chain drive.

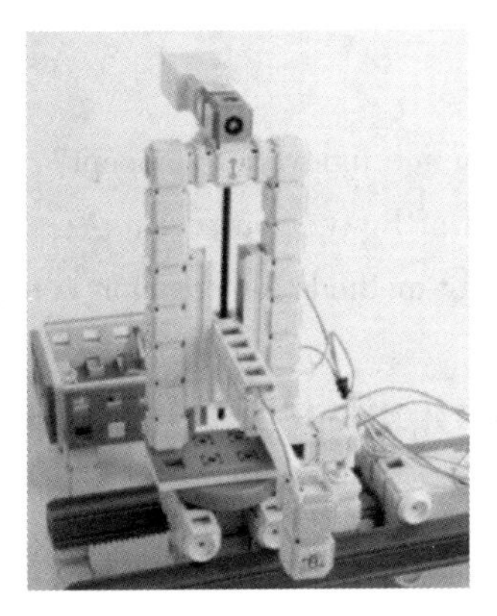

Figure 1-33 Screw drive with lead screw

6) Chain drive: When the parallel drive is not suitable for the belt drive and gear drive, and the two shafts are required to be parallel, the distance is far, the power is large, and the chain drive is often used.

The chain drive, which is composed of active sprocket, driven sprocket and chain, is wound on the sprocket and meshed with the sprocket. It is a transmission way to transfer the movement and power of the active sprocket with special tooth shape to the driven sprocket with special tooth shape through the chain. In the POWERON, there is a chain drive in the intelligent elevator project, which uses pulley and cotton thread simulation.

### 1.1.4 Evaluation

After completing the task, the task is evaluated, as shown in Table 1-1.

**Table 1-1 Task evaluation form**

| No. | Evaluation content | Achievement ratio | Self evaluation | Teacher evaluation |
|---|---|---|---|---|
| 1 | Cognition of engineering practice innovation | 10 | | |
| 2 | Functional characteristics of structural parts, connectors, transmission parts and electrical components | 30 | | |
| 3 | Common transmission | 25 | | |
| 4 | Construction method of structural parts and connectors, electrical components and transmission parts | 35 | | |

Through the study of the above content, students have a preliminary understanding of the use and role of these kits. There are rack drive, worm gear drive, belt drive, screw drive, chain drive and so on in the transmission mode. At the same time, the construction method of electrical components and common sensors is also introduced. Through learning, we have a preliminary overall understanding of the ability source innovation curriculum suite, and master the basic building methods and skills of common components, so as to lay the foundation for the next stage of project practice.

### 1.1.5 Development

(1) What is the engineering practice innovation concept?
(2) What does the classification of Power Source Kits?
(3) What does the basic building methods for common components of Energy Source Kits?

## Task 1.2 VJC Graphical Interactive Developing System

### 1.2.1 Introduction

In the POWERON, each functional part seems to be the human body, the POWERON controller seems to be the brain that directs the human body, and the program is like the thought in our brain. By using the VJC graphical interactive developing system, we can easily compile programs to achieve logical control of complex electromechanical systems. It can be seen that the VJC graphical interactive developing system has given the soul to the suite system we built, enabling it to complete the set tasks independently and autonomously.

### 1.2.2 Relevant Knowledge

VJC graphical interactive development system (hereinafter referred to as VJC) is a special software of ability source innovation course suite. It supports flow chart programming and interactive JC language programming. The written program can be downloaded to the controller and run directly. The flow chart is in the form of modular programming, close to human natural language (shown in Figure 1-34). The flow chart program is completely consistent with the standard flow chart, which will not cause students mis-understanding later.

#### 1.2.2.1 VJC4.3 Installation Process

Users can get VJC4.3 software on our website (http://vjc.xpartner.cn) or download from website directly. After downloading the software, follow installation tips to finish the installation. During the installation, you can selection installation address by yourself. Details are list as follows:

(1) When installing the software, the software will verify whether your computer has installed '.NET2.0' (or more advanced version). If '.NET2.0' is not installed, the software will install '.NET2.0' firstly. If '.NET2.0' already installed, you can directly install the VJC software.

(2) Default save path for VJC software is the root directory of partition C. Users can revise the save path by themselves.

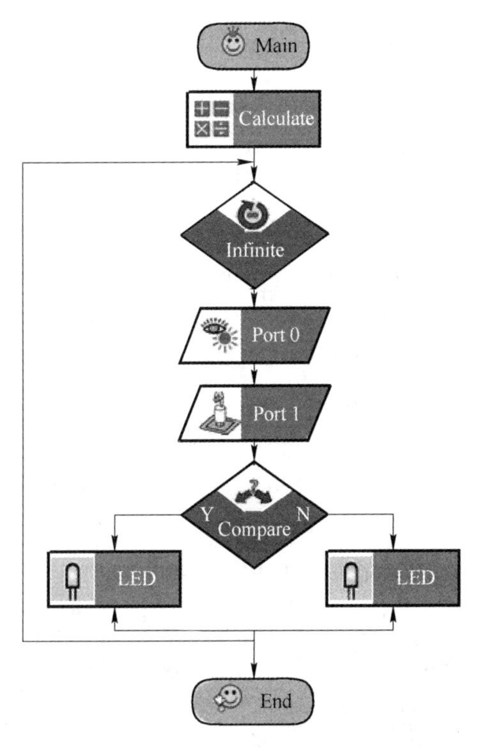


Figure 1-34 Flowchart grogram language

(3) After the installation, a short cutnamed 'VJC_4. 3_CH' will appear on the desktop and quick launch bar automatically. Double click the shortcut, then the software will be started.

Note: Suggested operating environment for VJC4. 3 is Win7.

### 1. 2. 2. 2 First Use

After successful installation, firstly entering VJC, the new interface will be displayed, as shown in Figure 1-35. The corresponding options are selected according to the program format you want to create. If you need to open an existing program, you can choose to open the menu for operation.

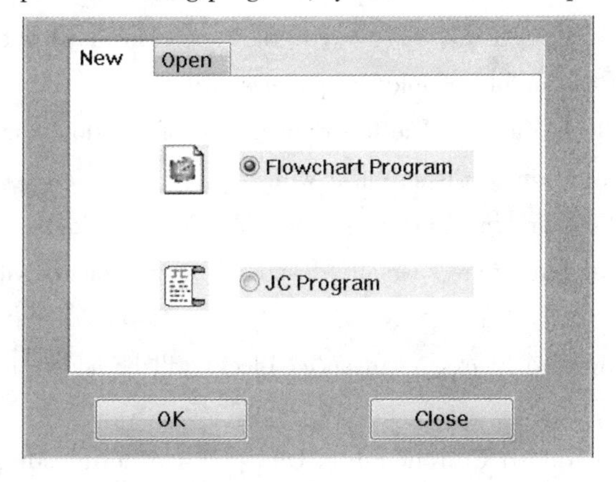


Figure 1-35 New program

### 1.2.2.3 Menu Bar and Toolbar

Menu bar and toolbar are very user-friendly, similar to your commonly used MS Office. In this manual, we will have a detailed intro of related items on the menubar and toolbar.

**Flow Chart Interface**

The selections of menu items under flow chart interface are as follows:

(1) 'File (F)→Export JC File (J)...': The current flow chart is saved as JC program.

(2) 'Edit (E)→Main Program (M)': The main program interface is backed to from subprogram interface. This function only works when the present interface is subprogram interface.

(3) 'Edit (E)→New Subprogram (N)...': A dialogue box will pop up for creating a new subprogram (shown in Figure 1-36). Users can create a 'System Subprogram' or call a subprogram from another program, and then name the new subprogram by filling the blank behind 'Subprogram Name'. Behind the 'Author Name', users can write down their own name for the convenience of the other users to find the author if they find out some problems.

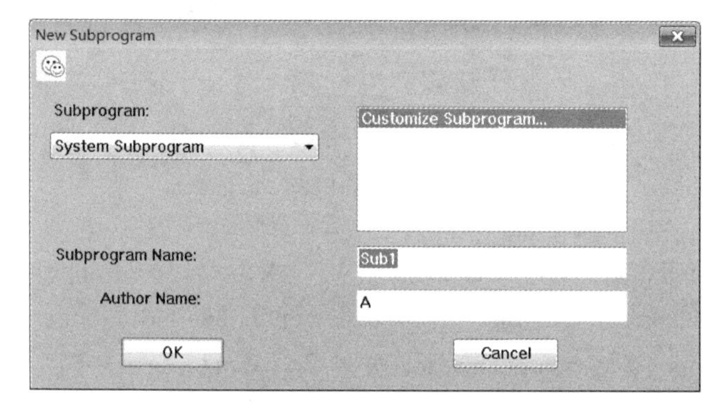


Figure 1-36 New subroutine

(4) 'Edit (E)→Delete Subprogram (D)...': Users can delete the subprogram they are editing. This function only works when the displayed interface is the subprogram interface. A subprogram couldn't be deleted incase that the subprogram has been used in a main program.

(5) 'View (V)→Show JC Code (C)': The right panel of the flow chart area is the zone for displaying JC code. You can unfold or fold the panel at will.

(6) 'View (V)→Block Mark': The top right corner of the flow chart area is the parameter prompt window. You can hide it or display it at will.

(7) 'View (V)→Line Tracing Blocks': The Line Tracing Blocks will be added on the left after being checked. These blocks are specially designed for line tracers with 5 grayscale sensors or 7 grayscale sensors.

(8) 'View (V)→Advanced Blocks': Advanced Blocks will be added on the left after being checked.

(9) 'Tools (T)→Download Current File (D)': When downloading the current file to the controller, you can use shortcut key 〈F6〉 to download current flow chart (〈F5〉 is for download-

ing JC program).

Note: USB cable need to be linked before downloading.

(10) 'Tools (T)→Screenshot': The image of current flow chart window is grabbed. When the function is activated, a pop-up dialogue box will suggest you to select a save path. If you want screenshots to be in a white background, you can remove the confirmation mark before 'Background' in the 'View'.

(11) 'Tools (T)→Save flow chart in real time': When this item is confirmed, the program in editing will be saved automatically. To make this function valid, you should save your file manually for one time. Automatic saving function can help you improve the security of the file you are editing in case of power outage or that you forget to save the file.

(12) 'Help (H)→Help Topics (H)': Users can call the help docs of the software by clicking the function key. Of course, users may also find updated information, which generally appear in the end of the texts.

(13) 'Help (H)→Check for Update (U)': VJC4.0 and more advanced versions support online update. This function helps users check whether there are updates, and then the update will proceed automatically. For updating online, the computer users using can get access to the internet.

**JC Code Interface**

The selections of menu items under JC code interface, as shown in Figure 1-37 are as follows:

(1) 'Edit (E)→Go to Line (G)': When users click this item, a dialogue box for line setting will pop up. Under JC code interface, each line of JC code has a serial number.

(2) 'View (V)→Interactive Window (I)': Interactive window refers to the window on the bottom side of the JC code interface. In general, compile error, cautions, progress errors, etc., will be displayed in the area. The position of the interactive window can be reset by dragging.

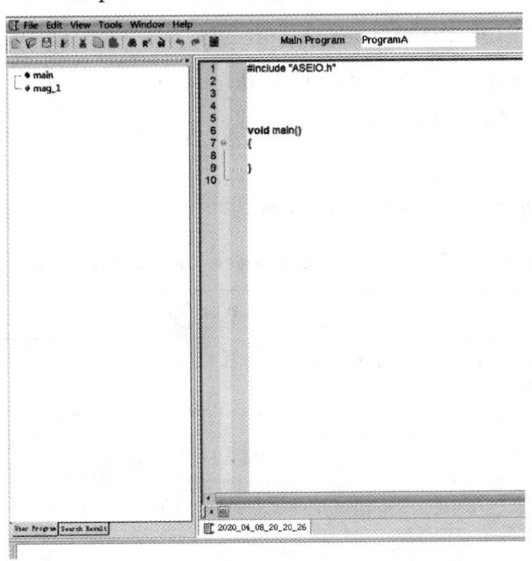


Figure 1-37 JC interface

(3) ‘View (V)→Multifunction Window’: The left side of JC code area is the default area for the multifunction window. Multifunction window can be divided into two sub windows. ‘User Program’ window displays the main program, subprogram, and variables of the current program. Users can easily position any of these items, such as a variable in the JC code area by double clicking the item in the multifunction window. The item in the JC code area is selected and then the mouse is right clicked, and users will see a pop-up menu list. Clicking ‘Find All’, users will find the results listed in the ‘Find in All’ window on the left side of the JC code area.

(4) ‘Tools (T)→Compile in real time’: When the function is confirmed to use, just-in-time compile result will be displayed in the ‘Interactive Window’ for better helping users find grammar errors or cautions.

(5) ‘Window (W)’: Under JC code interface, multiple program windows can be opened. Users can sort these program windows manually.

**Toolbar**

Figure 1-38 shows the toolbar under JC code interface.

Figure 1-38 Toolbar under JC code interface

Figure 1-39 shows the toolbar under flow chart interface.

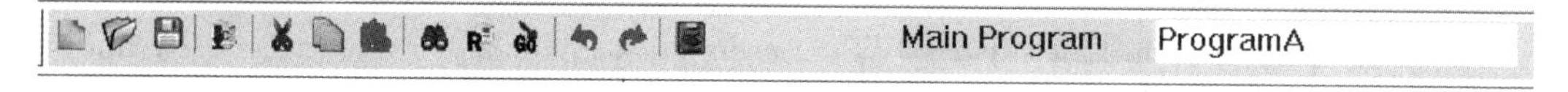


Figure 1-39 Toolbar under flow chart interface

When users move the mouse over an icon on a toolbar, the function of that icon is displayed. Ccon102 controller can save more programs in its storage space. To make programs more recognizable, users can define their names by ourselves. The content filled in after the ‘Main Program Name’ will be displayed on the screen of the Ccon102 controller as the program name. It should be noted that only English can be displayed.

**Shortcut Keys**

Table 1-2 shows the shortcut key in the program interface.

**Table 1-2 Shortcut keys under the program interface**

| | |
|---|---|
| 〈F2〉 | Full screen display |
| 〈Esc〉 | Quit full screen display mode. When more dialogue boxes are opened, the key works as cancel key |
| 〈F6〉 | Download flowchart program |
| 〈F1〉 | Open help document |
| 〈F9〉 | Ma make highlights on graphic blocks of the flow chart |
| 〈F12〉 | Switch between flow chart interface and JC code interface.<br>Please note that the switch won't convert flow chart to JC code |

Table 1-3 shows the switch between flow chart interface and JC code interface.

**Table 1-3 Some shortcut keys of code editor**

| 〈F2〉 | Look for marks |
|---|---|
| 〈Ctrl〉 + 〈+〉 | Enlarge JC code |
| 〈Ctrl〉 + 〈-〉 | Narrow JC code |
| 〈Ctrl〉 + 〈W〉 | Search all |
| 〈Ctrl〉 + 〈F3〉 | Search down |
| 〈Shift〉 + 〈F3〉 | Search up |
| 〈Ctrl〉 + 〈F〉 | Open the search dialogue box |
| 〈Ctrl〉 + 〈A〉 | Select all |
| 〈Ctrl〉 + 〈H〉 | Replace |
| 〈Ctrl〉 + 〈Z〉 | Repeat |
| 〈Ctrl〉 + 〈C〉 | Copy |
| 〈Ctrl〉 + 〈V〉 | Paste |
| 〈Ctrl〉 + 〈X〉 | Cut |
| 〈F5〉 | Download JC program |

### 1. 2. 2. 4 Program Download

The methods available for downloading the present flow chart Ccon102 program or JC program are as follows:

(1) Click 'Download Program' button on the toolbar.

(2) Click 'Tools Download Current File' in the menu bar.

(3) Press shortcut key 〈F5〉 to download current JC program, or 〈F6〉 to download current flow chart program.

Before downloading the program, make sure USB cable is linked. When Ccon102 controller is linked to the computer via USB cable, the controller will beep. If no beeps heard, the controller may not be well linked to the computer or the controller driver not be installed. When Ccon102 controller is linked to the computer via USB cable, the LCD screen of the controller will enter program interface automatically. In case that USB cable is not well connected, downloading dialogue box of VJC software will also pop up while downloading.

If grammar mistakes exist in the program, compile process will be terminated, and download process cannot be complete. In general, grammar errors of flow chart program appear in the 'Customize' block. As to JC program, grammar errors will be displayed in the 'Interactive Window'. Double clicking the error will lead users to the error in the program, and thus is convenient for error correction.

Please do not pull off the USB cable during downloading, because it may lead to machine crash. After downloading the program, please pull off the USB cable firstly, then run the program.

### 1.2.2.5 Software Online Upgrade

VJC supports online update. We will provide solutions for bugs found. The one-click update will help users get everything done. It's very easy and simple. On the other hand, users will also provide new functions, tutorial programs by online update. Click 'Help Update' in the menu bar, and the dialogue box as shown in Figure 1-40 will pop up. Make sure you have saved your program before closing for update.

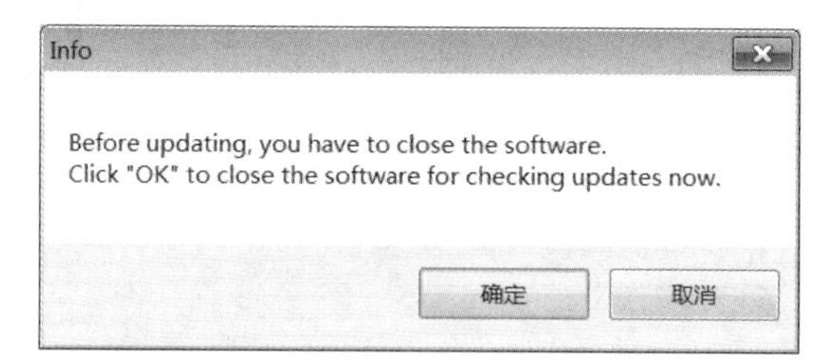


Figure 1-40 Update file prompt

Click 'OK' to open the upgrade management dialog box, as shown in Figure 1-41.

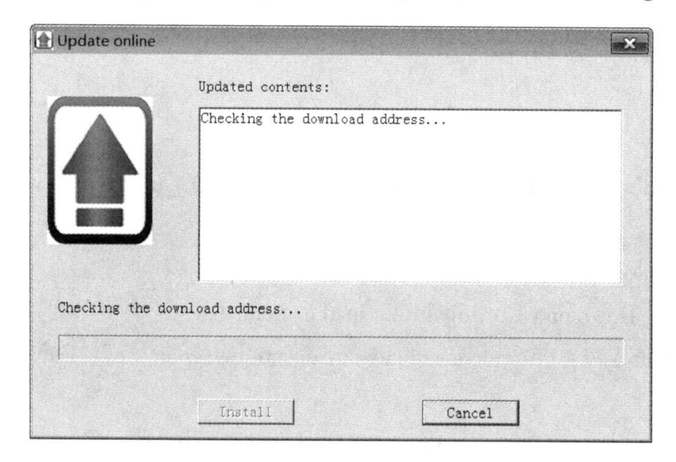


Figure 1-41 Check update

After confirming that the computer is connected to the network, click 'Check Update' button to start downloading the files that are not downloaded on the server. If there are files that need to be updated, there will be a prompt. Click 'Start Installation' to install new updates. After the installation, click 'Quit' to close the dialogue box.

### 1.2.2.6 Flow Chart Interface

**Introduction of Flow Chart Interface**

As shown in Figure 1-42, it is the flow chart programming interface. Block libraries are on the left panel, and programming area is the central area. Right panel displays JC code, which can be hidden, code generated automatically. Code cannot be revise manually, and it is convenient for learning C language grammar structure, and parameters of various blocks can be read here.

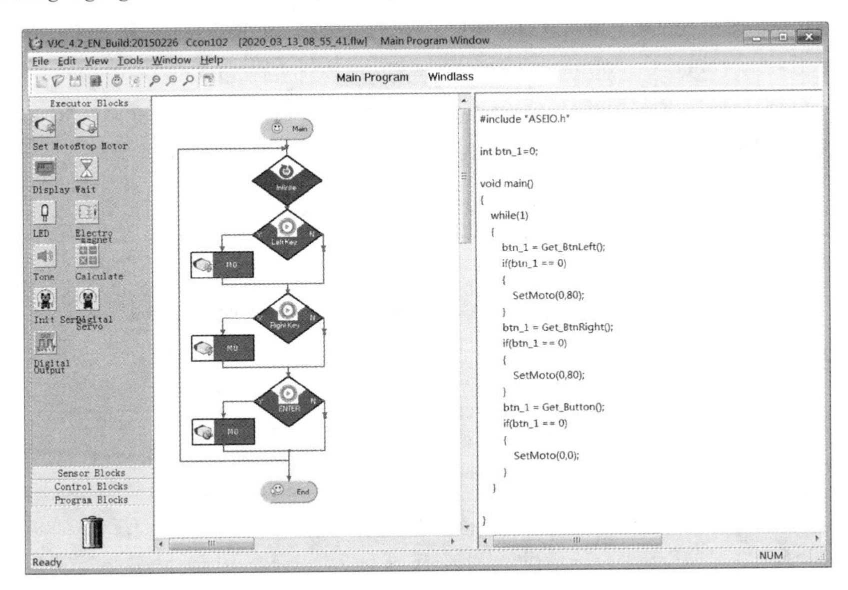


Figure 1-42  Flow chart interface

A robot mainly consists of below parts: a controller, sensors, actuators and user program. Sensors and actuators are linked to various ports on the controller, while user program runs in the controller. We can see a robot in this way: Users program read sensor values via controller ports, and then send commands to various actuators via related ports after various calculations. Then actuator will make certain motions after receiving these commands from the controller.

Ccon102 controller also has RS485 communication function, so it supports sensors, actuators and communication which are RS485 based.

In VJC 4.3, block libraries are reclassified. They include actuator blocks, sensor blocks, control blocks and program blocks. Besides, several block libraries that are not used frequently can also be called from the menu bar.

**Names and Functions of Graphic Blocks**

The compositions of graphic blocks are as follows:

(1) Actuator blocks: Actuator blocks include all electronic components that can perform certain motions in case that they receive certain commands from a related controller. Main commands from controller consist of DO (Digital Output) startup and shutdown, DC rotation, voice, display, etc.

(2) Sensor Blocks: Sensor blocks include all collection modules that can gather environment data for the controller. Controller collects environment data via functions such as AI, DO, counter, etc.

(3) Control Blocks: In user program, return value from each port generally has either of the below two functions: data storage or judgment making. And judgment making is used more frequently. As to VJC software, three judgment modes are available: 'While' sentence, 'If...

else...' sentence, and 'For' sentence.

If a program needs to make a judgment, there should be an object and reference value. The object for comparison is often the return value of a certain sensor or updated variable value. So each block that is capable of 'Reading' function in sensor blocks can be converted to a 'Condition Judgment' block. Table 1-4 shows the detailed list of control blocks.

**Table 1-4 Control blocks**

| No. | Icon | Name | Function |
|---|---|---|---|
| 1 |  | Multiple loop | The block works as 'For' sentence in C language. 'Loop Number' parameter in this block means how many times the block will loop. Exact loop number is specified by users |
| 2 |  | Infinite loop | The block works as 'While (1)' sentence in C language. It repeats the body inside the sentence infinitely |
| 3 |  | Condition loop | The block works as 'While' sentence in C language. Parameters are the conditions that can be set by users. As long as the loop condition is met, the loop will continue, or the loop will be terminated and the program will run below sentences |
| 4 |  | Compare | The block works as 'If...else...' sentence in C language. Parameters can be set by users. If the condition is met, the program will implement the 'Y' branch, or the program will implement the 'N' branch |

'Condition' mentioned above is an expression including three parts: left part, middle part and right part. Left and right part can be formulas, variables or numerical values, while the middle part can be '==', '!=', '>', '<', '>=', '<=' etc. Return value of such condition sentence only has two possibilities, which are 0 and 1. If return value is 0, the condition is not met; If return value is 1, then the condition is met.

When setting condition loop and condition judgment, users can find 'Condition 1' and 'Condition 2', as shown in Figure 1-43. Usually 'Condition 1' is used, and 'Condition 2' will become valid if 'Valid' option on tab 'Condition 2' is selected. After 'Valid' on tab 'Condition 2' is selected, other parameters for 'Condition 2' will be activated as follows:

1) Only when 'Valid' selected will parameters for 'Condition 2' become activated.

2) Logic relation means the relation between 'Condition 1' and 'Condition 2'. Three logic relations are available, which are 'AND', 'OR' and 'NOT' (for C language, '&&', '||' and '&&!' instead). Parameters for 'AND', 'OR' and 'NOT' can be formulas or values. Three results are available for the calculation:

① 'Condition 1' and 'Condition 2': When both conditions are met, result is 1, or result is 0.

② 'Condition 1' or 'Condition 2': Result will be 1 if one condition is met at least, or result is 0.

③ ‘Condition 1’ not ‘Condition 2’: If return values for both conditions are different, result is 1, otherwise result is 0.

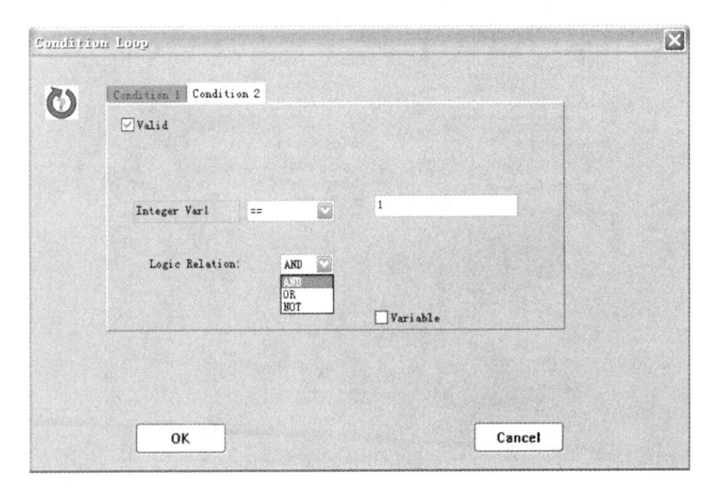


Figure 1-43 Conditional expression

(4) Program blocks: Table 1-5 shows the detailed list of program blocks.

**Table 1-5 Program blocks program blocks**

| No. | Icon | Name | Function |
|---|---|---|---|
| 1 | | New subprogram | This block can be used to build a new subprogram. Users can also reference a subprogram of another program. Subprogram can be taken as a package of many blocks for realizing a certain function. Subprogram can be used easily in later operations and can keep the main program relatively brief |
| 2 | | Subprogram return | This block is used to end a certain subprogram and keep the flow chart of the subprogram complete. There is no practical function for the block |
| 3 | | End block | This block is used to end the main program and keep the flow chart of the main program brief. There is practical function for this block |

(5) Variables Box: For graphic blocks, users may find ‘Variable’ option or sensor variable option. After clicking that option, users will find a pop-up dialogue box as shown in Figure 1-44, which is the variables box.

Variables box integrates all variables used in flow charts. Users can switch types of variables by clicking icons of different sensors or others in the picture. Then users can choose a serial number from 10 options above icons on the picture.

## 1. 2. 3 Implementation

### 1. 2. 3. 1 Task Description

Requirements for building a rotating workbench are as follows:

(1) The workbench will stop its rotation for 3s once its fan blade moves to a second station,

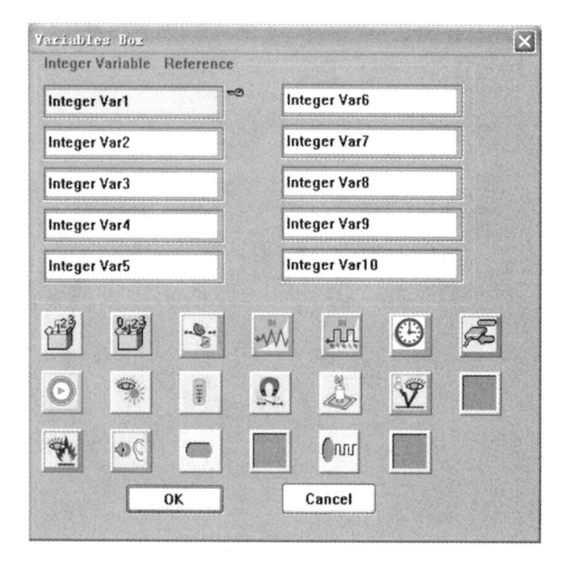


Figure 1-44 Variables box

and the red light is turned on simultaneously.

(2) After 3s, the controller will beep for 0.5s to inform people, and the workbench will start rotating again.

(3) When the rotation starts rotating again, the red light is turned off at the same time.

### 1.2.3.2 Structure Construction

Figure 1-45 shows the model of rotating workbench, and Figure 1-46 to Figure 1-49 show the steps of model of setting up rotating workbench respectively.

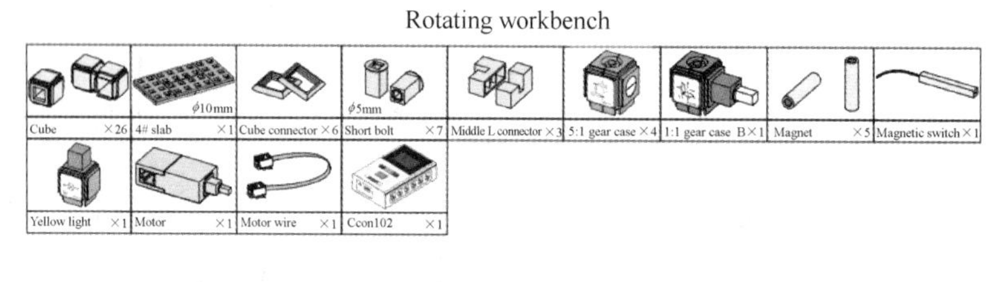


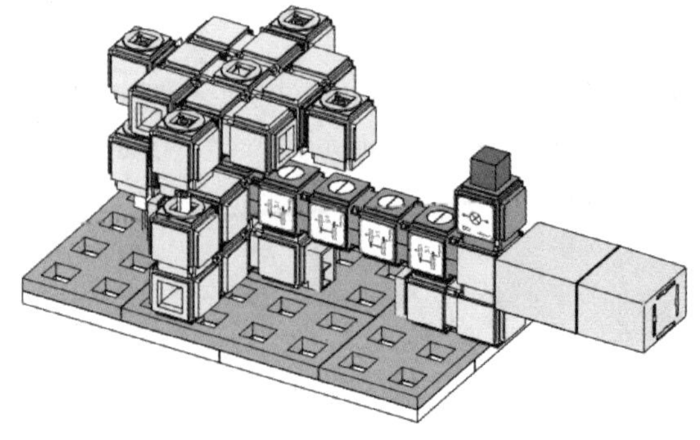

Figure 1-45 Rotating workbench

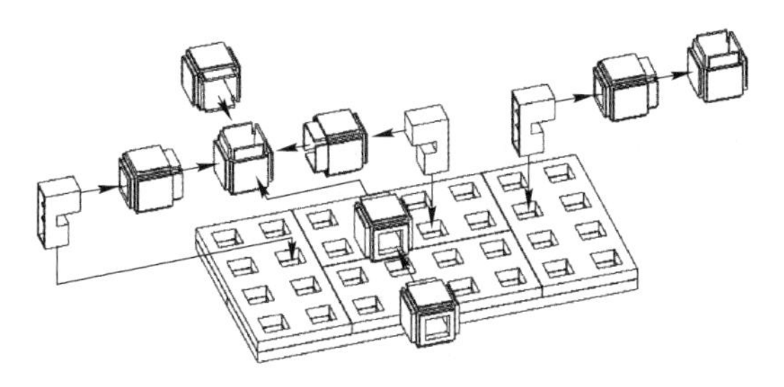

Figure 1-46　Step of setting up rotating workbench (1)

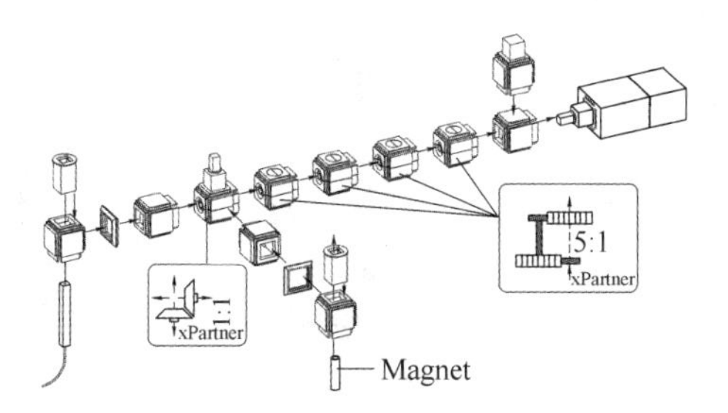


Figure 1-47　Step of setting up rotating workbench (2)

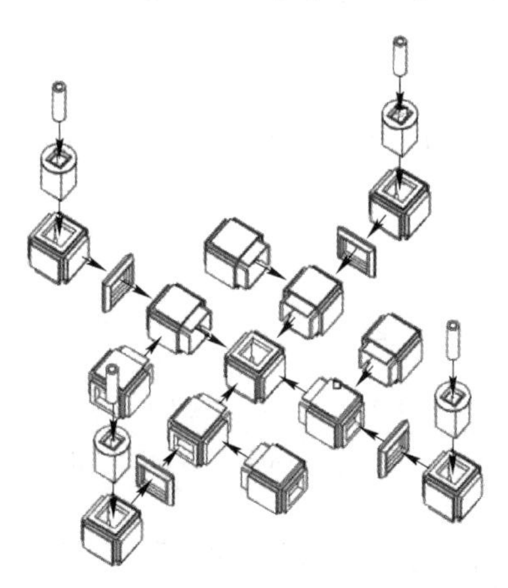

Figure 1-48　Step of setting up rotating workbench (3)

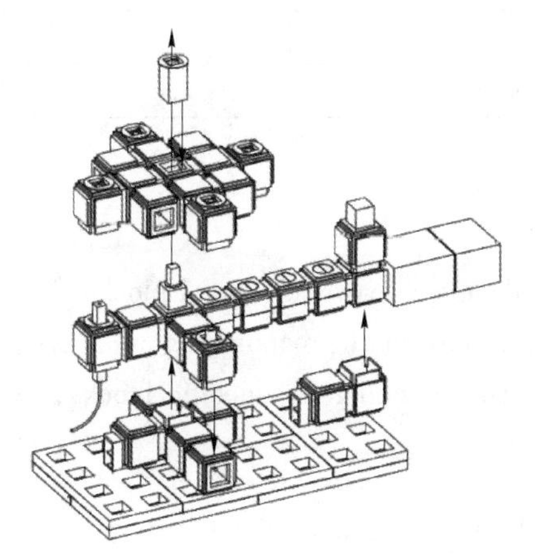

Figure 1-49　Step of setting up rotating workbench (4)

### 1.2.3.3 Program Design

When magnetic sensor finds a magnet of the fan blade, the return value of the sensor changes to 1 from 0. The rotating table procedure is shown in Figure 1-50.

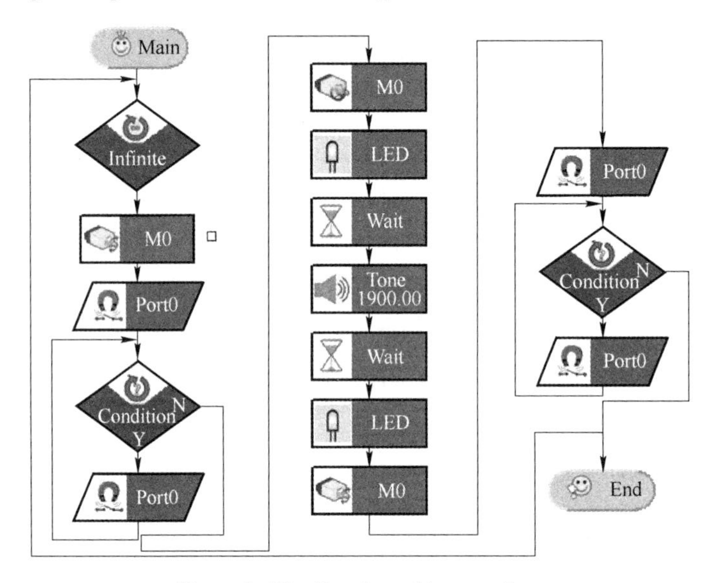


Figure 1-50 Rotating table procedure

## 1.2.4 Evaluation

The task evaluation is shown in Table 1-6.

**Table 1-6 Task evaluation**

| No. | Evaluation content | Achievement ratio | Self evaluation | Teacher evaluation |
|---|---|---|---|---|
| 1 | Working principle of magnetic sensor | 10 | | |
| 2 | Principle of conditional cycle module | 10 | | |
| 3 | Construction of rotary table | 30 | | |
| 4 | Programming of rotary table | 30 | | |
| 5 | Debugging of rotary table | 20 | | |

## 1.2.5 Development

For advanced users, we have developed several special blocks. To use these blocks, users should list them on the left panel of the flow chart programming area. Users can find these special blocks in 'Tools' of the main menu. After clicking the specific blocks, they will be listed on the panel left to the flow chart programming area.

### 1.2.5.1 Line Following Blocks

Line Following Blocks are the blocks library specially added for line tracers with grayscale sensors. It is required that the grayscale sensors should be lined in a row in the front of the line tracer. The number of grayscale sensors can be 5. The line following blocks is shown in Figure 1-51.

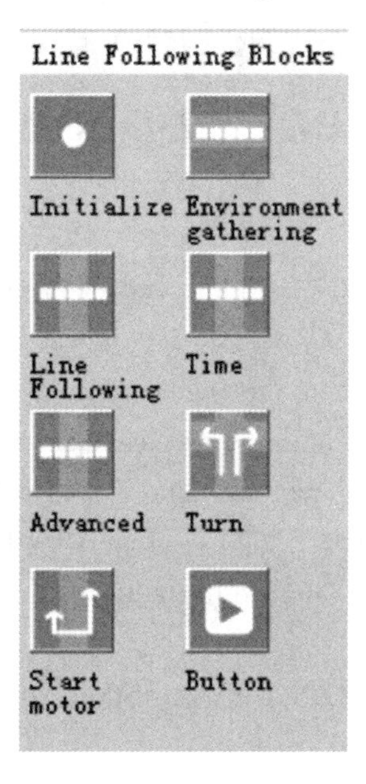


Figure 1-51　Line following blocks

### 1.2.5.2　Advanced Module Library

There're also some not-often-used functions stored in 'Advanced Blocks'. Blocks in this block library will be added in accordance with the update of functions and hardware, which are as follows:

(1) Coding motor: The module is designed for the encoder motor of RS485 communication, which contains two parameters of 'Motor ID' and 'Motor Speed'. One encoder motor module can control up to four motors at the same time. If users need to control more encoder motors, they can use multiple encoder motor modules.

(2) Steering angle: There is a 'Digital Steering Gear' module in the actuator module library, which makes the steering gear turn to the corresponding angle, while the 'Steering Gear Angle' module reads the current angle of the digital steering gear.

(3) Digital tube: The module designed for the digital tube of RS485 communication can control the digital value displayed by the digital tube. It should be noted that the nixie tube can only display 1-bit of 0~9 value, so when using the 'Reference Variable' function, the referenced variable should not exceed this range.

(4) Read EEPROM: Read the EEPROM value of a certain address inside the controller. The key parameter of EEPROM is its address, which ranges from 0 to 31, generally critical values of programs are stored in EEPROM.

(5) Write EEPROM: Revise the EEPROM value of a certain address inside the controller. For Ccon102 controller, the operation can be done by key operation of the controller.

# Project 2 Practice of Innovation Project Application

This project contains five typical tasks: building and debugging automatic door control systems, intelligent elevator control systems, AGV systems, industrial robot control systems, and CNC systems from simple to complex details of the engineering practice using the POWERON. This project introduces in detail the process of building and debugging a typical task of engineering practice innovation and VJC software programming, using the POWERON from simple to complex. Through these tasks, students can be proficient in the use of the POWERON and the programming and debugging process of the program, and improve the students' innovative ability and practical ability.

## Task 2.1 Build and Debug Automatic Door Control System

### 2.1.1 Introduction

The theoretical understanding of automatic doors should be an extension of the concept of using doors. Single, double or multiple doors, which use various signals to control automatic opening and closing, are called automatic doors. It is a complete set of device that uses electric power to drive the door to open and close, and the control system sends instructions to drive the door to open and close through a motor and a reduction drive system. Door engineering is a comprehensive fringe industry, which integrates building technology, electronic control technology, mechanical design and manufacturing technology, computer technology, and building decoration technology. It will have broad development space.

### 2.1.2 Related Knowledge

#### 2.1.2.1 Principle of Automatic Door

Automatic door is a door system that uses advanced induction technology to control the electromechanical actuator through the control system to automatically open and close the door. When a person or other moving target enters the sensor detection working range, the door leaf will automatically open; When a person or other moving target leaves the sensing area, the door leaf will automatically close.

#### 2.1.2.2 Alassification of Automatic Door

**Automatic Sliding Doors**

Automatic sliding doors are mainly used in company, government agencies, restaurants, banks, hospitals, office buildings, shopping malls and workshop channel of the factory. The advantages of the sliding door are low cost of daily maintenance, low price of accessories, no occupation of longitudinal space, and wide appearance. The disadvantage is that it needs a large horizontal space, the whole project supporting construction should be done during the installation, the maintenance is inconvenient, and the absolute closed space cannot be formed when the door is opened.

**Automatic Swing Doors**

Automatic swing doors are mainly used in commercial offices, entrances to stylish residential buildings, medical access, elderly apartments, toilets, disabled access, and fire doors or places with small horizontal spaces. The main advantage is that it is easy to install and maintain. It can be installed on the original door body with short construction time. The cost of single door is lower than that of sliding door. The cost of a single-opening door is lower than that of a sliding door. Since a double-opening door requires two sets of equipment, the cost of the door is higher than a sliding door, and an absolute closed space cannot be formed when the door is opened.

**Other Special Doors**

Other special doors include household doors, automatic folding doors, automatic revolving doors, automatic curved doors, security doors, fire doors, sound doors, radiation doors, mute doors, bank vault doors, etc.

### 2. 1. 2. 3 Structure of Automatic Doors

The structure of automatic doors include:

(1) Main controller: Master controller is the command center of the automatic door. It controls the operation of motor or electric lock system by sending appropriate instructions from the internal large-scale integrated circuit programmed with instructions. It also adjusts the opening speed, opening degree and other parameters of the door. Figure 2-1 shows the dedicated master controller of the automatic door.

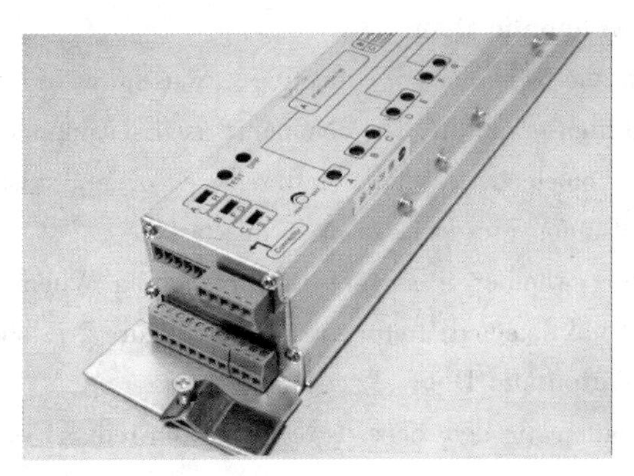

Figure 2-1 Dedicated controller in automatic door

(2) Inductive detector: It is in charge of collecting external signals. When a moving object comes into view, it, like an eye, sends a pulse signal to the master controller. The inductive detector outside view is shown in Figure 2-2.

Figure 2-2 Inductive detector outside view

(3) Power motor: It provides driving force to open and close the door. At the same time, it controls the door to slide rapidly or slowly.

(4) Door track: Functionally like rail, it restricts the door sling wheeling system to run toward a specific direction.

(5) Door sling wheeling system: It holds the sliding door and puts the door in motion under pulling force.

(6) Synchronous belt: It transmits the actuating force from the motor to pull the door sling wheeling system.

(7) Lower guiding system: At the lower part of the door there is a guiding and positioning system which prevents the door from swaying back and forth in motion.

### 2.1.2.4 Core Technology of Automatic Doors

**Sensing Technology of Automatic Door**

Sensor technology refers to the technology of collecting various forms of information with high precision, high efficiency and high reliability. The commonly used sensing technology of automatic door includes travel switch, photoelectric switch, infrared sensor and visual sensor. The magnetic sensor was used instead of automatic door sensor in this task.

Digital sensor, detecting whether a magnet is approaching. When magnet detected, return value is 1; When magnet not detected, return value is 0. Figure 2-3 shows the magnetic switch.

**Driving Technology of Automatic Door**

The driving device of an automatic door consists of motor and reducer or motor and hydraulic system. It actuates the door to open and close following the programmed instructions. From the driving device to the door motion, there must be a transmission mechanism, and reducer is a part of the

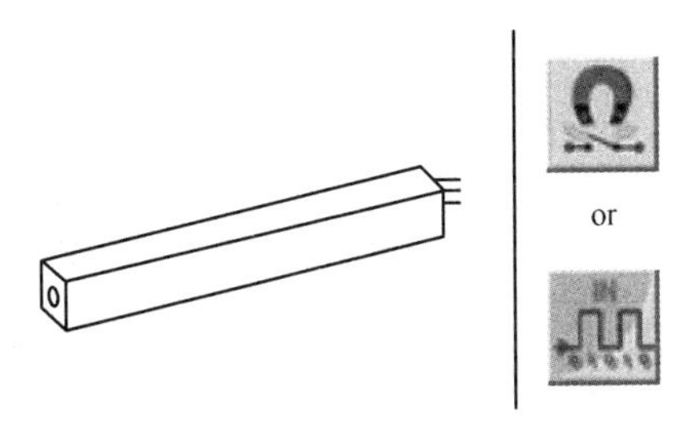


Figure 2-3 Magnetic switch

transmission mechanism. Now, reducer and motor have been integrated together, and collectively called gear motor. Gear motor contributes to simplification of equipment, which can be easily standardized and become smaller. In addition to reducer, the transmission mechanism must be designed according to specific needs, for which, gear drive, chain drive or high-efficient synchronous belt drive can be used.

In the POWERON, a DC motor (shown in Figure 2-4) is used. Its shape has been processed for the purpose that the motor can be easily connected with other component. Through a reduction gear case with the reduction ratio of 5 : 1, it has the function of speed reduction. The specific technical parameters are listed in Table 2-1.

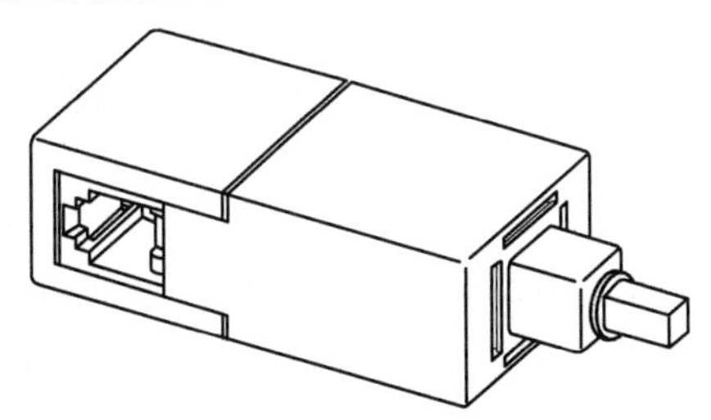

Figure 2-4 POWERON DC motor

**Table 2-1 Technical parameters of capacity source DC motor**

| | |
|---|---|
| Working voltage/V | 6.5~11.2 |
| No-load speed/r · min$^{-1}$ | >5800 |
| Torque at maximum efficiency/g · cm | 163 |

## Control Technology of Automatic Door

Usually, PLC controller (shown in Figure 2-5) is used in automatic door. PLC controller has the advantages of powerful interference immunity, high reliability, simple control system structure, universal use, easy programming and easy realization. The common PLC controller brands are Mitsubishi, Siemens, etc.

A dedicated controller is shown in Figure 2-6. Usually, the popular 32-bit digital signal processor (DSP) is used as control center and the excellent communication and digital decoding technology are introduced, so the performance and reliability of automatic door are greatly improved.

The controller of automatic door has the function that supports the communications of handheld

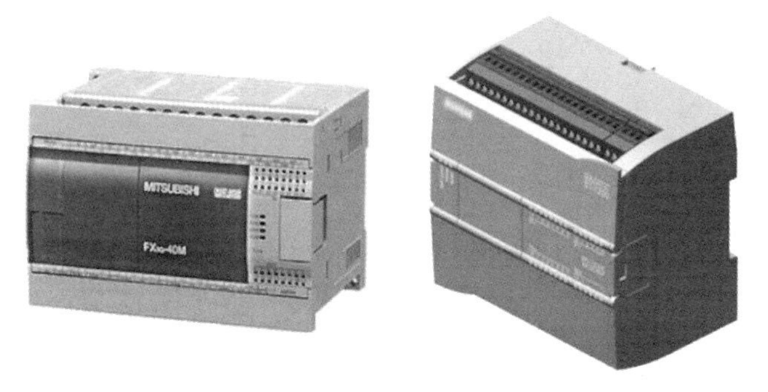


Figure 2-5 PLC controller

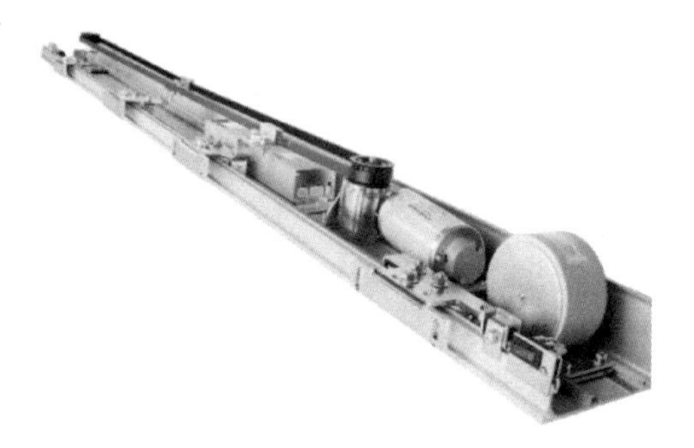

Figure 2-6 Special controller for automatic door

PDA and other intelligent terminals, conducive to easier and safer commissioning and maintenance of automatic door. All parameters of automatic door can be displayed vectorially, operation status can be clearly reflected by the PDA parameters, and control programs can be upgraded. Handheld PDA is shown in Figure 2-7.

Figure 2-7 Intelligent terminals handheld PDA

In the building of POWERON automatic door, the POWERON controller (shown in Figure 2-8) is used as control unit. It has 4-channel motor interfaces (numbered DCO DC3 respectively) and 12-channel 10 interfaces (numbers 1/O 0-10 11 respectively), of which I/O interfaces are A/D multiplex ones. After being programmed, the controller makes possible the various basic functions of the automatic door.

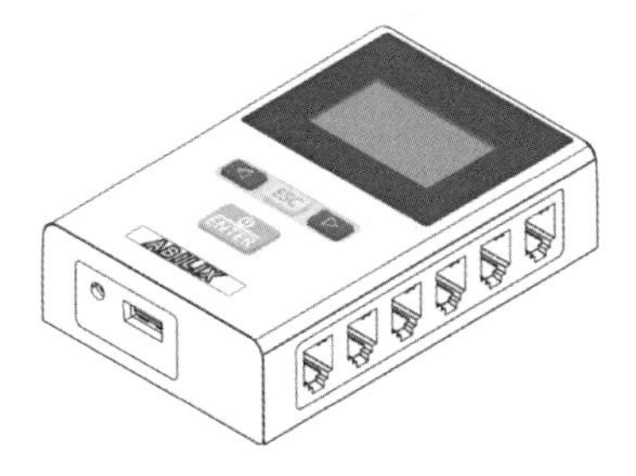

Figure 2-8　POWERON controller

## 2.1.3　Implementation

Select proper type of automatic door, and complete automatic door design. It is required that the door will be opened when automatic door controlling system detect the opening order, and the door will be closed when detecting the closing order, with three colors of lights used to respectively show the status of 'Closing, No Access', 'Moving, Caution' and 'Opening, Passable'.

### 2.1.3.1　Project Design

The project design is as follows:

(1) Automatic door takes the form of moving door.

(2) Automatic door is driven by DC motor. The the gear rack drives the door opening and closing after the gear slow down.

### 2.1.3.2　Material Preparation

The list of materials for supporting projects is shown in Figure 2-9.

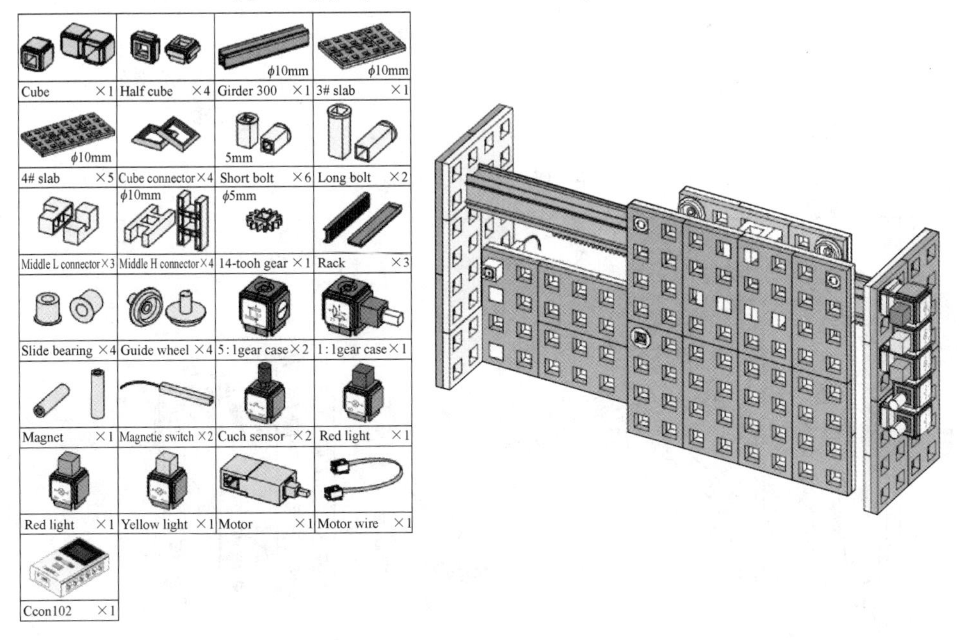


Figure 2-9　Automatic door

## 2. 1. 3. 3 Hands-on Construction

The models of hands-on construction are shown in Figures 2-10 to 2-16 respectively.

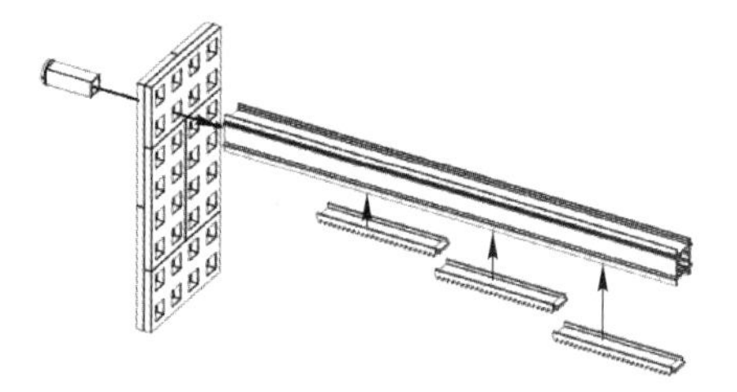

Figure 2-10 Install door travelling rail

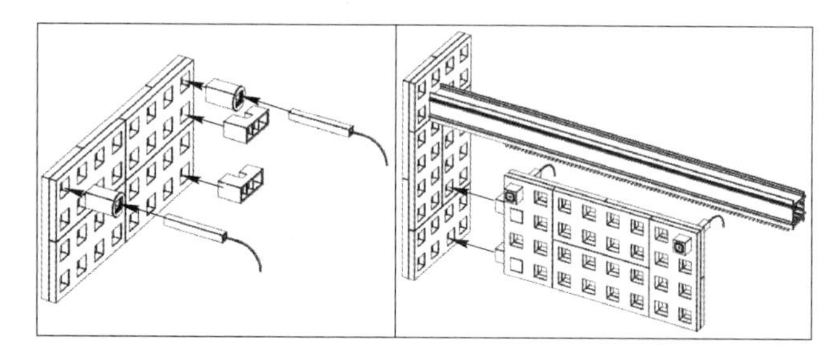

Figure 2-11 Install sensors and install the fixed bracket

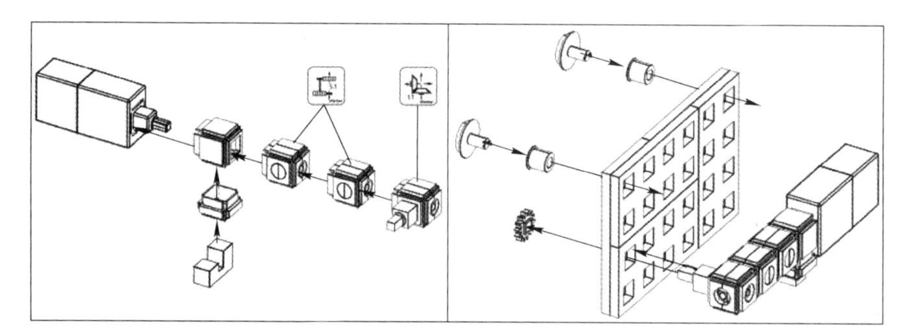

Figure 2-12 Install driving system

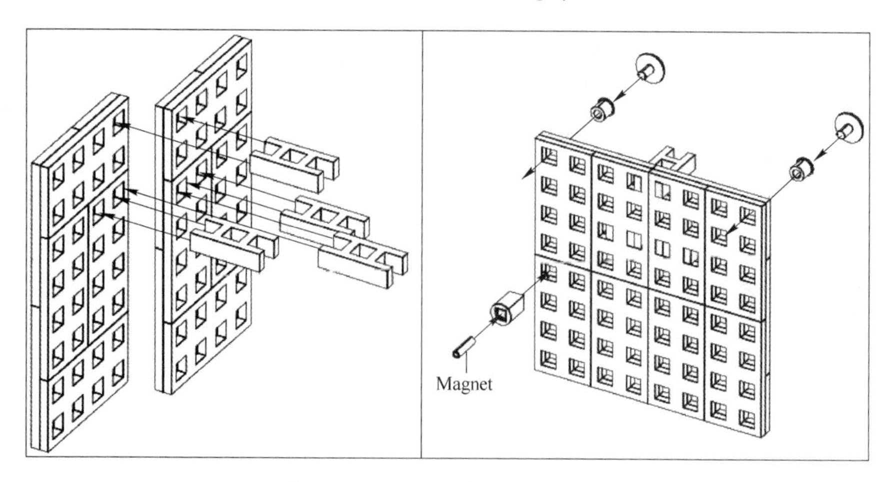


Figure 2-13 Install oling wheels

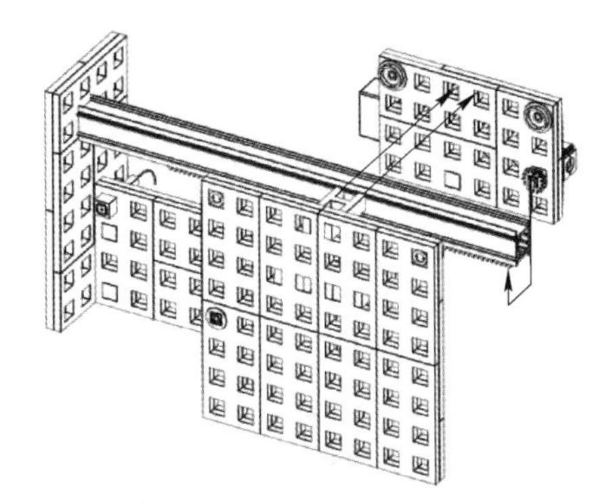

Figure 2-14 Install doors

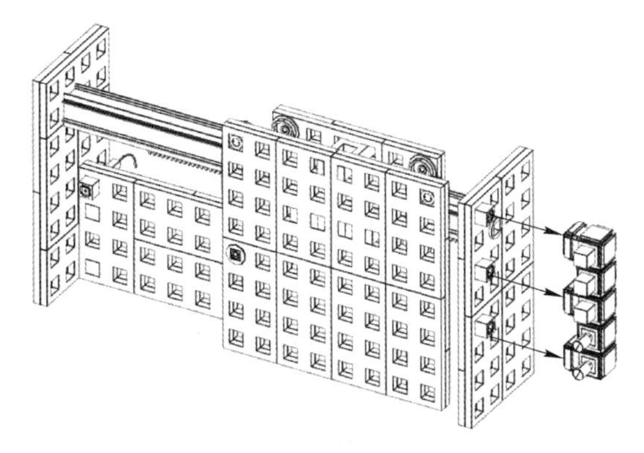

Figure 2-15 Install button and indicator

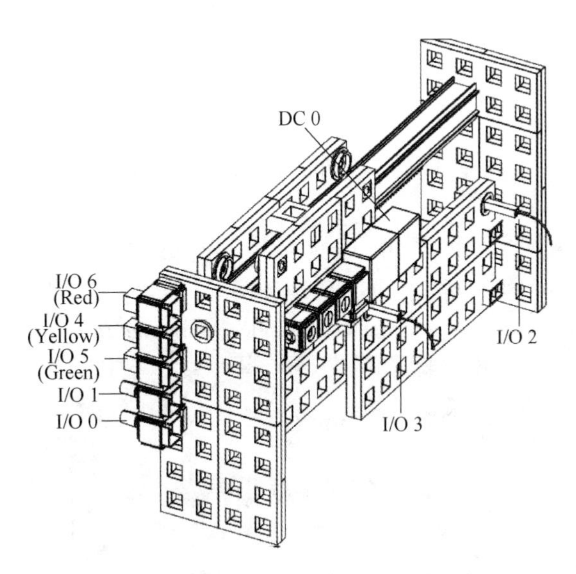


Figure 2-16 Electrical system wiring

#### 2. 1. 3. 4 Program Design

The controlling flow chart of automatic door is shown as Figure 2-17.

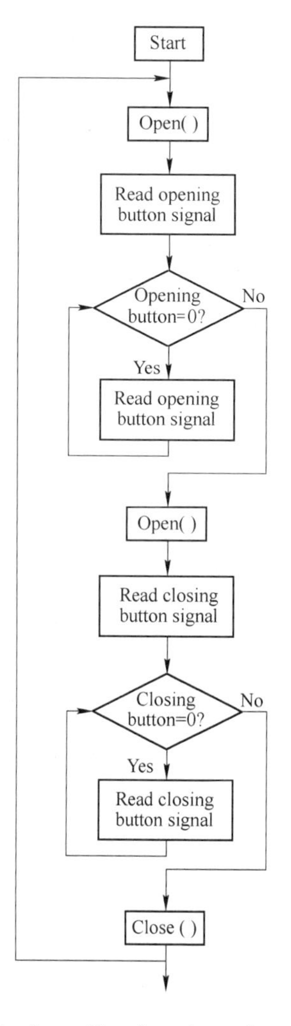


Figure 2-17 Controlling flow chart of automatic door

The new project dialog box is shown in Figure 2-18, the flowchart programming interface shown in Figure 2-19, the new built subprogram dialog box shown in Figure 2-20, the flowchart interface of subprogram shown in Figure 2-21, and the opening door subprogram shown in Table 2-2.

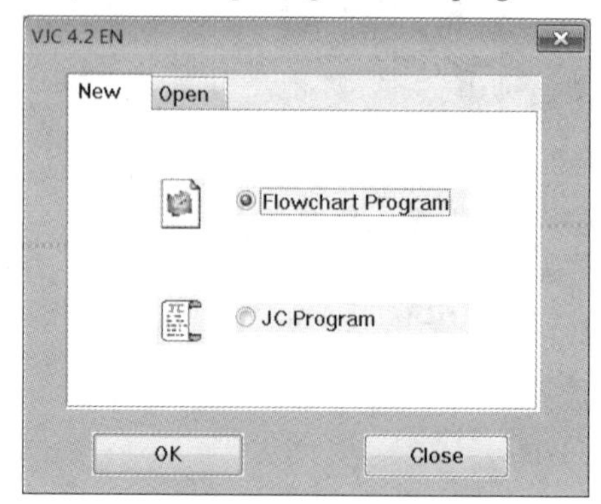


Figure 2-18 New project dialog box

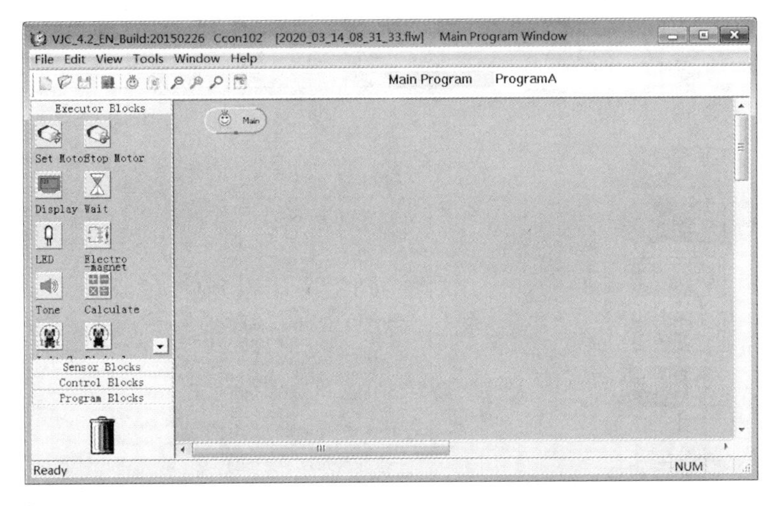


Figure 2-19 Flowchart programming interface

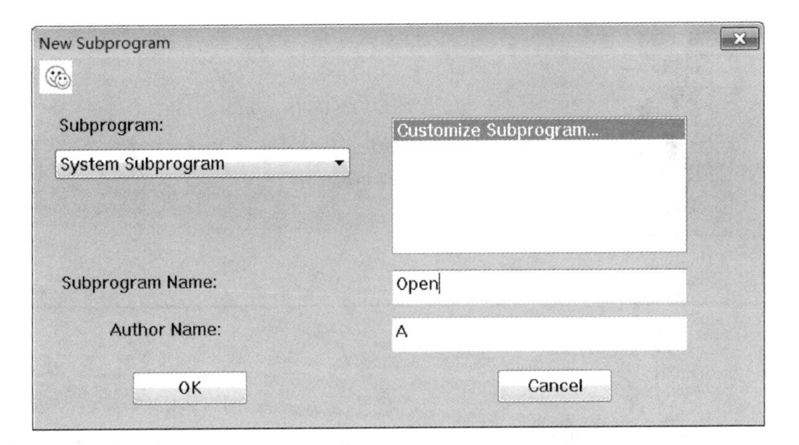


Figure 2-20 New built subprogram dialog box

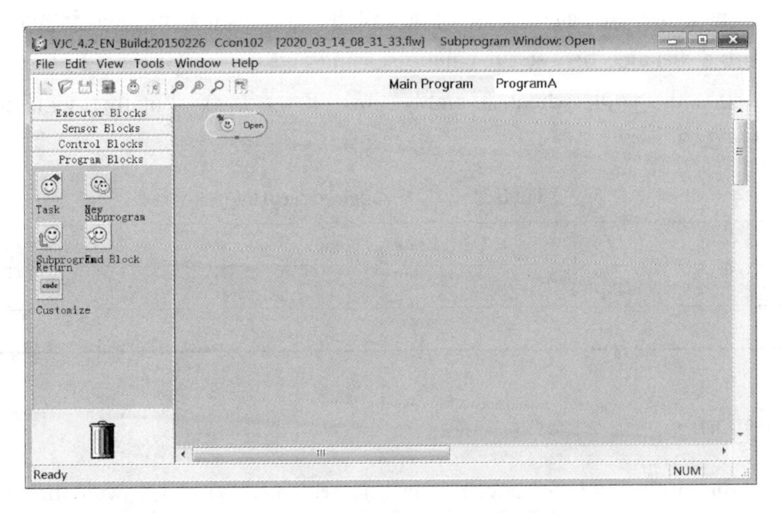


Figure 2-21 Flowchart interface of subprogram

**Table 2-2 Opening door subprogram**

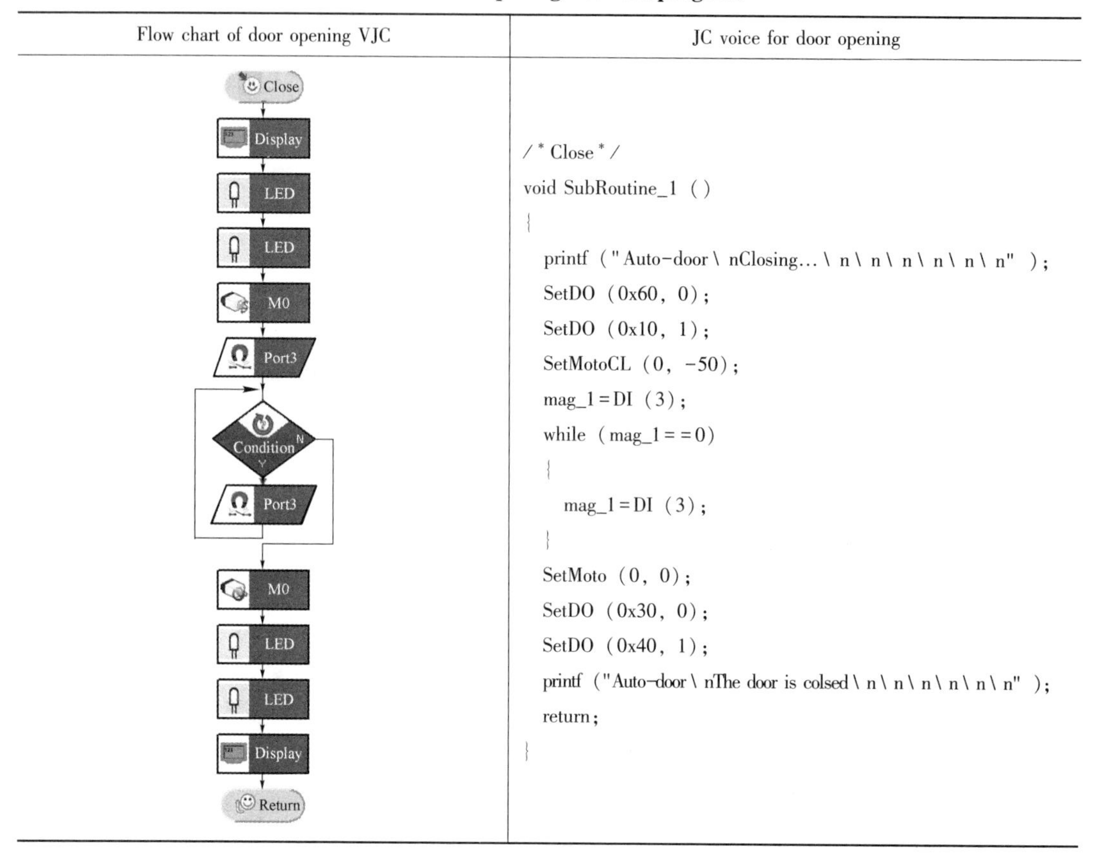

| Flow chart of door opening VJC | JC voice for door opening |
| --- | --- |



```
/* Close */
void SubRoutine_1 ()
{
  printf ("Auto-door\nClosing...\n\n\n\n\n");
  SetDO (0x60, 0);
  SetDO (0x10, 1);
  SetMotoCL (0, -50);
  mag_1 = DI (3);
  while (mag_1 == 0)
  {
    mag_1 = DI (3);
  }
  SetMoto (0, 0);
  SetDO (0x30, 0);
  SetDO (0x40, 1);
  printf ("Auto-door\nThe door is colsed\n\n\n\n\n");
  return;
}
```

### 2.1.3.5 Debugging Records

Test each port in turn by using the VJC's online testing function, to ensure the ports correspond correctly and the components work properly. When motor moves forward, it comes the closing action. If there's no response on magnetic sensing switch, the testing distance is too far generally, and could adjust the distance between magnetic sensing and magnet. If the 12 gear below the rack slips, check whether the guide wheel assembly slips off the beam. Finally, write the debugging records into Table 2-3.

**Table 2-3 Debugging records**

| Testing iterm | Testing process record | Testing Person |
| --- | --- | --- |
| | | |
| | | |

## 2.1.4 Evaluation

After completing the task, the automatic door task is evaluated. The task evaluation table is shown in Table 2-4.

Table 2-4　Task evaluation

| No. | Evaluation content | Achievement ratio | Self evaluation | Teacher evaluation |
|---|---|---|---|---|
| 1 | Principle of condition module | 10 | | |
| 2 | Principle of subprogram module | 10 | | |
| 3 | Construction of automatic door | 30 | | |
| 4 | Automatic door programming | 30 | | |
| 5 | Automatic door debugging | 20 | | |

## 2. 1. 5　Development

Can we add several components to further rich the function of automatic door? And think about how to realize it.

**Think and Practice**

(1) Can we increase the width of the door and use two sliding door?

(2) Can we increase alarm function? For example, when the door opens over 1min, the yellow light will flash to alarm.

(3) Can the automatic door immediately stop closing if it tests someone being going to go into or out of door during the closing process of the door?

(4) Can it count the amount of people passing through the door?

(5) Can it further separately count the amount of people going into an out of the door?

**Design Two Sliding Doors by Yourself**

(1) Design two parallel sliding doors by yourself.

(2) Use a timer to set time. When the door opens over 1min, the signal is sent, and the yellow light flash is let. And think how to make it flash.

(3) Add testing in closing subprogram, when it detects someone passing, stop output immediately.

(4) Add a sensor to test the number of people going into and out of the door.

(5) Add a sensor, judge whether person want to go into or out of the door according to the action sequence of two sensor (shown in Figure 2-22).

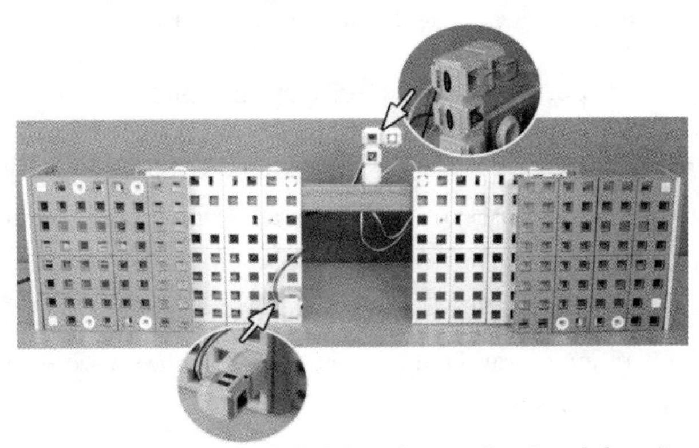

Figure 2-22　Two parallel sliding doors with infrared detection

# Task 2.2 Build and Debug Intelligent Elevator Control System

## 2.2.1 Introduction

A long time ago, people began to use the original lifting tools to transport people and goods, and most of them used human or animal power as the driving force. At the beginning of the 19th century, with the development of the industrial revolution, steam engine became an important source of power. Elevator has been out for more than 100 years. From the earliest simple, unsafe and uncomfortable elevator to today, it has experienced numerous improvements and improvements, and its technical development is endless. In the 21st century, with the contradiction between population and usable land area further intensified, multi-purpose and full-function high-rise tower buildings will be vigorously developed, and ultra-high speed elevator will continue to be the research direction.

## 2.2.2 Related Knowledge

### 2.2.2.1 Intelligent Elevator

The equipment for competition is of high simulation and is proportionally micro-designed on the basis of actual elevators. They are mainly composed of mechanical and electrical systems fitted with independent standard electric control systems; The mechanical system is composed of driving system, elevator car and counterweight device, guiding system, floor door and elevator door, door open-close system, and mechanical safety system. The electrical control system is mainly composed of parts for pulling control, use and operation, shaft information collection, safety protection, etc. The control system comprises programmable controller, frequency converter, sensor, motor driving and low voltage electrics. In addition, they are also equipped with video monitor, video display, fire extinguisher and telephones. The functions of button control, signal control, centralized selection control, man-machine interaction and so on can be realized, and intelligent group control, remote control and fault setting and diagnostic inspection can be checked, which represent the currently main stream of elevator technology in the world.

### 2.2.2.2 Definition and Classification of Intelligent Elevator

**Definition of Elevator**

Elevators are of vertical lifting mechanism with motors as driving force, equipped with a box like cars and used in multistory buildings for carrying people and goods, and they are fixed lifting equipment serving for specified floors. There are other elevators in use, such as those of step type, with footboards fixed on a running crawler track, commonly known as escalators.

**Classification of the Intelligent Elevator**

The intelligent elevator can be classified from different angles. Since the basis for classification differ from each other, the contents related are different, which are as follows:

(1) Classification according to their uses: In this way, the intelligent elevator can be classified into 9 types: passenger elevator, goods elevator, medical elevator, lumber elevator, sight-seeing elevator, vehicle elevator, ship elevator, construction site elevator and special elevator.

(2) Classification according to the way of driving: In this way, the elevator can be divided into several groups: AC elevator, DC elevator, hydraulic elevator, gear and gear rack elevator, and screw elevator.

(3) Classification according to the way of control: In this way, the elevator can be divided into 8 types: handle on-off switch control insidecar, button control, signal control, centralized selection control, lower centralized (selection) control, parallel control, group control and intelligent group control.

### 2.2.2.3 Main Brands of Elevator in the World

Common elevator brands include OTIS, Schindler, KONE, GUANGRI, Mitsubishi, FUJI Elevator, HITACHI, TOSHIBA, ThyssenKrupp, etc.

### 2.2.2.4 Structure and Composition of Intelligent Elevator

An elevator is a kind of electro-mechanical product. Its mechanical part is just like man's body; its electric part like man's nerve, and its control part like man's brain. The mechanical part and the electrical part can be operated and closely coordinated through control systems, to make sure the elevator's normal running. Although there are many kinds of intelligent elevators, most of the intelligent elevators currently in use are electric driven types and are of steel rope dragging structure, as shown in Figure 2-23.

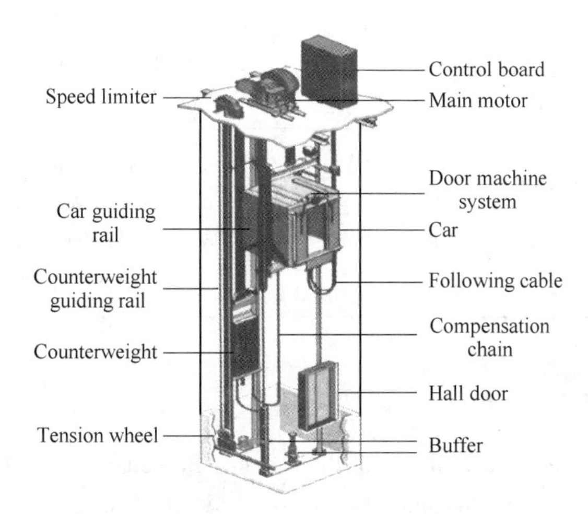


Figure 2-23 Structure of elevator

In terms of the elevators' space positions, there are four parts: building attached machine house, shaft, space for carrying passengers and goods (car); places for passengers or goods getting in or out of the car (floor stop), i.e. machine house, shaft, car and floor stop.

In terms of the function of each part of the elevator, the elevator can be divided into eight parts: dragging system, guiding system, car system, door system, weight balancing system, electric dragging system, electric control system and safety protection system. Functions of each part of elevator is shown in Table 2-5.

**Table 2-5 Functions of each part of elevator**

| System | Function | Main parts and installations |
|---|---|---|
| Dragging system | Output and transmit power, drive elevators | Dragging machine, steel dragging rope, guiding wheel and diversion sheave |
| Guiding system | Limit freedom of car and counterweight | Guiding rail and its support |
| Car system | For carrying passengers and goods | Car and its support |
| Door system | Entries and exits for passengers and goods; the car and floor doors must be closed while moving, and opened only upon arriving | Car door, hall door, door machine and door lock |
| Weight balancing system | Help in balancing the weight of car and eliminate the influence of the length of dragging | Counterweight and compensation chain |
| Electric dragging system | Supply power and control the speed of elevator | Electric motor, power supply system, speed feedback device and speed regulator device |
| Electric control system | Operate and control elevator's running | Control board, leveling unit, operation box, calling box and controller |
| Safety protection system | Ensure the safe operation of elevator and prevent any accident endangering human life | Speed limiter, safety clamp, buffer, end stop safety installation, over speed protection, open and reversed phase protector, upper and lower limit protector and door interlock |

## 2.2.3 Implementation

Simulate an elevator system for a building of three floors. Passengers on each floor of them can enjoy the 'Calling' service. And when the elevator reaches the floor a light indicator should be on.

### 2.2.3.1 Project Design

The Project design including system, function and simulation project is shown in Table 2-6.

**Table 2-6 Project design**

| System | Function | Simulation project |
| --- | --- | --- |
| Dragging System | Output and transmit power drive elevators | Simulation with cotton thread |
| Guidance system | Limit the freedom of car and counterweight | Simulation of guide wheel |
| Car | Components for transporting passengers and goods | Plate and beam construction |
| Electric drive system | Provide power and speed control for elevator | Motor and gearbox simulation |
| electric control system | Operation and control of elevator | Switch, magnetic sensor, etc. |

### 2. 2. 3. 2 Material Preparation

The list of materials of materials for supporting projects are shown in Figure 2-24.

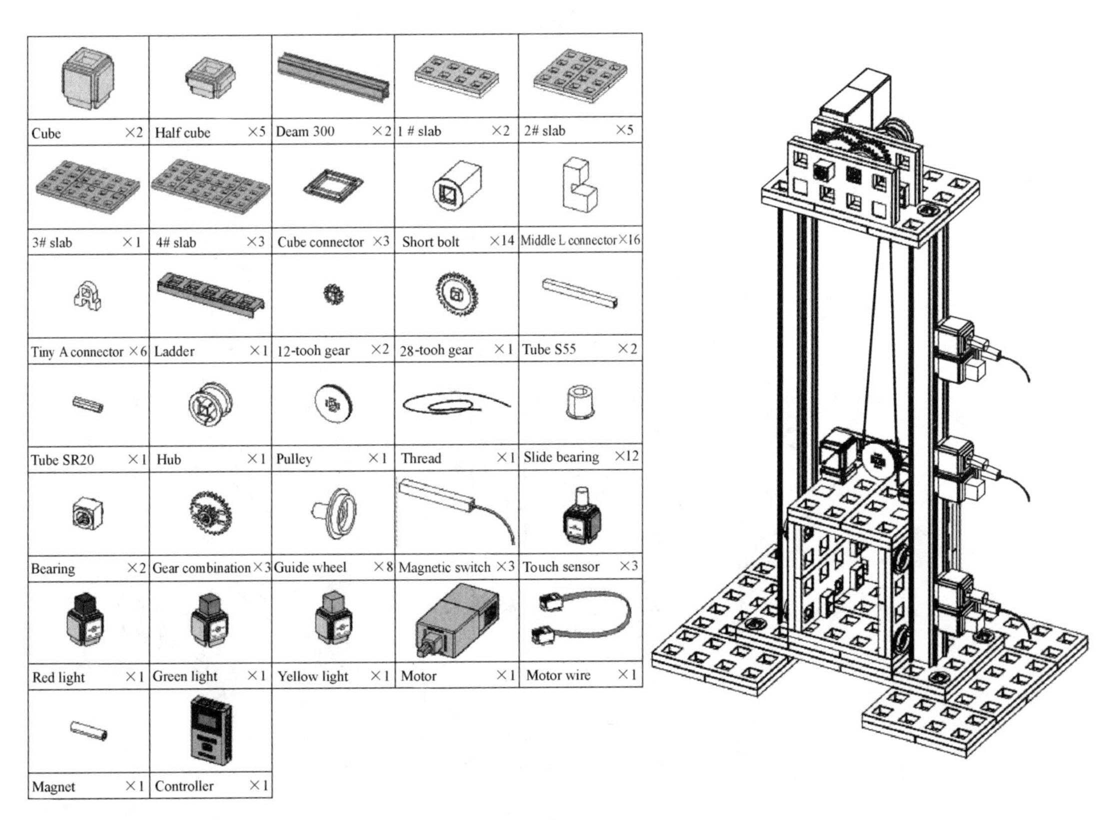


Figure 2-24 Intelligent elevator

### 2. 2. 3. 3 Hands-on Construction

The models construction of a intelligent elevator are shown in Figures 2-25 to 2-30 respectively.

### 2. 2. 3. 4 Program Design

The flow chart of the intelligent elevator is shown in Figure 2-31.

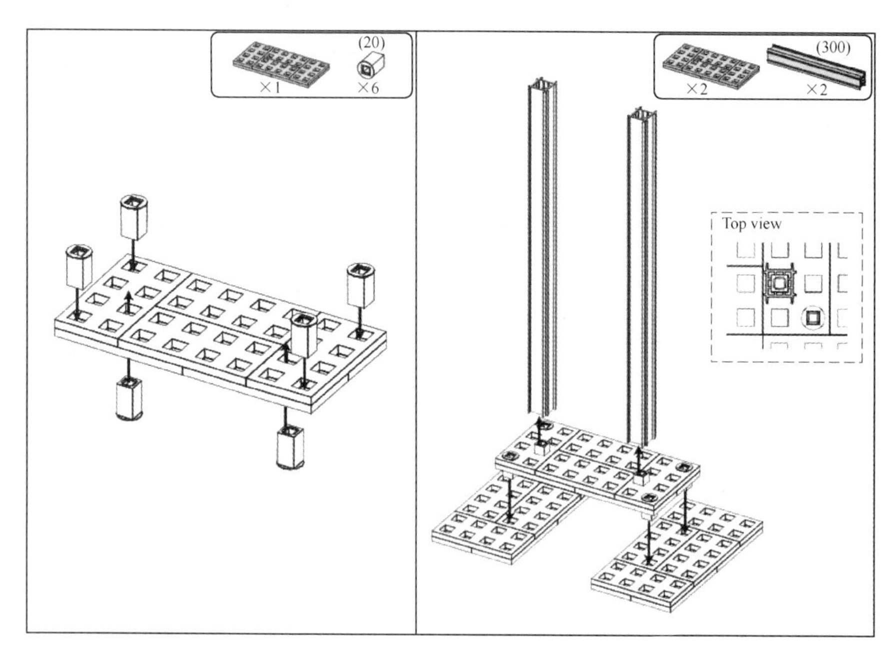

Figure 2-25 Construction of the support

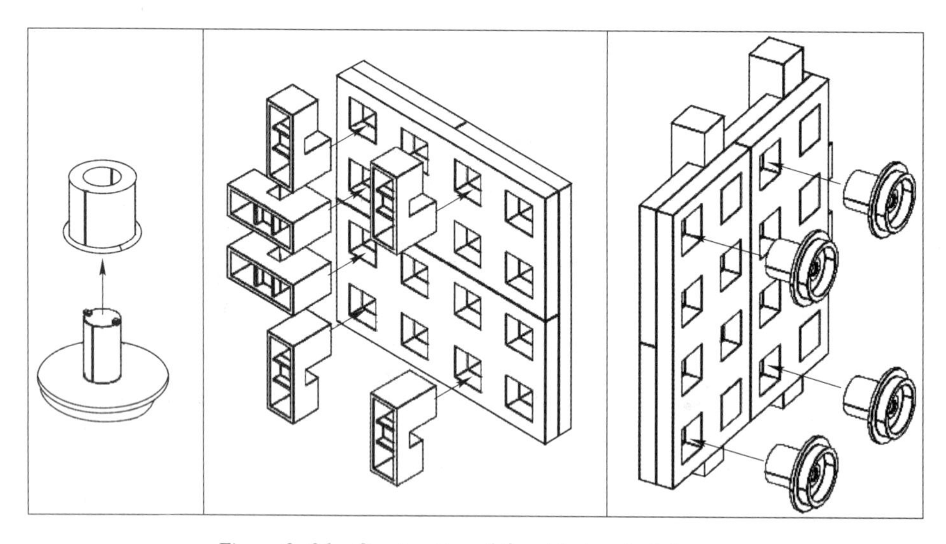
Figure 2-26 Construction of the side board of the car

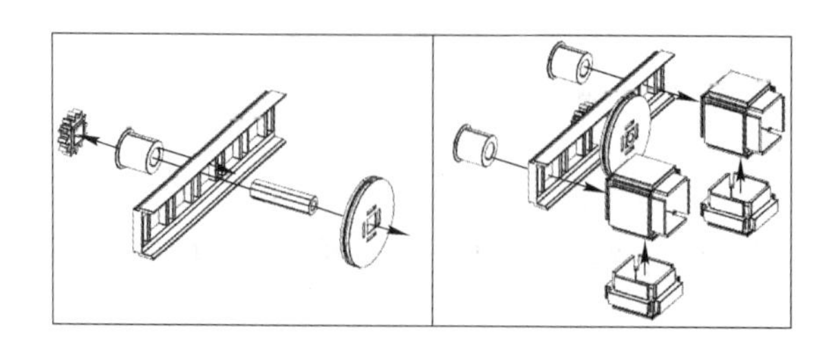
Figure 2-27 Construction of the dragging mechanism for elevator

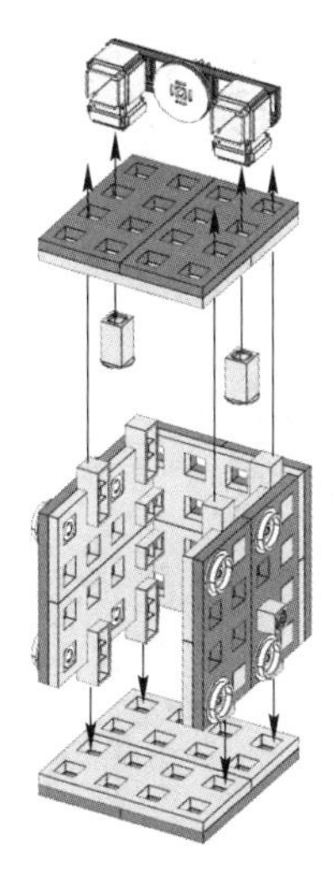

Figure 2-28  Construction of the elevator's car unit

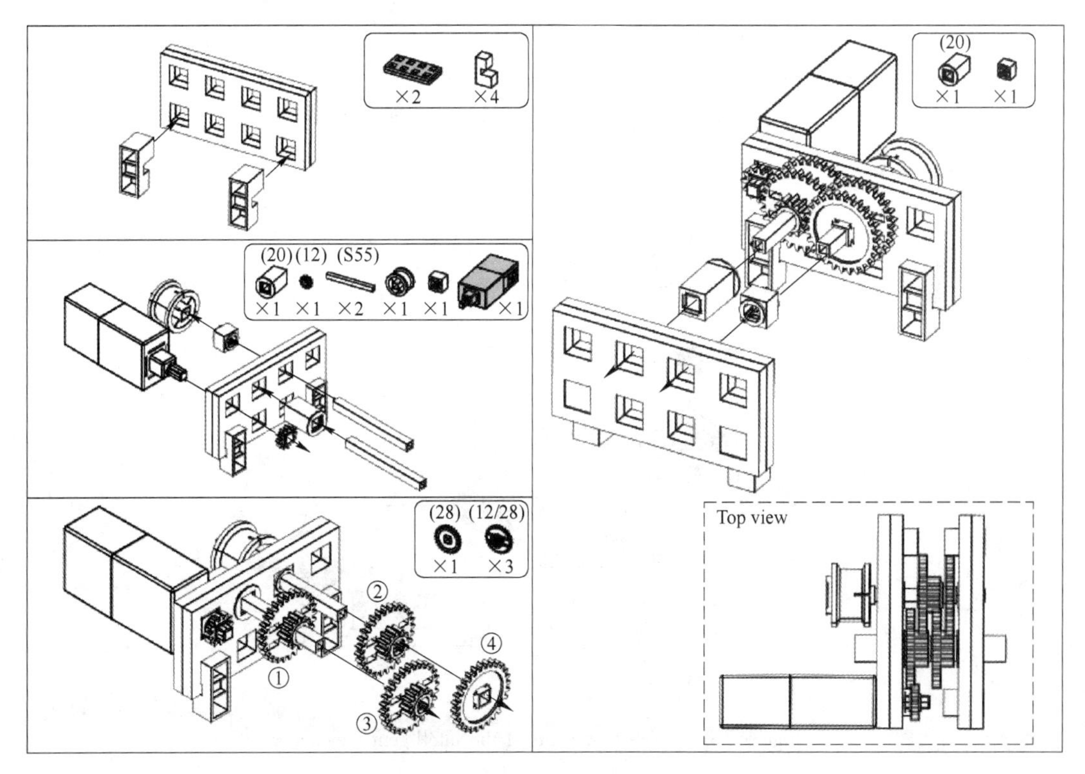


Figure 2-29  Construction of the driving system for elevator

### 2. 2. 3. 5  Debugging Records

The debugging records are shown in Table 2-7.

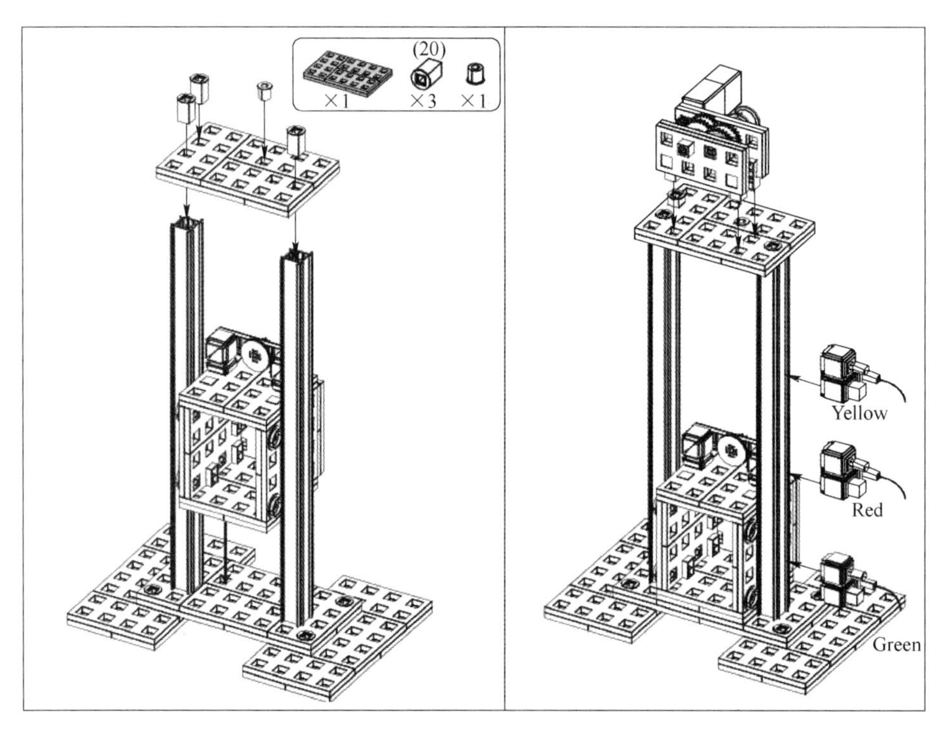


Figure 2-30  The all parts into an elevator

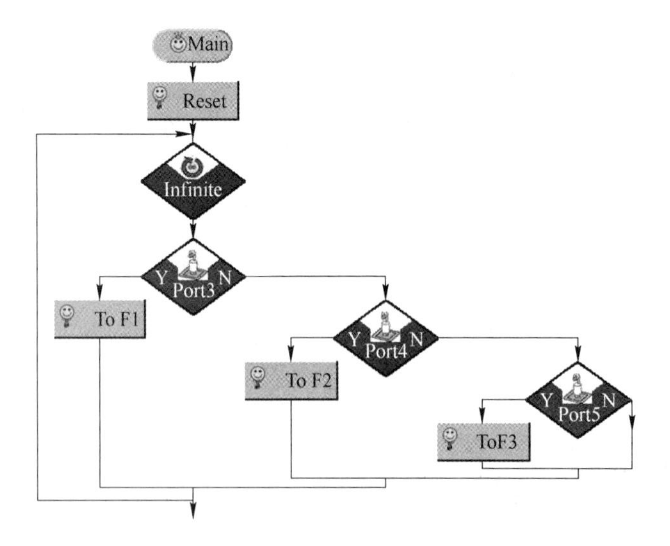


Figure 2-31  The flow chart of the intelligent elevator

**Table 2-7  Debugging records form**

| Testing iterm | Testing process record | Testing person |
|---|---|---|
| | | |
| | | |

### 2. 2. 4 Evaluation

The task evaluation is shown in Table 2-8.

**Table 2-8 Task evaluation**

| No. | Evaluation content | Achievement ratio | Selfevaluation | Teacher evaluation |
|---|---|---|---|---|
| 1 | Application of magnetic sensor | 10 | | |
| 2 | Application of subprogram module | 10 | | |
| 3 | Construction of elevator | 30 | | |
| 4 | Elevator programming | 30 | | |
| 5 | Elevator commissioning | 20 | | |

### 2. 2. 5 Development

(1) How to achieve a four-story elevator structure?

(2) Whether the floor to be reached can be selectod in the car.

(3) How to dispatch control for multiple elevators in parallel?

## Task 2. 3 Build and Debug AGV Trolley Control System

### 2. 3. 1 Introduction

The world's first automated guided vehicle (AGV) was successfully developed in the early 1950s. It is a traction trolley system that can be easily connected to other logistics systems automatically and significantly increase labor productivity, and greatly improve the degree of automation of loading, unloading and handling. In 1954, an electromagnetically guided automatic guided vehicle was developed. Figure 2-32 shows the picture of the car. This task completed the AGV car design and program design, focusing on fostering independent innovation capabilities.

Figure 2-32 AGV physical picture

### 2.3.2 Related Knowledge

#### 2.3.2.1 Definition of AGV Trolley

The AGV trolley is an automatic guided vehicle that is an industrial vehicle that loads goods by automatic or manual means. It moves on the set route or pulls a truck to a specified location, and then loads the goods automatically or manually. AGV is a battery-operated industrial vehicle with automatic operation.

#### 2.3.2.2 Classification of AGV Trolley

According to different guiding principles, automatic guided trolleys can be divided as follows:

(1) Externally guided: A guidance information medium (such as a wire, a tape, a ribbon, etc.), is set on the vehicle's running route, and the guidance information (such as frequency, magnetic field intensity, light intensity, etc.) of the line medium is received by the guidance sensor on the vehicle. The vehicle is controlled to follow the correct route.

(2) Self-guided: Free path guidance, the coordinate positioning principle is used, and the coordinate information of the running route is set on the vehicle in advance. When the vehicle is running, the actual vehicle position coordinates are measured in real time, and then the two are compared to control the vehicle's guidance operation.

#### 2.3.2.3 Structure and Composition of AGV Trolley

AGV car, as shown in Figure 2-33, is usually composed of four subsystems: car body system, vehicle control system, walking device, and safety and assistance system, which are as follows:

(1) Car body system: Including the chassis, frame, housing, driving device, steering mechanism and control room, etc., the car body system is the body of the AGV, and it has the basic characteristics of electric vehicles. The frame is usually a steel structure, which requires a certain strength and stiffness. The drive device is composed of wheels, reducer, brake, motor and governor, etc. It is a servo-driven speed control system. The drive system can be controlled by computer or manual. It can drive AGV and has speed control and braking capabilities. Through the steering mechanism, the AGV can realize all-round movements forward, backward, longitudinal, lateral, oblique and turning.

(2) Vehicle control system: The on-board control system is the core of the AGV, and is gen-

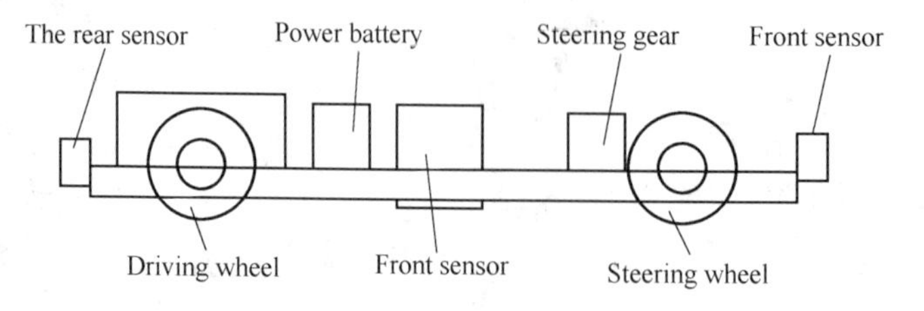


Figure 2-33 Composition of AGV trolley

erally composed of a monitoring system, a navigation system, a communication system, a motor driver, and a control panel.

The guidance system is an important part of AGV. The most basic technical requirement is to reliably guide the AGV from point A to point B, avoiding known obstacles. With different guidance methods, the guidance system has different components. There are currently 10 common guidance methods: ground chain traction guidance, electromagnetic induction guidance, tape guidance, inertial guidance, infrared guidance, laser guidance, optical guidance, teaching guidance, visual guidance, and GPS (Global Positioning System) guide.

(3) Traveling device: The traveling device is generally composed of a driving wheel, a driven wheel and a steering mechanism, also including three-wheel, four-wheel, six-wheel and multi-wheel, etc. The three-wheel structure generally adopts front-wheel steering and driving, and adopts two-wheel driving, four-wheel, six-wheel differential steering or independent steering. In order to improve the positioning accuracy, the drive and steering motors use DC servo motors.

(4) Safety and auxiliary systems: In order to avoid AGV collision during operation, the safety of personnel and other devices have been protected. AGVs have safety measures such as obstacle detection, collision avoidance, siren, warning and emergency stop. Generally, AGV adopts multi-level hardware and software security monitoring measures.

#### 2.3.2.4 Sensors and Principles

**Grayscal Sensor**

This task requires the use of sensors for guidance. The following describes the grayscale sensors used in the POWERON. The gray-scale sensor uses different colors to detect light with different degrees of reflection. The photoresistor detects the color depth of the principle that the light returned by different detection surfaces has different resistance values. It is used to distinguish black from other colors when the ambient light interference is not very serious. It also has a relatively wide operating voltage range, and it can still work normally in the case of large power supply voltage fluctuations. The sensor outputs a continuous analog signal, so it can easily determine the reflectance of an object through an A/D converter or a simple comparator, which is a practical robot line-tracking sensor. Gray sensor is shown in Figure 2-34.

Figure 2-34 Gray sensor

The grayscale sensor is an analog sensor with a light-emitting diode and a photoresistor installed on the same surface. The gray-scale sensor uses different colors to detect light with different degrees of reflection. The photoresistor detects the color depth of the principle that the light returned by different detection surfaces has different resistance values. Within the effective detection distance, the light emitting diode emits white light and irradiates the detection surface. The detection surface reflects part of the light. The photoresistor detects the intensity of this light and converts it into a signal that the robot can recognize.

**Adjustment Method**

Here is no signal indicator on the grayscale sensor, but it is equipped with an analog size adjuster that detects the color return. When you want to detect a given color, you can place the transmitting/receiving head at the given color and cooperate with the regulator to call up the appropriate return analog quantity. The methods are as follows:

The adjuster counterclockwise is rotated to increase the analog value. The adjuster clockwise is turned to decrease the analog value, and you can adjust it until you need the value. If an accurate analog quantity is needed, it can be displayed on the LCD screen by a program, and the accurate analog quantity can be called up with the regulator.

Note: When turning the adjuster with a screwdriver, do not turn it too fast and do not turn it too hard to prevent it from being damaged. When you find that it cannot be turned, stop it immediately.

**Precautions**

The precautions for grayscale sensors are as follows:

(1) The material of the detection surface will also cause a difference in its return value.

(2) The intensity of external light has a great influence on it, which will directly affect the detection effect. When detecting specific items, pay attention to packaging sensors to avoid interference from external light.

(3) According to its working principle, the light sensor determines the color depth of the detection surface with the light intensity reflected from the detection surface. Therefore, the accuracy of the measurement is directly related to the distance from the sensor to the detection surface. The vibration of the body while the robot is in motion will also affect its measurement accuracy.

### 2.3.3 Implementation

The next task implementation link uses the POWERON to simulate the AGV car and programs the simulation operation to design an AGV car.

#### 2.3.3.1 Projet Design

As shown in Figure 2-35, the AGV trolley adopts two-wheel differential drive, the rear wheels are active, and the front wheels are guided. In the form of optical guidance, two gray sensors are installed on the left and right of the car.

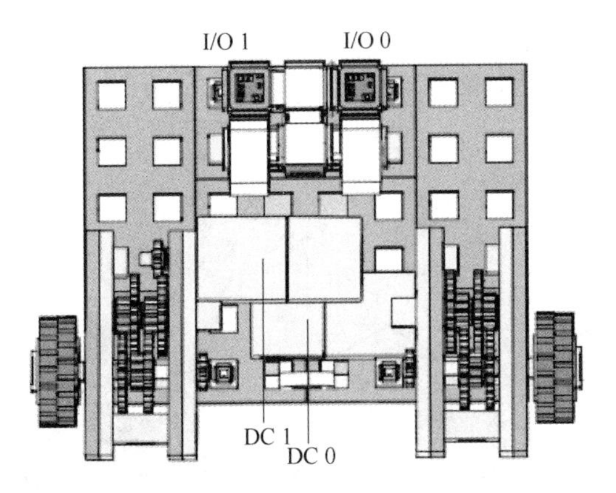


Figure 2-35　AGV trolley

### 2. 3. 3. 2　Material Preparation

The material of designing AGV trolley is shown in Figure 2-36.

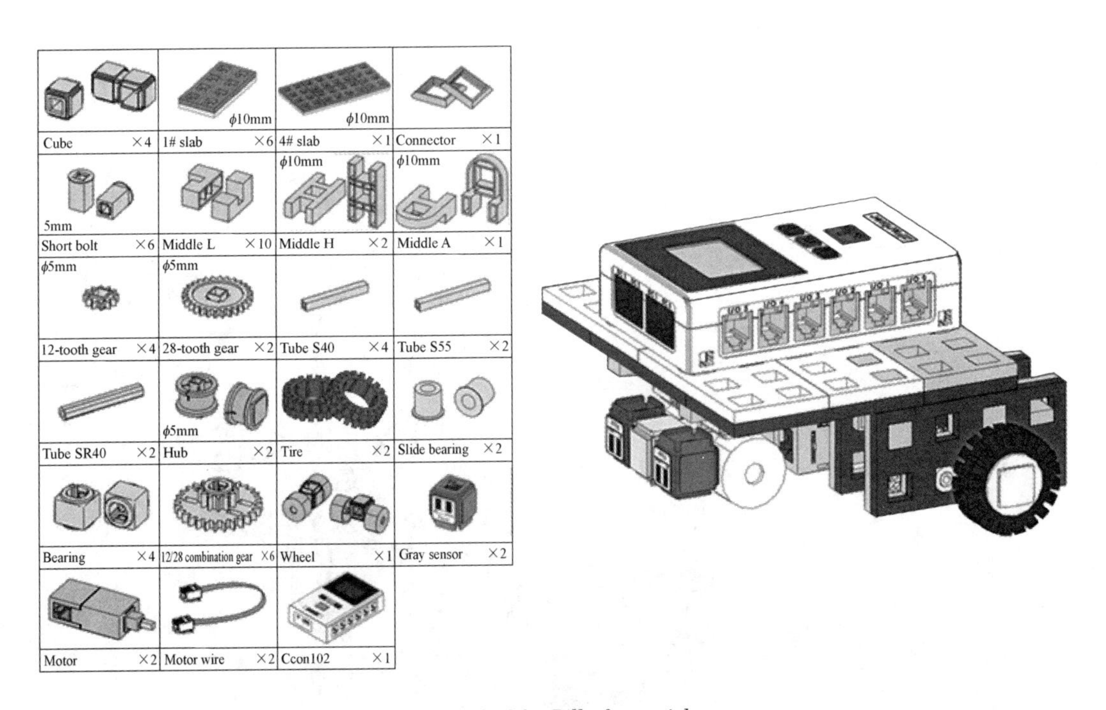


Figure 2-36　Bill of materials

### 2. 3. 3. 3　Hands-on Construction

Next, some key steps are introduced in the construction of the trolley to help students complete the hardware splicing of the trolley. And the models of key steps are shown in Figures 2-37 to 2-40 respectively.

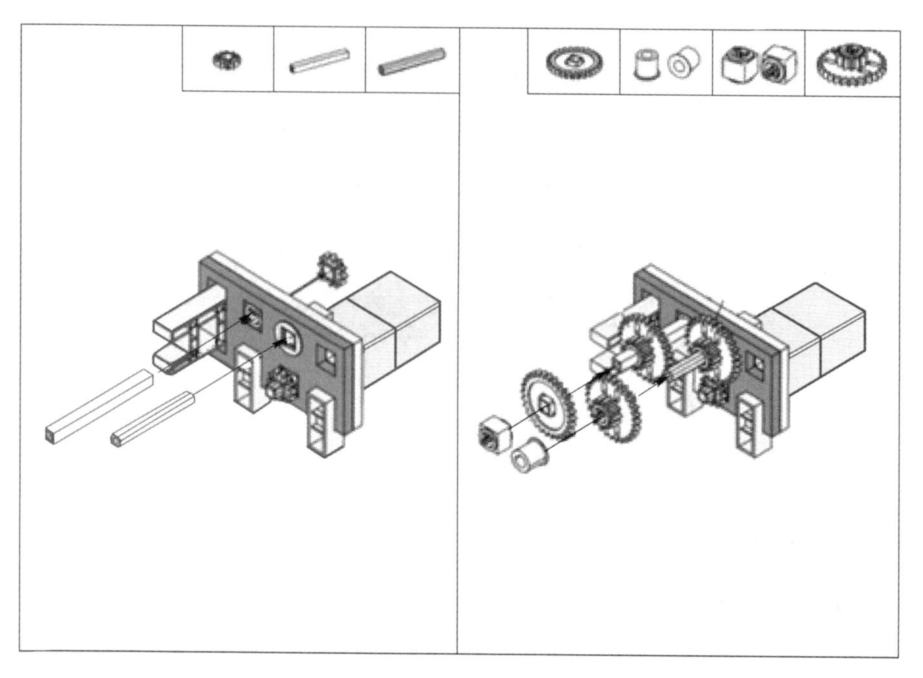

Figure 2-37　Gear and motor construction

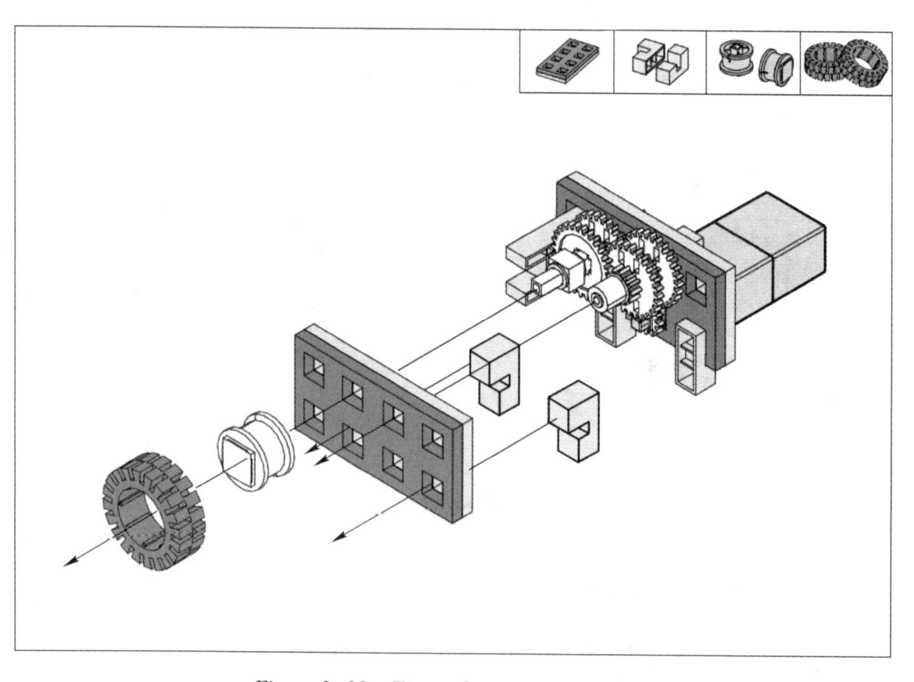

Figure 2-38　Tire and motor construction

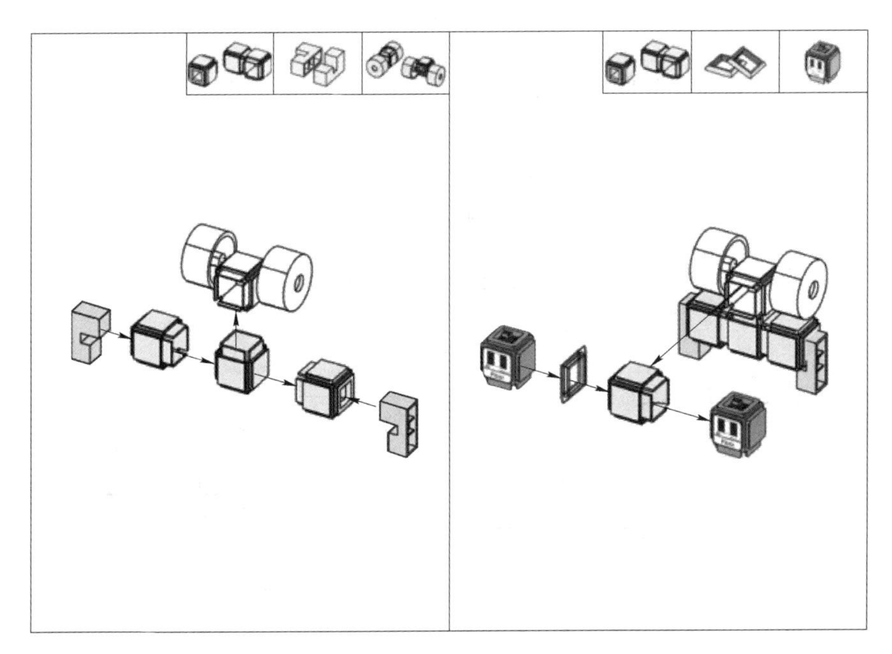

Figure 2-39　Grayscale sensor setup

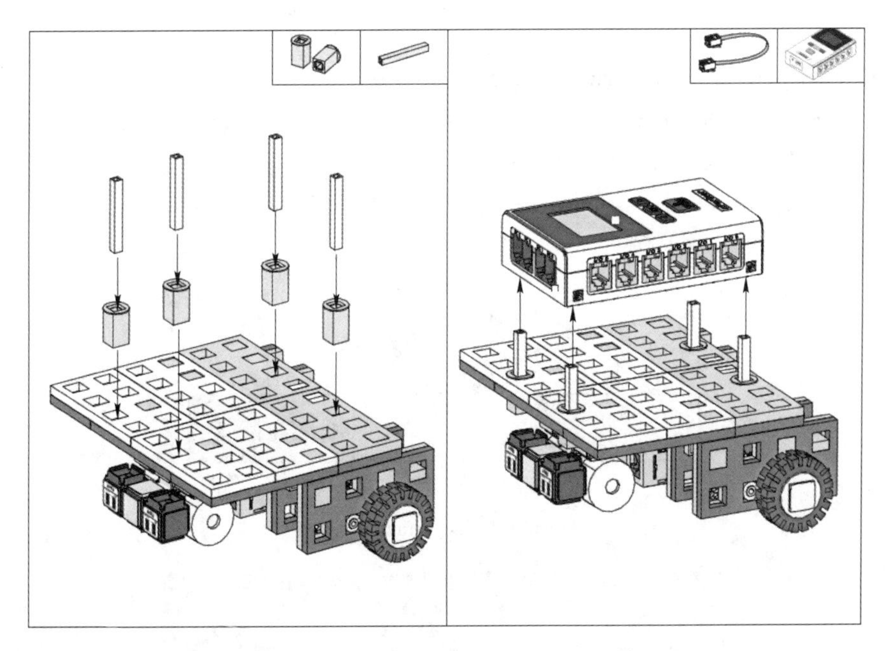

Figure 2-40　AGV controller setup

### 2.3.3.4　Program Design

**Control Requirements**

AGV car control requirements are autonomously guided car that can move along the black line on the whiteboard to achieve the guidance function. The program design block diagram is shown in Figure 2-41.

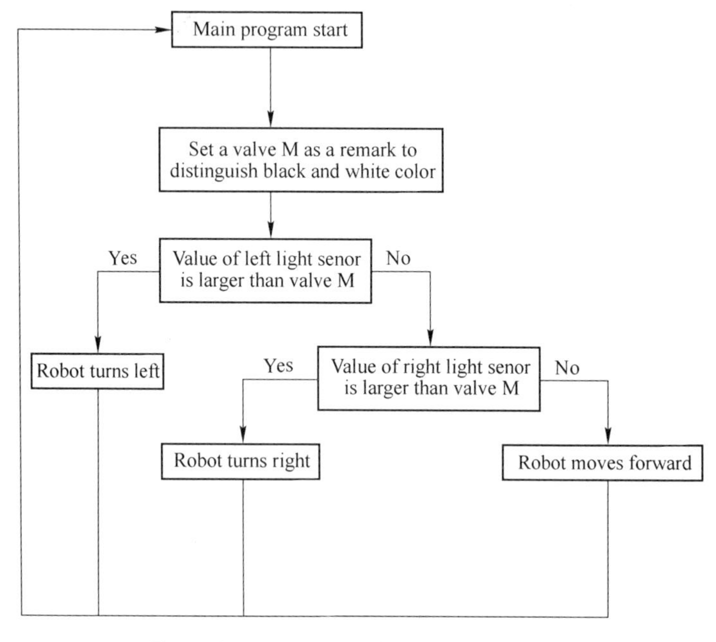


Figure 2-41 Program design block diagram

## Usage and Characteristics of Gray Sensor

Using the grayscale detection module in the software, you can make a judgment based on the grayscale value. The judgment of the black and white limit is completed, and the steering and speed of the motor are setted, as shown in Figures 2-42 to 2-44.

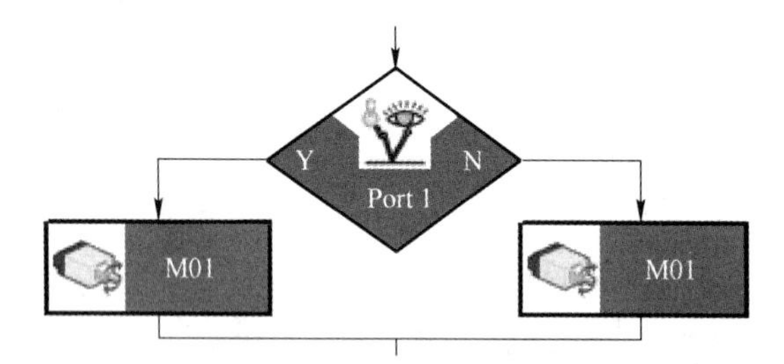


Figure 2-42 Gray sensor

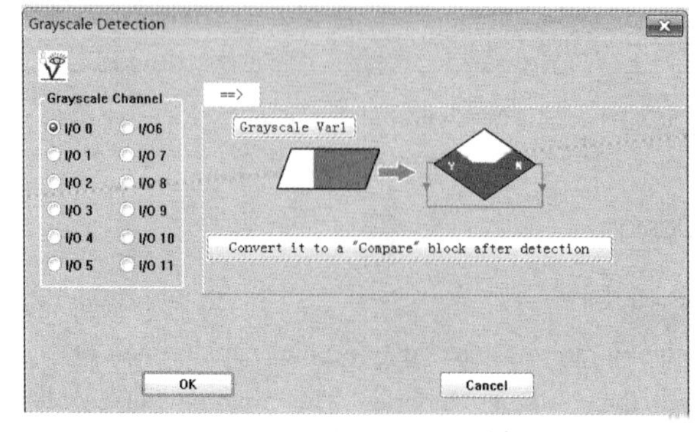


Figure 2-43 Grayscale sensor debugging

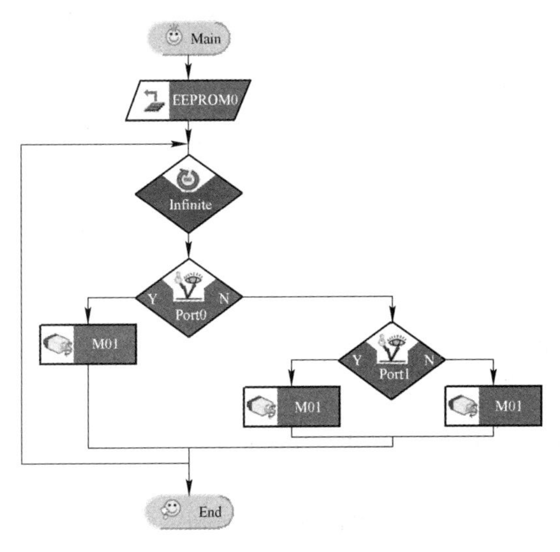

Figure 2-44　Reference procedure diagram

### 2.3.3.5　Commissioning Records

Test the DC0 and DC1 ports of the motor, pay attention to the positive and negative rotation of the motor, place the grayscale sensor on the black and white positions, and read the value. Then obtain the critical values for the black and white judgment of the left and right grayscale, and assign them to the two Integer variables. Finally, write the datas into the table, as shown in Table 2-9.

**Table 2-9　Commissioning records**

| Testing iterm | Testing process record | Testing person |
|---|---|---|
| | | |
| | | |

## 2.3.4　Evaluation

The table of task evaluation is shown in Table 2-10.

**Table 2-10　Task evaluation**

| No. | Evaluation content | Achievement ratio | Selfevaluation | Teacher evaluation |
|---|---|---|---|---|
| 1 | AGV guiding principle | 10 | | |
| 2 | Principle of grayscale sensor | 10 | | |
| 3 | Construction of AGV trolley | 30 | | |
| 4 | AGV car programming | 30 | | |
| 5. | Commissioning of AGV trolley | 20 | | |

By knowing AGV, the application, definition and classification of AGV are understood. The structure and common components of typical AGVs are also understood, and the basic composition of AGVs are commonly used. The AGV detection system and control system are analyzed, the AGV automatic guidance principle is explained, the AGV power system is described, familiar with the control and communication methods and safety measures of the AGV, the ability source kit is used to build and debug the AGV car control system, and the grayscale sensor construction and programming methods, the use and precautions of motors, gears and other kit parts are mastered.

### 2.3.5 Development

(1) Try to make a different auto-guided car, find information and see more information about auto-guided car.

(2) Check the information to understand common AGV models and more of AGV's core technology.

(3) Master the engineering work method through AGV construction, and think about what other types of AGVs can be built using a source of innovation curriculum package.

(4) If there is an obstacle in front of the AGV guide path, how can the AGV detect it and respond accordingly?

(5) Can you add the function of handling goods to the AGV trolley?

(6) What is the role of the internal gear of the trolley? and how do your calaulate the transmission ratio?

## Task 2.4 Build and Debug Industrial Robot Control System

### 2.4.1 Introduction

In the industrial production line, a robot is needed to sort the two color bottles from the production line, and put them on different production lines according to two colors. To build an industrial robot using the POWERON, it is required that the industrial robot has one degree of freedom of rotation and two degrees of freedom of straight line. The sorting bottles can be moved from one place to another.

### 2.4.2 Related Knowledge

The industrial robot is the most widely used automatic mechanical device in the field of robot. It can be used in industrial welding, industrial assembly, industrial handling and other fields. Although their shapes are different, they have a common feature, which can receive instructions and accurately locate a point on the three-dimensional (or two-dimensional) space for operation.

#### 2.4.2.1 Definition and Classification

**Definition**

The definition of industrial robot put forward by US. RIA is that industrial robot is a special device used to carry materials, parts, tools and other programmable multi-functional manipulator, or to complete various work tasks by calling different programs.

In 1987, the International Organization for Standardization (ISO) defined that industrial robot is a programmable manipulator with automatic operation and mobile function, which can complete all kinds of work.

According to the national standard GB/T 12643—1990 in China, industrial robot is a kind of operating machine which can be controlled automatically, reprogrammed repeatedly, multi-function and multi-degree of freedom. It can carry materials, workpieces or operating tools to complete various operations.

No matter which definition, it emphasizes four characteristics of robot:

(1) Bionic characteristics: imitating human body movements.

(2) Flexible features: wide adaptability to operation.

(3) Intelligent features: having the ability to perceive the outside world.

(4) Auto feature: auto complete tasks.

**Classification**

Industrial robots can be divided into different categories according to different classification standards:

(1) According to the movement form of the robot, it can be divided into rectangular coordinate industrial robot, cylindrical coordinate industrial robot, spherical coordinate industrial robot, multi joint industrial robot, plan joint industrial robot and parallel industrial robot, as shown in Figure 2-45.

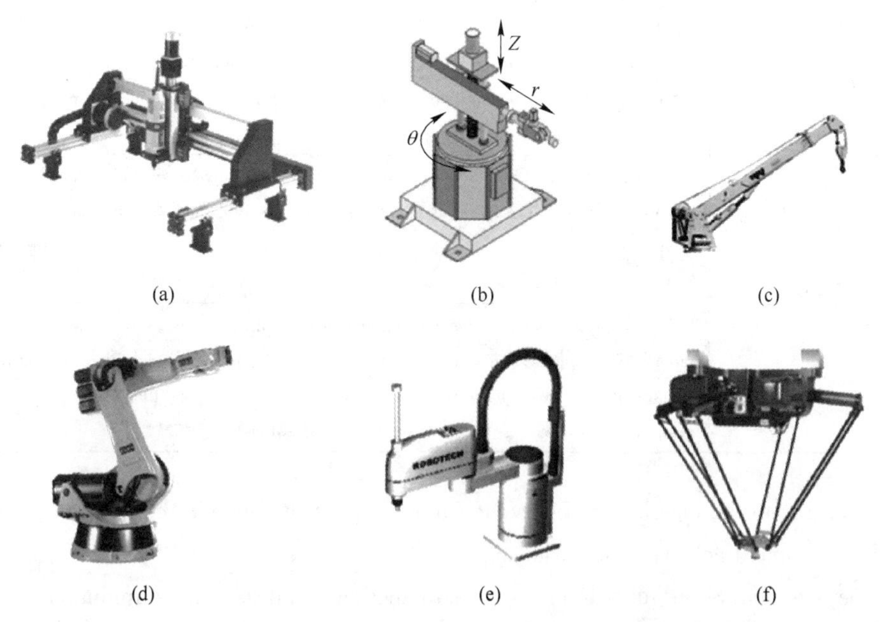


Figure 2-45 Industrial robots classified by movement form

(a) Rectangular coordinate; (b) Cylindrical coordinate; (c) Spherical coordinate; (d) Multi-joint; (e) Plan-joint; (f) Parallel

(2) According to the input information, it can be divided into operation manipulator, fixed program industrial robot, programmable industrial robot, program-controlled industrial robot, teaching industrial robot and intelligent industrial robot, as shown in Table 2-11.

**Table 2-11 Classification by input information**

| Classification | Characteristic |
|---|---|
| Operating manipulator | A manipulator with several degrees of freedom operated directly by the operator |
| Fixed program robot | A manipulator that performs a given task step by step in a predetermined sequence, condition, and location |
| Programmable robot | It is basically the same as the fixed program robot, but its working order and other information are easy to modify |
| Program controlled robot | Its task instruction is provided by computer program to robot, similar to CNC |
| Teaching robot | The action taught by a person can be repeated according to the information stored in the memory device, and the teaching action can be repeated automatically |
| Intelligent robot | Sensors are used to sense the changes of working environment or working conditions, and complete corresponding tasks with the help of their own decision-making ability |

(3) According to the driving mode, it can be divided into hydraulic industrial robot, electric industrial robot and pneumatic industrial robot, as shown in Table 2-12.

**Table 2-12 Classification by driving mode**

| Classification | Characteristic |
|---|---|
| Hydraulic industrial robot | The hydraulic industrial robot has a large capacity of grasping, which can reach thousands of Newtons. This kind of industrial robot has compact structure, stable transmission and sensitive action, but it has high requirements for sealing and is not suitable for use in high or low temperature environment |
| Electric Industrial Robot | At present, the most widely used industrial robot, not only because of the variety of motors, but also because of the use of a variety of flexible control methods. The motor directly drives each shaft, or drives after decelerating through a device such as a reducer, so the structure is very compact and simple |
| Pneumatic industrial robot | Using compressed air to drive the manipulator has the advantages of convenient air source, rapid action, simple structure, low cost and no pollution; the disadvantage is that the air has compressibility, resulting in poor stability of working speed, and the grasping force of such industrial robots is small, generally only tens of Newtons |

(4) According to the classification of motion track, it can be divided into point type industrial robot and continuous track type industrial robot.

Point type is to control the robot from one pose to another, and its path is unlimited; According to the programmed pose and speed, continuous track is, to move the robot on the specified track.

Generally, the industrial manipulator, which belongs to the intelligent, continuous track, multi joint industrial robot and the end tool, is mostly pneumatic or electric control device.

**Core Parameters**

The core parameters are as follows:

(1) Degree of freedom refers to the number of independent coordinate axis motion when the robot determines the motion, which should not include the opening and closing degrees of freedom of the end tool.

In the industrial robot system, one degree of freedom needs to be driven by a motor. It needs 6 degrees of freedom to describe the position and attitude of an object in three-dimensional space. However, the degree of freedom of industrial robot is designed according to its application, which may be less than 6 degrees of freedom or greater than 6 degrees of freedom.

(2) Mechanical origin is the reference point in the mechanical coordinate system shared by all degrees of freedom of industrial robots.

(3) Work origin, the reference point of industrial robot workspace.

(4) Positioning accuracy, the difference between the actual arrival position of robot and the target position.

(5) Repetitive positioning accuracy, the ability of the robot to repeatedly locate the position and the same target, is expressed by the dispersion degree of the actual position value.

(6) Working range refers to the collection of all points that can be reached at the end of the robot arm, and generally refers to the working area where the end tool is not installed.

(7) Rated load refers to the maximum allowable load at the mechanical interface of wrist within the specified performance range.

### 2.4.2.2 Structural Composition

In terms of architecture, industrial robots are divided into 'Three Major Parts and Six Systems', which are a unified whole. The three parts refer to the mechanical part for realizing various actions, the sensing part for sensing internal and external information, and the control part for controlling the robot to complete various actions. The six systems include driving system, mechanical structure system, robot environment interaction system, sensing system, human-computer interaction system and control system.

The mechanical structure driving system includes the power device and the driving mechanism to make the actuator produce the corresponding action.

The system of industrial robot consists of four parts: body, arm, wrist and end actuator, as shown in Figure 2-46. Some robots are even equipped with walking guide rails on the base. Most industrial robots have 3~6 degrees of freedom, of which the wrist usually has 1~3 degrees of freedom.

Robot environment interaction system realizes the connection and communication between robot and external devices.

The sensing system is composed of internal sensors and external sensors to detect its movement position and working state, such as position, force, vision and other sensors.

The human-computer interaction system is the unit of communication and coordination between human and robot.

The control system sends command signals to the six systems according to the input program and collects information from them.

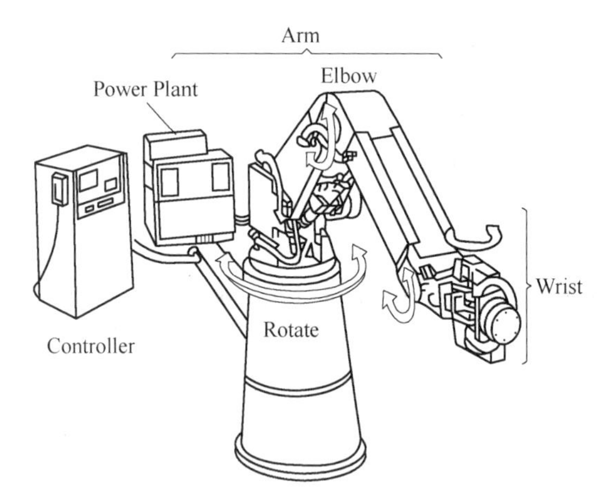


Figure 2-46 Structure of industrial robot

### 2. 4. 2. 3 Core Technology

The core technology of industrial robot mainly includes sensing technology, driving technology, transmission technology and programming technology.

**Sensing Technology of Industrial Robot**

The industrial robot with sensors can better cooperate with the operating person and carry out appropriate operation. Industrial robots often use sensors, including vision sensor, tactile sensors, proximity sensors and force sensors and so on.

Vision sensor is mainly used for position compensation of parts or workpieces, identification and confirmation of parts. Vision sensor is usually installed at the end of industrial robot for local vision or installed at the periphery of industrial robot system for global vision, as shown in Figure 2-47.

Tactile and proximity sensors are generally fixed at the tool to compensate the position error of parts or workpieces and prevent collision, as shown in Figures 2-48 and 2-49.

Figure 2-47 Vision sensor

Figure 2-48 Tactile sensor

The force sensor is generally installed at the wrist, and it is generally used in precision assembly or flanging operations requiring force control, as shown in Figure 2-50.

Figure 2-49 Proximity sensor

Figure 2-50 Force sensor

In this project task, the position detection device used is a magnetic sensor and a rotation counter, which are used together. The magnetic sensor is used to detect the reset state of the equipment and the rotation counter is used to detect the position of the equipment in operation.

**Driving Technology**

There are three common driving modes of industrial robots: hydraulic drive, pneumatic drive and electric drive. The most common one is electric drive technology. The key component of electric drive technology is motor.

The motor is the executive component of the robot drive system. The commonly used motors are step motor, DC or AC servo motor.

AC servo motor is the most widely used motor in robot. The internal rotor is a permanent magnet. The three-phase electric field of U/V/W controlled by the driver forms an electromagnetic field. The rotor rotates under this magnetic field. At the same time, the encoder provided by the motor feeds back the signal to the driver. The driver compares the feedback value with the target value and adjusts the rotation angle of the rotor. The accuracy of servo motor depends on the accuracy of encoder.

**Transmission Technology**

The transmission mechanism is used to transfer the mechanical energy from the prime mover to the joints or other working parts, so as to realize various necessary motions of the robot. There are several kinds of drives commonly used in industrial robots: gear transmission, screw drive, and belt and chain drive.

In addition to the above three main transmission forms, such as hydraulic transmission, pneumatic transmission, linkage mechanism or cam mechanism, are also used in industrial robots.

**Programming Technology**

There are three different ways of industrial robot programming as follows:

(1) Teaching programming: Teaching programming is a widely used programming method for industrial robots at present. According to the needs of tasks, the robot end tools are moved to the required positions and postures. Then each postures, together with the running speed and parameters, are recorded and stored, and the robot and persons can reproduce according to the teaching postures. There are two teaching programming methods: hand teaching and teaching box teaching.

At present, most robots are programmed by teaching. Teaching mode is a mature technology, which is easy to be mastered by the person, and it can be carried out with simple equipment and control devices. The teaching process is very fast. After teaching, it can be applied immediately. If necessary, the process can be repeated many times. In some systems, it can also be used for different speed reproduction during teaching.

The button installed on the control box can drive the robot to operate in the required order. In the teaching box, each joint has a pair of buttons, which control the movement of the joint in two directions respectively, and sometimes provide additional maximum allowable speed control. Although in order to achieve the highest efficiency, people always hope that the robot can achieve multi joint synthetic motion, but it is difficult to move multiple joints at the same time in the way of teaching box teaching.

(2) Robot language: Robot language, which provides a universal means of communication between human and robot, is a special language. It uses symbols to describe the robot motion, which is similar to the ordinary computer programming language, like C language. For example, the RAPID language of ABB robot and the AS language of Kawasaki robot.

(3) Off line programming: In three-dimensional model, equipment, environment and workpiece are built in the computer, and the robot is programmed in such a virtual environment. The off-line programming (OLP) system of robot is an extension of robot language programming. It makes full use of the achievements of computer graphics, establishes the model of robot and its working environment, and then uses some planning algorithms to program the control and operation of the graphics in the off-line situation.

### 2.4.3 Implementation

In the task implementation phase, the POWERON components are used to simulate the industrial robot and program to simulate the operation. The design of an industrial robot model is shown in Figure 2-51.

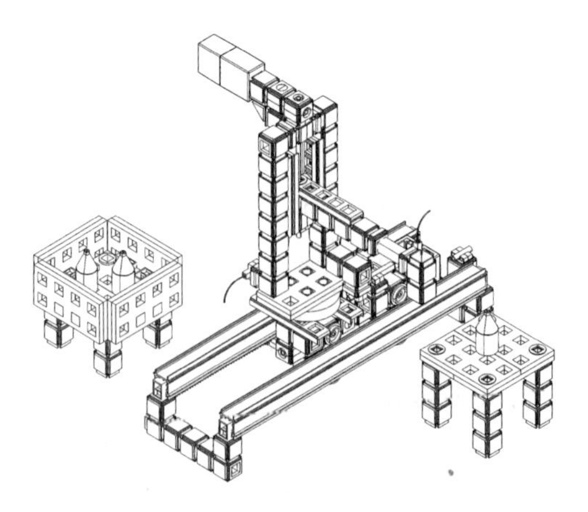

Figure 2-51 Industrial robot model

This industrial robot needs to be able to lift, translate and rotate. Industrial robots move color bottles from place A to designated place B through space movement.

### 2.4.3.1 Projet Design

In the drive system (shown in Figure 2-51), the translation of industrial robot is realized by motor 0, decelerating through gear set, and by meshing gear and rack, turning the rotation movement into translation movement. The rotation movement is realized by motor 1 through gear set. Lifting movement is realized by motor 2 driving lead screw. The robot grabs the color bottle by driving the electromagnet, and grabs the color bottle with iron block by magnetic force.

The mechanical structure system is realized by the assembly of various structural parts, such as cube, beam 300, 1# slab, etc.

The sensing system is based on the rotation counter to calculate the robot position and control the direction and speed of each motor. The detection of the reset position of each degree is based on the detection of magnetic sensor.

The human-computer interaction system is realized by the controller. The operator controls the machine through the current state of the controller displayed on the screen.

### 2.4.3.2 Material Preparation

Figure 2-52 shows the list of materials for supporting projects.

Figure 2-52 Bill of materials

## 2.4.3.3 Hands-on Construction

Next, some key steps are learnt in the process of building industrial robots, as shown in Figures 2-53 to 2-55, to help students complete the hardware splicing of industrial robots.

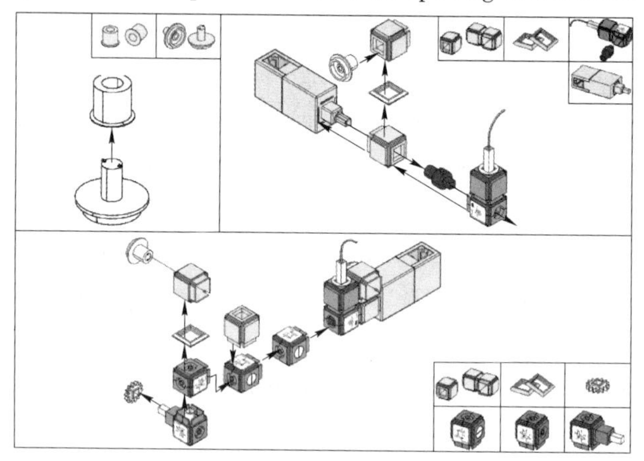

Figure 2-53 Robot with revolution counter and reducer

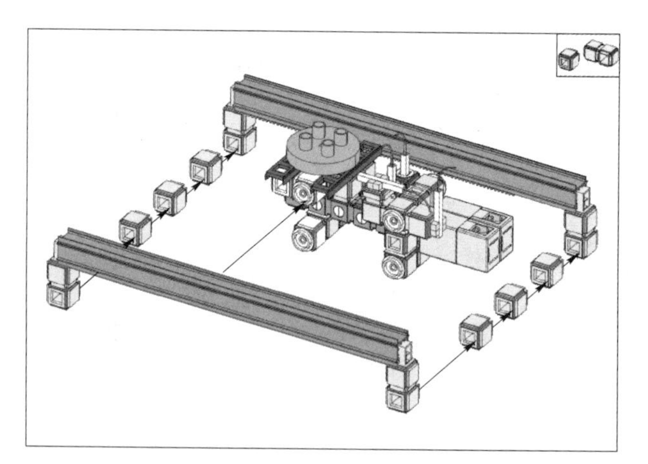

Figure 2-54 Overlap of translation and rotation mechanism on beam

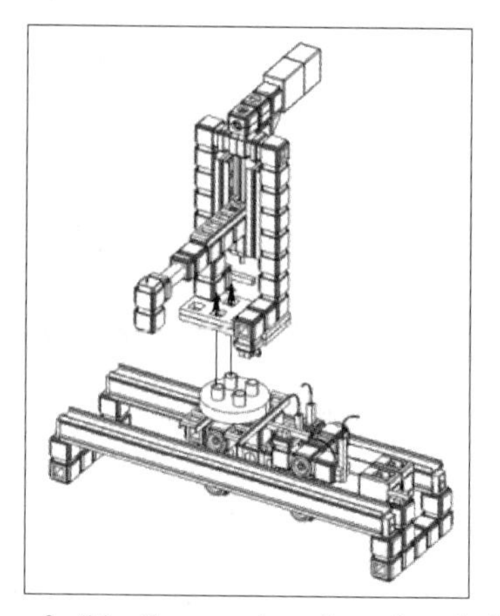

Figure 2-55 Construction of rotating platform

Finally, the input and output ports of industrial robots are shown in Figures 2-56 and 2-57.

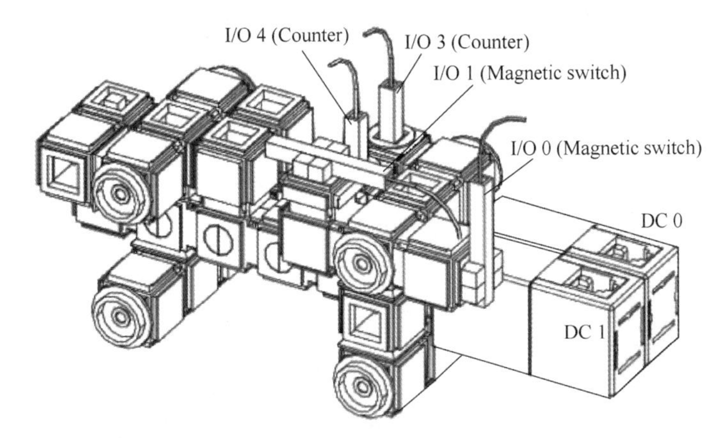


Figure 2-56 Signal interface of translation and rotation

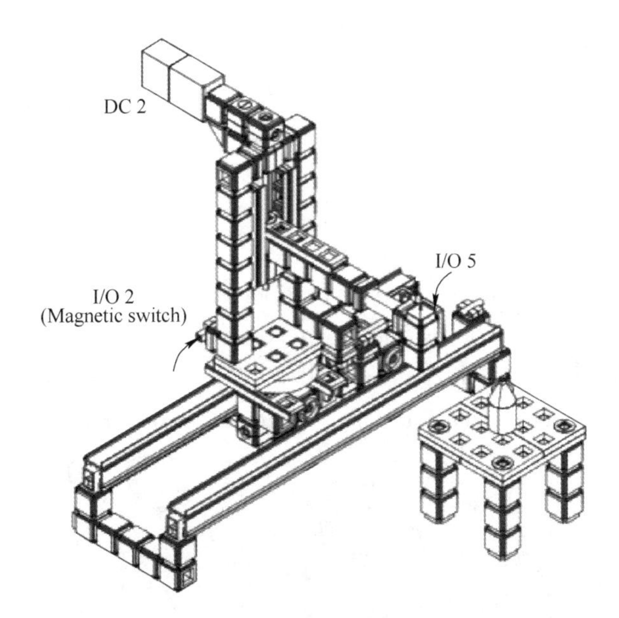


Figure 2-57 Signal interface between lifting and grabbing

### 2.4.3.4 Program Design

**Control Requirements**

The control requirements are as follows:

After the program is running, three degrees of freedom of lifting, translation and rotation will first move to the reset position. At this time, the electromagnet should be just above the color bottle to be handled. The working sequence is that: the electromagnet descends and sucks up the color bottle, and then lifts it up. The color bottle is moved to the upper part of the target by rotating and moving. The electromagnet descends and puts down the color bottle, and the reset action is executed to enter the next cycle, as shown in Figure 2-58.

**Usage and Characteristics of Revolution Counter**

The rotation counter in the POWERON consists of a magnetic sensor and a drive shaft with magnets, as shown in Figure 2-59.

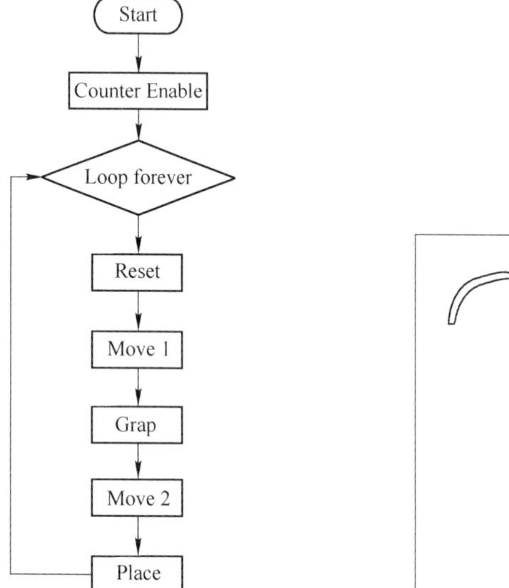


Figure 2-58 Block diagram of program design

Figure 2-59 Structure of rotation counter

The rotation counter module in the software has four functions: counter start, clear, get and stop. The counter can be read for logical judgment.

According to the conditions set in Figures 2-60 and 2-61, it means that when the count value of channel 4 greater than 2000, motor 1 stops, otherwise motor 1 rotates.

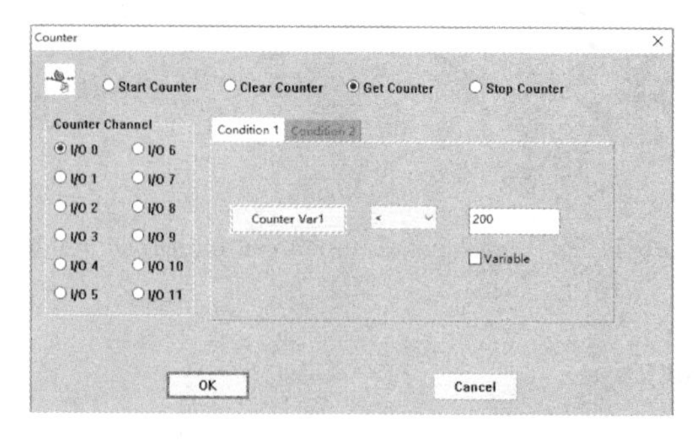


Figure 2-60 The function of rotation counter module

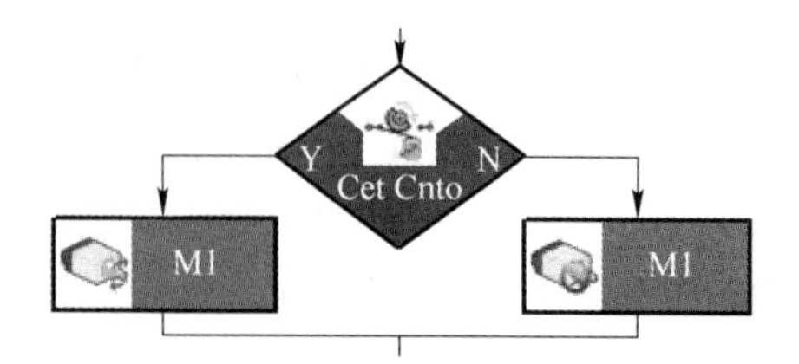


Figure 2-61 Rotating counter module condition judgment

In the industrial robot, the system can control the rotation and speed of the motor according to the rotation counter, and the position of each attitude is calculated according to the rotation counter.

#### 2.4.3.5 Debugging Records

In the process of debugging, the 'Motor' interfacing on the controller is used to test first, so as to ensure the correct direction of the motor drive, and ensure that each degree of freedom can move smoothly.

Finally, the datas is written into the Table 2-13.

**Table 2-13 Debugging records**

| Testing iterm | Testing process record | Testing Person |
| --- | --- | --- |
| | | |
| | | |

### 2.4.4 Evaluation

The task evaluation is shown in Table 2-14.

**Table 2-14 Task evaluation**

| No. | Evaluation content | Achievement ratio | Selfevaluation | Teacher evaluation |
| --- | --- | --- | --- | --- |
| 1 | Definition and classification | 20 | | |
| 2 | Structure composition | 10 | | |
| 3 | Core technology | 20 | | |
| 4 | Construction | 10 | | |
| 5 | Program debugging | 40 | | |

There are four characteristics: bionic feature, flexible feature, intelligent feature and automatic feature. Industrial robots are generally composed of six systems: drive system, mechanical structure system, robot human environment interaction system, perception system, human-computer interaction system and control system. The key technologies include sensor technology, drive technology, transmission technology and programming technology.

Through the construction of industrial robots, the use skills and engineering simulation process are further consolidated, which is conducive to training students' innovative thinking and ability, and strengthening students' engineering consciousness and team consciousness.

### 2.4.5 Development

(1) Calculate the gear ratio of the rotating part in this model.

(2) Realize an industrial manipulator with more rotational degrees of freedom, and realize the grab and placement of objects.

(3) Build a planar joint industrial robot using the POWERON.

# Task 2.5 Build and Debug CNC Machine Control System

## 2.5.1 Introduction

There is a CNC milling machine that needs to be retrofitted. Firstly, a simulation CNC milling machine model is built with the POWERON components to understand the basic situation of the CNC milling machine. The simulation requirements are that the CNC milling machine needs three feed motion (X, Y, Z) and a main motion, and the position of the three feed motion needs to be detected.

## 2.5.2 Related Knowledge

The common CNC machine tools includes CNC lathe, CNC milling machine and machining center.

CNC lathe is one of the most widely used CNC machine tools, accounting for about 25% of the total number of CNC machine tools. CNC lathe is mainly used for turning. Generally, it can process the rotating surface, such as inner and outer cylinder surface, cone surface, forming rotating surface and thread surface. In CNC lathe, it can also process high-precision curved surface and section thread. Figure 2-62 shows horizontal and vertical CNC lathes respectively, and Figure 2-63 shows the actual figure of the cylindrical surface being processed by the CNC lathe. The parts shown in Figure 2-64 are processed by CNC lathe.

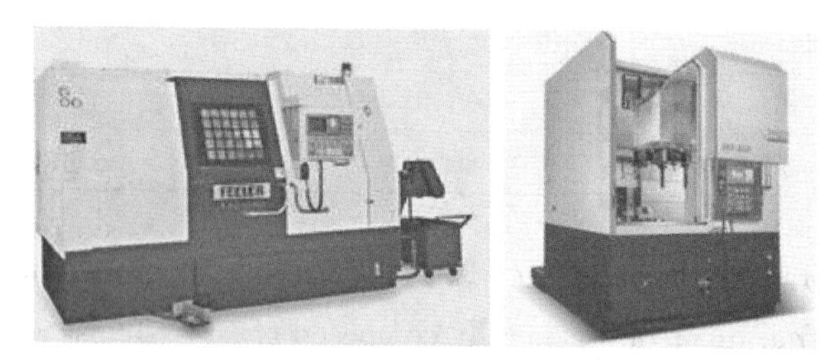

Figure 2-62 Horizontal and vertical CNC lathe

Figure 2-63 Lathe processing

Figure 2-64 CNC turning parts

CNC milling machine is a kind of milling machine controlled by computer. It can be used to machine the plane contour which is composed of two geometric elements of straight line and circular arc. The plane contour which is composed of non-circular curve can be directly processed by approximation method, and the three-dimensional curved surface and space curve can also be processed, such as blade and propeller. Figure 2-65 shows the typical CNC milling machine, and Figure 2-66 shows the parts processed by CNC milling machine. In addition, milling machine has the function of hole processing.

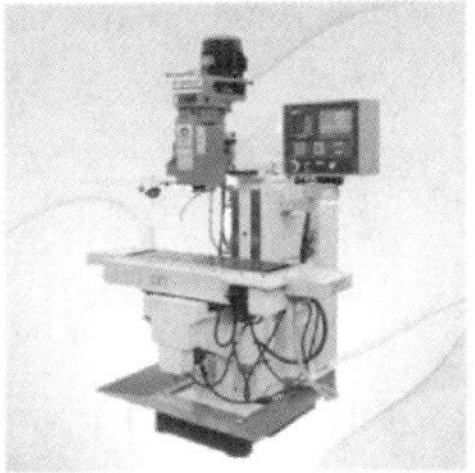

Figure 2-65　Typical CNC milling machine

Figure 2-66　Parts processed by CNC milling machine

Machining center is one of the most widely used CNC machine tools with the highest output in the world. The machining center is equipped with a tool magazine and an automatic tool changing device. When machining, the workpiece can be automatically processed by milling, boring, drilling, reaming, tapping and other processes on each machining surface of the workpiece after being clamped once. As for the batch workpiece with medium processing difficulty, its efficiency is 5~10 times of that of ordinary equipment. Especially it can complete many processing tasks that ordinary equipment can not complete, and it is more suitable for single piece processing with complex shape and high precision requirements, or small and medium batch production and multi variety production. Figure 2-67 shows the CNC machining center, and Figure 2-68 shows the parts processed by the CNC milling machine.

Figure 2-67　CNC machining center

Figure 2-68 Parts processed by CNC machining center

### 2.5.2.1 Definition and Classification

**Definition**

CNC machine tool is the abbreviation of computer numerical control (CNC) machine tool. It is an automatic machine tool equipped with program control system. The control system can logically process the program with control code or other symbol instructions, and translate it into code, so that the machine tool can act and process parts.

**Classification**

The classifications of CNC machine are as follows:

(1) According to the process use, CNC machine tools can be divided into metal cutting, metal forming and special machining machine tools.

(2) According to the driving mode, it can be divided into point control, linear control and contour control machine tools.

Point position control machine tool is characterized by that the moving parts of the machine tool. It can only realize the precise movement from one position to another, and no machining process is carried out in the process of movement and positioning. The typical point control CNC machine tools include CNC drilling machine, etc.

Linear control machine tool is characterized by that, not only the machine tool moving parts can achieve a coordinate position to another coordinate position of the precise movement and positioning, but also the machine tool can achieve parallel to the coordinate axis of the linear feed movement or control two coordinate axes to achieve oblique feed movement. Typical linears control CNC machine tools, such as early CNC lathe.

The contour control machine tool is characterized by that the moving parts of the machine tool can realize the linkage control of two coordinate axes in the same pair. It not only needs to control the coordinate position of the starting point and the end point of the moving parts of the machine tool, but also needs to control the speed and displacement of each point in the whole processing process. For the other words, it needs to control the small moving track to process the parts into straight lines, curves or curved surfaces in the plane or in the space. The typical contour controlled CNC machines includes machining centers, etc.

(3) According to the control mode, it can be divided into open-loop control, semi closed-loop control and closed-loop control, as shown in Figure 2-69.

Open loop control is the control mode without position feedback device.

Semi closed loop control refers to the open-loop control servo electric several axes equipped with angular displacement detection device, which indirectly detects the displacement of the moving parts by detecting the rotation angle of the same service motor and feeds it back to the comparator of the numerical control device, comparing it with the input command, and controlling the moving parts with the difference value.

Closed loop control is to directly install a linear or rotary detection device at the corresponding position of the final moving parts of the machine tool. The displacement or angular displacement value is feed directly measured back to the comparator of the numerical control device for comparison with the input command displacement, and the difference is used to control the moving parts, so that the moving parts move strictly according to the actual demand displacement.

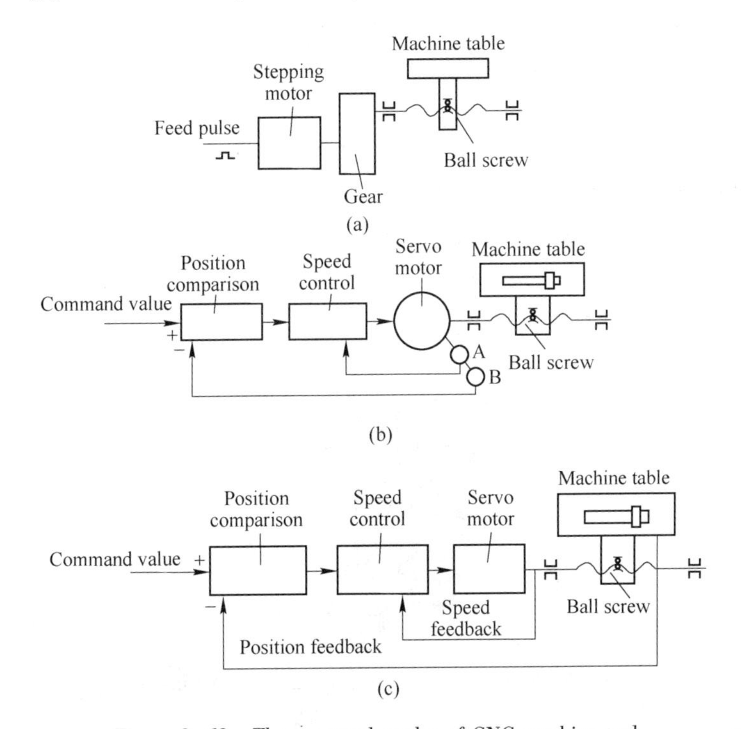


Figure 2-69 Three control modes of CNC machine tool

(a) Open-loop coutrol CNC machine tool; (b) Semi closed loop coutrol CNC machine tool;

(c) Full closed-loop control CNC machine tool

### 2.5.2.2 Structural Composition

CNC machine tool is mainly composed of CNC system, drive system, auxiliary control device, machine body, etc.

**Numerical Control System**

CNC system is the core and central link of CNC machine tools, which includes hardware (printed circuit board, display, etc.) and corresponding software, which is used to input the digital part program, and complete the storage of input information, data transformation, interpolation operation and various control functions.

**Driving System**

The drive system is the driving part of the actuator of CNC machine tools, which includes the main drive unit, feed drive unit, spindle motor and feed motor. Under the control of the numerical control device, it realizes the spindle and feed drive through the electric or electro-hydraulic servo system. When several feeders are linked, positioning, straight line, plane curve and space curve can be processed.

**Auxiliary Controls Device**

Auxiliary control device refers to some necessary supporting parts of CNC machine tools to ensure the operation of CNC machine tools, such as cooling, chip removal, lubrication, lighting, monitoring, etc. The device includes hydraulic and pneumatic device, chip removal device, exchange table, CNC turntable and CNC dividing head, as well as cutter and monitoring and testing device.

**Machine Tool Ontology**

The machine body is the main body of the CNC machine tool, which is composed of the basic large parts of the organic bed (such as bed and base) and the moving parts (such as workbench, bed saddle, spindle, etc.). Figure 2-70 shows the structure of vertical CNC milling machine.

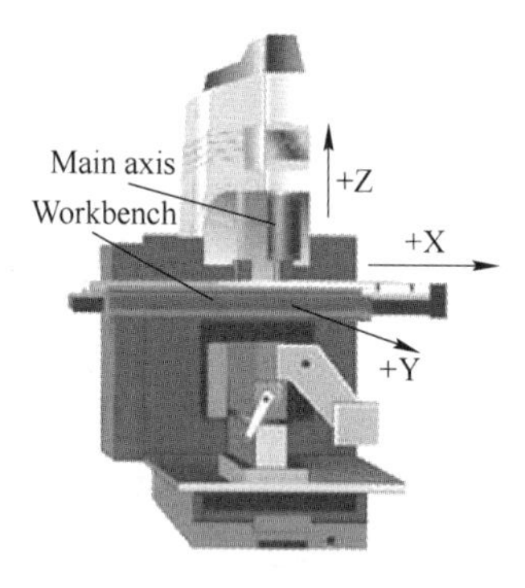


Figure 2-70 Structure of vertical CNC milling machine

### 2.5.2.3 Core Technology

There are four core technologies in CNC machine tools: machine tool transmission technology, CNC system, servo drive system and machine tool detection device.

**Transmission technology**

The drive system of CNC machine tool includes spindle drive and feed drive, which can be divided into main motion and feed motion. Both main motion and feed motion have drive parts.

(1) The main movement of CNC machine tools: The main motion of CNC machine tool refers to the transmission motion of chip production. For example, the spindle on CNC lathe drives the rotary motion of workpiece, and the spindle on vertical CNC milling machine drives the rotary mo-

tion of milling cutter. The main motion of CNC machine tool is driven by the main motion motor, which usually needs to be converted into the main motion of machine tool with the help of transmission components. The transmission methods often used include:

1) Speed regulating motor: The direct use of speed regulating motor greatly simplifies the structure of spindle box and spindle, and effectively improves the rigidity of spindle components. But the output torque of the spindle is small, and the motor heating has a great influence on the accuracy of the spindle.

2) Variable speed gear: Variable speed gear is one of the most widely used methods in large and medium-sized CNC machine tools. Through a few pairs of gears decelerating, the output torque is increased to meet the requirements of the spindle for the output torque characteristics. Some small-scale CNC machine tools also adopt this transmission mode to obtain the torque required for strong chip cutting.

3) Belt drive: Belt drive is mainly used in small CNC machine tools, which can avoid the vibration and noise caused by gear drive, but it can only use the main shaft which meets the requirements of torque characteristics.

(2) Feed motion of CNC machine tool: Typical CNC machine tool usually uses closed-loop control feed system, which is composed of position comparison, amplifying elements, driving units, mechanical transmission devices and detection feedback elements. The feed system receives the speed command provided by the control system for each moving coordinate axis, and outputs the drive system signal to drive the feed motor to rotate by adjusting the speed and torque, so as to realize the movement of the coordinate axis of the machine tool. At the same time, it receives the speed feedback signal to implement the speed closed-loop control. The mechanical transmission device is to change the rotary motion of the driving source (feed motor) into the linear motion of each coordinate axis of the worktable. The feed transmission device includes:

1) Gear: Gear transmission is a kind of mechanical transmission widely used. Almost all kinds of machine tool transmission devices have gear transmission. Through gear transmission, the output of servo motor with high speed and low torque is changed into the input of actuator with low speed and high torque.

2) Synchronous toothed belt: Synchronous toothed belt drive is a new type of belt drive. They uses the tooth shape of the toothed belt and the teeth of the belt wheel to mesh in turn to transmit the motion and power, so they have the advantages of belt drive, gear drive and chain drive. And there is no relative sliding. The average transmission is more accurate, the transmission accuracy is high, and the toothed belt has high strength, small thickness and light weight. So it can be used for high-speed transmission. The toothed belt is widely used in CNC machine tools because of its small load and high transmission efficiency.

3) Ball screw nut pair: In order to improve the sensitivity of the feed system, positioning accuracy and prevent creeping, it is necessary to reduce the friction of the feed system of CNC machine tools and the difference of static and dynamic friction coefficient. Therefore, ball screw pairs are often used in linear motion mechanisms with short stroke.

**CNC System**

CNC system of CNC machine tool is the core part of CNC machine tool, which include computer system, servo drive device, position detection device, PLC and interface circuit, as shown in Figure 2-71.

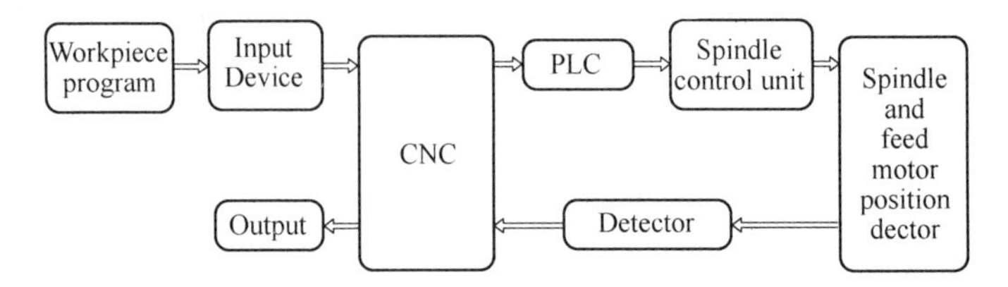


Figure 2-71 Composition of computer numerical control system

From the point of view of automatic control, CNC system is a kind of position control system. In essence, it is an automatic control system which takes the position of multiple executive parts as the control object and makes them coordinate the movement. It is a computer control system equipped with a special operating system.

From the external characteristics, CNC system is composed of hardware and software.

The hardware structure of CNC system is divided into computer basic system, equipment support layer and equipment layer. Computer basic system refers to computer system, display equipment, input/output devices, etc. Equipment support layer refers to the human-computer interaction programming system and motion control equipment between computer and machine tool.

In essence, CNC system software is a special operating system with real-time and multi task, which consists of CNC machine management software and CNC control software. It is the living soul of CNC machine tool system. With the support of hardware, the software of CNC system reasonably organizes and manages all works of the whole system. All kinds of CNC functions are realized, and CNC machine tools are enabled to process in an orderly manner according to the requirements of operators. The hardware and software of the CNC system constitute the CNC system platform.

**Servo Drive System**

There are two kinds of servo drive systems for CNC machine tools: One is the feed servo system, which controls the cutting feed motion of each coordinate axis of the machine tool; The other is the spindle servo system, which controls the cutting motion of the spindle, mainly the rotation motion.

Servo drive system can be divided into open-loop servo system, closed-loop servo system and half closed-loop servo system according to the control principle. According to the use of the actuator, it can be divided into stepper motor drive system, DC servo motor and AC servo motor; According to the nature of control quantity, it can be divided into position servo system and speed servo system. The motors commonly used in servo system include stepping motor, DC servo motor and AC servo motor.

**Detection Device**

The detection device of CNC system is mainly divided into displacement detection device. The following is mainly about position detection.

The position detection device used in the servo system of CNC machine tools is divided into two types: linear type and rotary type. The linear position detection device is used to detect the linear displacement of moving parts, and the rotary position detection device is used to detect the rotational displacement of rotating parts.

In the closed-loop system, the main function of position detection is to detect the displacement, and send out feedback signal to compare with the command signal sent by the numerical control device. If there is a deviation, the executive part is controlled after amplification to move towards the direction of eliminating the deviation until the deviation is equal to zero.

Common position detection devices include grating position detection device and photoelectric encoder, as shown in Figures 2-72 and 2-73.

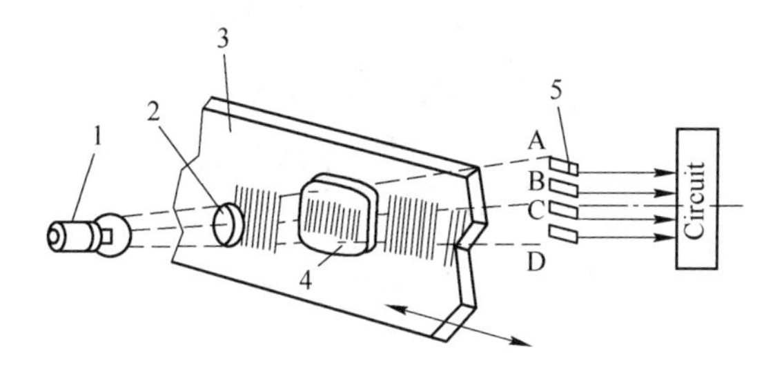


Figure 2-72 Grating detection principle

1—Light source; 2—Lens; 3—Ruler grating; 4—Indicating grating; 5—Optoelectronic element

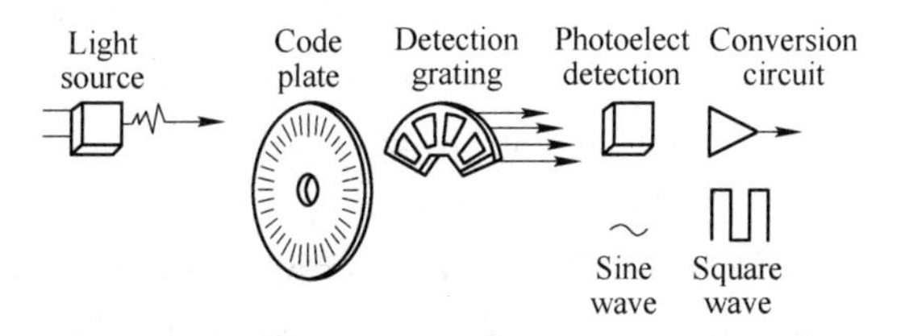


Figure 2-73 Principle of photoelectric encoder

Grating has been used in CNC machine tools as testing device for decades. It is used to detect length, angle, speed, acceleration, vibration, creep, etc. It is a kind of testing device used more in the closed-loop system of CNC machine tools.

The photoelectric encoder is used to measure the angular displacement of rotating motion. It uses the photoelectric principle to change the mechanical angular displacement into a pulse electrical signal. And it is a widely used angular displacement sensor.

In this project task, the position detection device used is a magnetic sensor and a rotation counter, which are used together. The magnetic sensor is used to detect the reset state of the equipment and the rotation counter is used to detect the position of the equipment in operation.

### 2.5.3 Implementation

In the task implementation phase, the POWERON components are used to simulate the CNC milling machine and program to simulate the operation. The design of a CNC milling machine model is shown in Figure 2-74. This controlled milling machine requires three feed movements of X, Y and Z. One main movement, and the positions of three feed movements need to be detected.

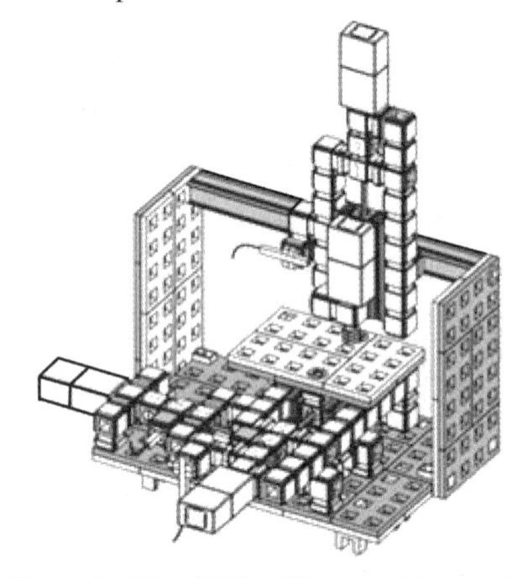

Figure 2-74 CNC milling machine model

#### 2.5.3.1 Projet Design

CNC system is the core of CNC machine tool, which is used to input the digital part program. In this model, the CNC system is ABILIX controller, which can control X, Y and Z by operating the controller.

The drive system is the driving part of the actuator of CNC machine tools, including the spindle drive and the feed drive. In this model, the motor drive mode is adopted, the No. 0 motor drives the lead screw, transforms the rotation movement into translation, and realizes the X-axis movement of the CNC milling machine; The No. 1 motor uses the same way to realize the Y-direction movement of the CNC milling machine; The No. 2 motor also uses the same way to realize the Z-direction movement of the CNC milling machine, and the above three motor drives realize the feed drive of the CNC machine. The No. 3 motor realizes the spindle drive of the numerical control.

Auxiliary control system is a system that uses some necessary supporting parts to ensure the operation of CNC machine tools. In this model, three magnetic sensors are used to detect the reset position of the X, Y and Z of the feed system, and two rotation counters are used to detect the position of X and Y.

The machine body is the main body of the CNC machine tool, which is composed of the basic large parts of the organic bed (such as bed and underlay) and various moving parts (such as workbench, bed saddle, spindle, etc.). In this model, the milling machine body is composed of all kinds of structural parts, such as cube, beam 240, 4# slab, etc.

#### 2.5.3.2 Material Preparation

Figure 2-75 shows the list of materials for this projects.

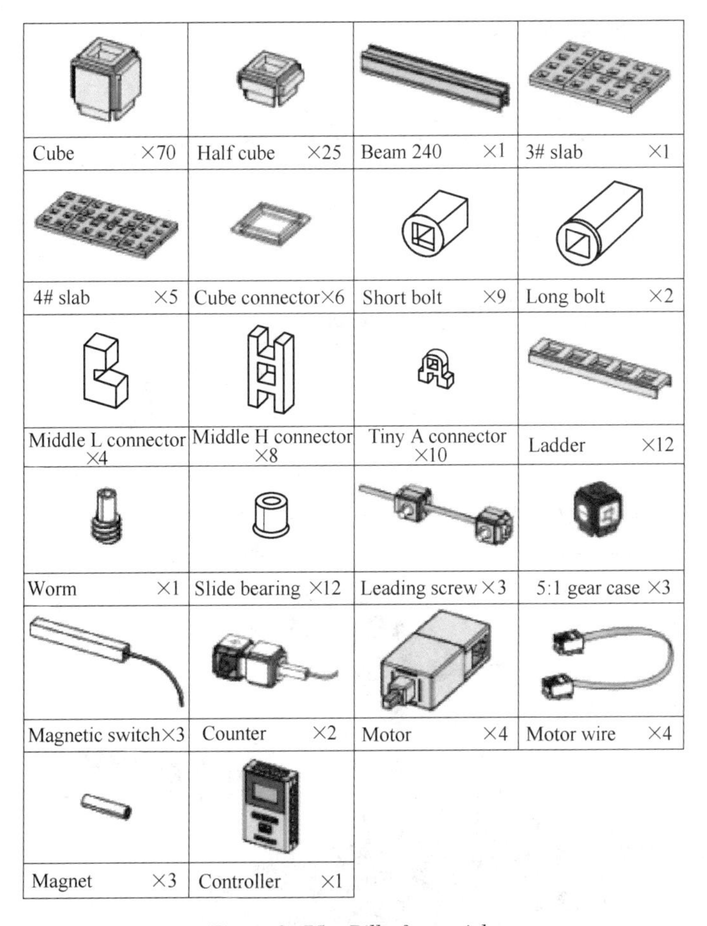


Figure 2-75 Bill of materials

### 2. 5. 3. 3 Hands on Construction

Next, some key steps are learnt in the process of building industrial robots, as shown in Figures 2-76 to 2-79 to help students complete the hardware splicing of CNC milling machine.

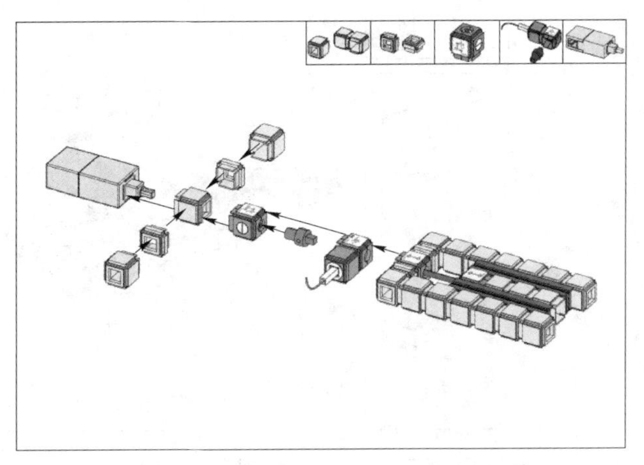

Figure 2-76 X-axis feed drive splicing of CNC milling machine

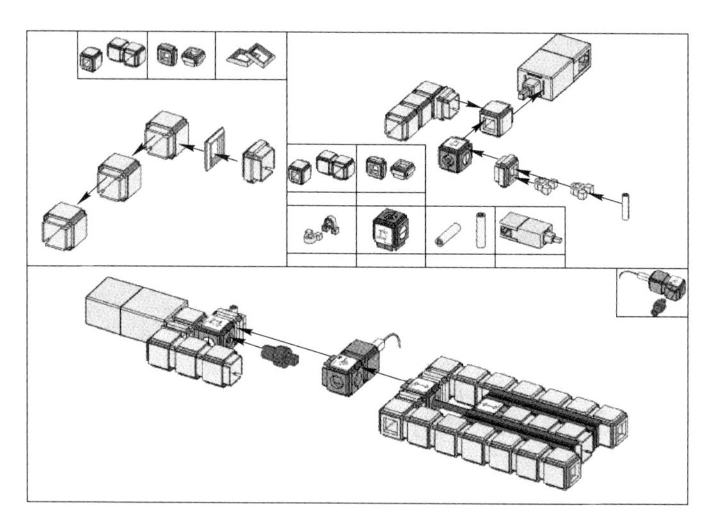

Figure 2-77　Y-axis feed drive splicing of CNC milling machine

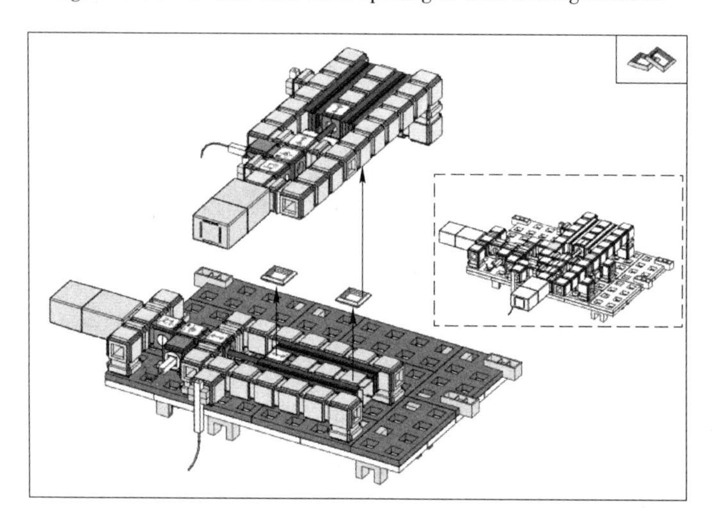

Figure 2-78　X/Y-axis feed drive splicing of CNC milling machine

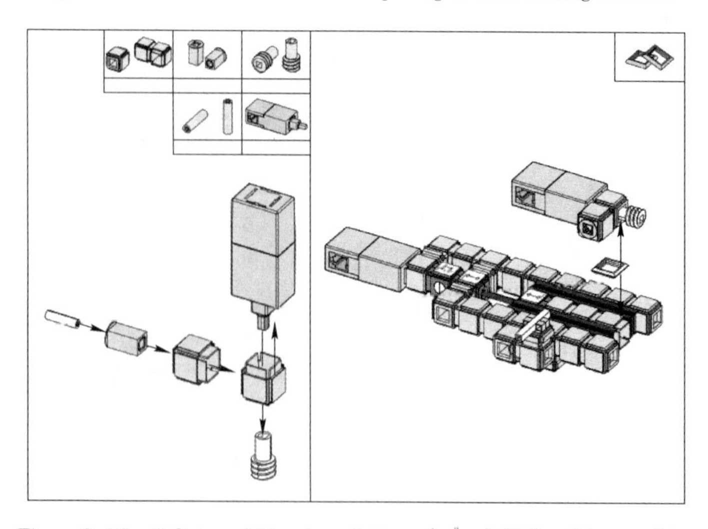

Figure 2-79　Splicing of Z-axis and main shaft of CNC milling machine

Finally, the input and output ports of CNC machine tools are shown in Figures 2-80 and 2-81.

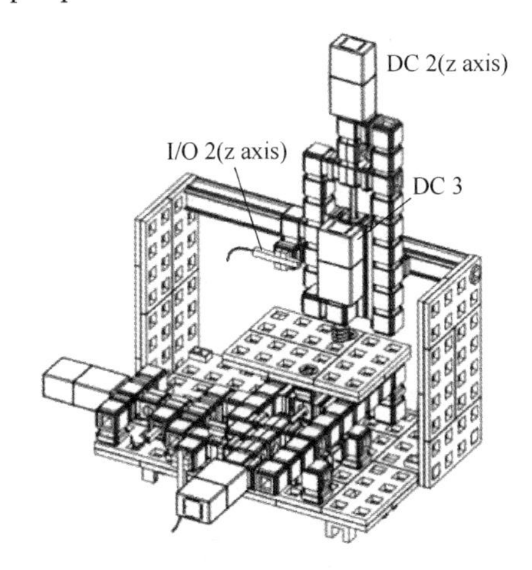


Figure 2-80 Signal interface between Z and main shaft

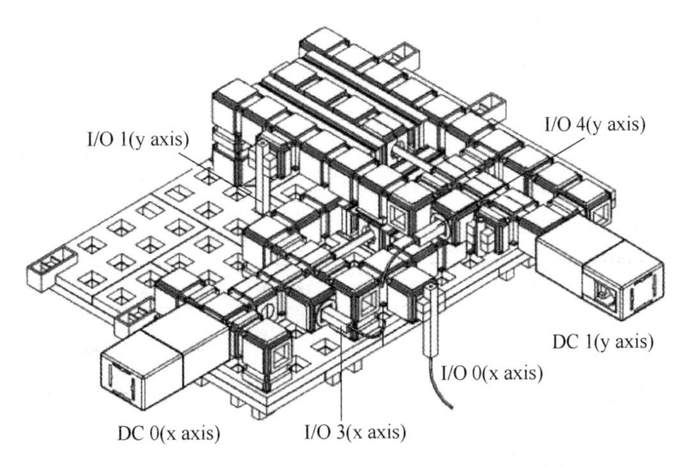


Figure 2-81 Signal interface between X and Y

### 2.5.3.4 Program Design

The control requirements of CNC milling machine are as follows:

(1) The three lead screws are defined as three axes of rectangular coordinate system, which are X, Y and Z axes from the bottom to the top. The reset position is the origin and the extension direction is positive.

(2) After the program is running, reset is performed first (Z is lifted first. X and Y can move at the same time). After the reset is completed, the system will beep to prompt to add materials, and then process the first point. This routine only processes one point, as shown in Figure 2-82.

### 2.5.3.5 Debugging Records

In the process of debugging, the 'Motor' interface on the controller is used to test first, so as to

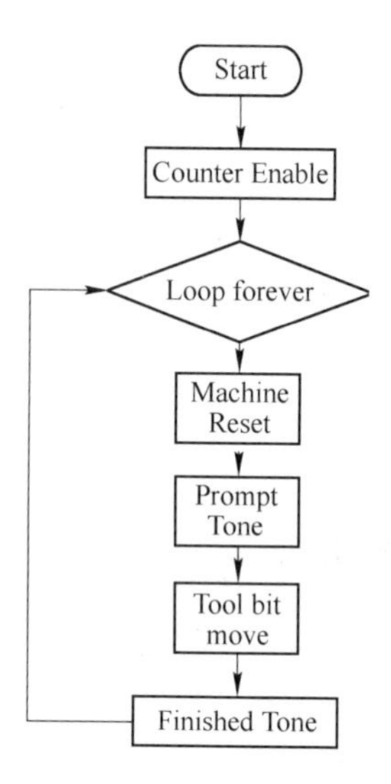


Figure 2-82 Program design block diagram

ensure the correct direction of the motor drive, and to ensure that each degree of freedom can move smoothly.

According to the specific task, program and debug the system, and record the debugging process in Table 2-15.

**Table 2-15 Debugging records**

| Testing iterm | Testing process record | Testing Person |
|---|---|---|
| | | |
| | | |

## 2. 5. 4 Evaluation

The table of the task evaluation is shown in Table 2-16.

**Table 2-16 Task evaluation**

| No. | Evaluation content | Achievement ratio | Self evaluation | Teacher evaluation |
|---|---|---|---|---|
| 1 | Definition and classification | 20 | | |
| 2 | Structure | 10 | | |
| 3 | Core technology | 20 | | |
| 4 | Construction | 10 | | |
| 5 | Program debugging | 40 | | |

CNC machine tool is mainly composed of CNC system, servo drive system, inspection feedback device, auxiliary control device and machine tool body. Transmission, CNC system, drive technology, sensor technology and electrical technology are the five core technologies of CNC machine tools. The transmission of CNC machine tools mainly includes screw drive, synchronous belt drive and gear drive. Numerical control system is the center of numerical control machine tool, which will receive all the functional instructions for decoding, operation, and then orderly issued a variety of control instructions and a variety of machine tool function control instructions, until the end of the movement and function. The main motion drive and the feed motion drive are controlled by different motors respectively. The sensor mainly realizes the feedback to the machine speed and the position.

By using the POWERON to build the CNC machine tool system, we learned the project scheme design of engineering practice innovation, material preparation, component installation, programming and debugging, and further became familiar with the application of the ABILIX controller and its VJC development system. The main technical links of engineering realization of CNC machine tool system are mastered.

### 2.5.5 Development

(1) Change the cutter head into a brush, drive and control the movement of X, Y, Z reasonably through the movement, and write simple patterns such as triangles and pentagram stars.

(2) Think about how to control the X, Y (components can be added) and write complex alphabet, such as 'A'.

(3) Think about how to display the movement speed of X, Y and Z by adding devices and appropriate algorithms.

(4) Think about how to transform this CNC machine into a CNC engraving machine.

# Project 3 Knowledge Operation of Industrial Robots

As early as thousands of years ago, the concept of robots had appeared. However, robot technology, especially industrial robot technology, has been developed rapidly until modern times. Robots must more firstly to replace people to work. There are many ways to set a robot in motion, for example, manually controlling it with the FlexPendant, automatically controlling it by writing a RAPID program, or controlling robots with external signals. For beginners, operating a robot manually is the base of learning it.

## Task 3.1 Preliminary Knowledge of Industrial Robots

### 3.1.1 Introduction

The industrial robot is a kind of typical device of mechatronics, which is a multi-joint manipulator or multi-degree of freedom robot facing industrial fields. The emergence of industrial robots is a milestone in the use of machinery to promote social development. Through the study of this task, we will know the history development, application, characteristics, typical structures and safety precautions of industrial robots.

### 3.1.2 Related Knowledge

Industrial robot is a general term for robots used in industrial production environment. In 1954, George. G. Devol first proposed the concept of industrial robot and applied for a patent; In 1959, the first industrial robot was born in the United States, creating a new era of robot development. After 60 years of development, industrial robots have made great changes in performance and utility. As the structure of modern industrial robot becomes more and more reasonable, the control more and more advanced and the function more and more powerful. Modern industrial robots are gradually developping toward the direction of walking ability, multiple perception ability and strong adaptive ability to working environment.

At present, industrial robots are used in various fields of industrial production more and more widely, in which automobile manufacturing industry, electronic and electrical industry, metal products and manufacturing industry are the main application fields. According to the functions and uses, industrial robots can be roughly divided into four categories: machining robots, assembling robots, handling robots and packaging robots. Machining robots are directly used in the processing of industrial products. Assembling robots can combine different parts or materials into components or finished products. Handling robots usually means the ones which are engaged in moving objects. Packaging robots are used for classification, packaging and palletizing of goods.

At present, Japan and the European Union are the main production bases of industrial robots in the world. The main enterprises include FANUC, YASKAWA, ABB, KUKA and so on, which are known as the ‘Four Families’ of industrial robots. This book is based on ABB industrial robots in four families.

The most remarkable characteristics of industrial robots are as follows:

(1) Programmable: The further development of production automation is flexible start-up. Industrial robots can be reprogrammed according changes of their working environment, so they play a good role in flexible manufacturing of small batch and multi-varieties with balanced and high efficiency, which have become an important part of the flexible manufacturing system.

(2) Personification: The mechanical structure of industrial robots includes waist, big arm, small arm, wrist, claw and other parts similar to human and the control part mainly depending on computer. In addition, intelligent industrial robots also have many human-like ‘Biosensors’, such as skin contact sensors, force sensors, load sensors, visual sensors, acoustic sensors, language functions and so on. Sensors improve the adaptability of industrial robots to the surrounding environment.

(3) Versatility: In addition to the ones specially designed, general industrial robots are of good versatility in different tasks. For example, different work can be accomplished by changing the end manipulator, such as claw, tool, etc.

### 3.1.3 Implementation

#### 3.1.3.1 Typical Structures of Industrial Robots

**Cartesian-coordinate Industrial Robot**

The cartesian-coordinate industrial robot (shown in Figure 3-1) generally includes two or three degrees of freedom movement between which space angle is a right angle. It can realize automatic control and repeated programming. All movements can be executed according to program. The cartesian coordinate robot can work in bad environment, which is easy to operate and maintain.

**Planar Joint Industrial Robot**

The planar joint industrial robot (shown in Figure 3-2), known as SCARA, is a form of cylindrical coordinate robots. SCARA contains three rotating joints, whose axes are parallel to each other, to locate and orient in the plane. Meanwhile, it has a moving joint to complete vertical movement of end manipulator. SCARA industrial robot has the advantages of high precision, wide range of movement, simple coordinate calculation, light structure, fast response speed and small load, which are mainly used in electronic, sorting and other fields.

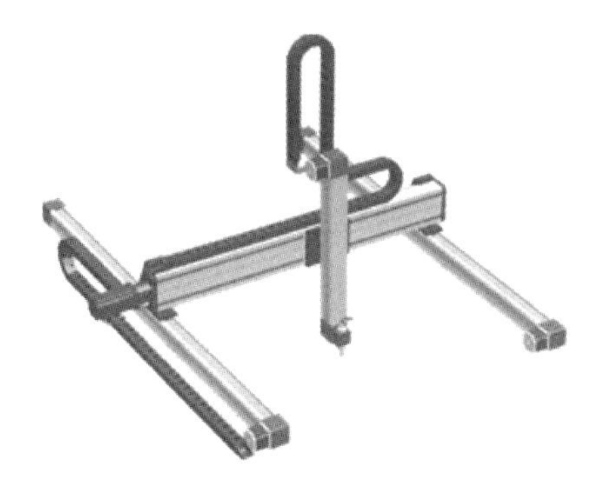

Figure 3-1　Cartesian-coordinate industrial robot

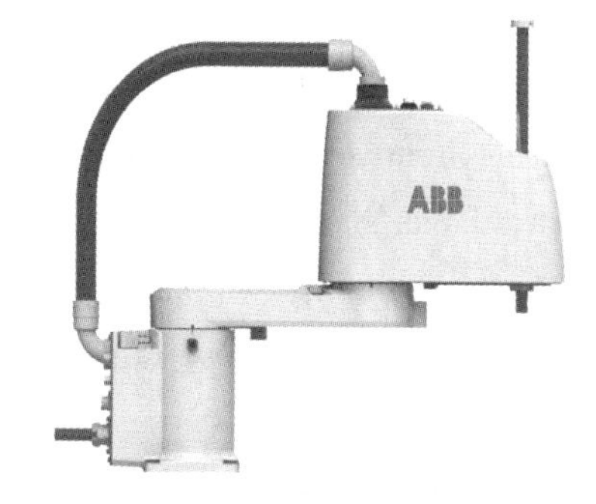


Figure 3-2　Planar joint industrial robot

**Parallel Industrial Robot**

Parallel industrial robot (shown in Figure 3-3), known as Delta, belongs to high-speed and light load industrial robot. Generally, it captures the target object through teaching programming or vision system, determines the position of tool center point (TCP) by three parallel servo axes and implements transportation, machining and other operations of target object. Delta robot is mainly used in machining and assembling of food, medicine, electronic products and so on.

**Serial Industrial Robot**

Serial industrial robot (shown in Figure 3-4) has four or more rotating axes, of which 6-axis robots are the most common form. Similar to human arms, it is of high degrees of freedom and high production efficiency, which can be programmed freely. The robot can replace people to complete complex work that is harmful to people's health, such as loading, unloading, painting, surface treatment, testing, measurement, arc welding, spot welding, packaging, assembly, chip machine, fixing, special assembly operation, forging, casting and other fields.

This book is written on basis of serial industrial robots.

**Cooperative Industrial Robot**

With traditional industrial robots replacing monotonous, repetitive and dangerous work gradually, cooperative industrial robot (shown in Figure 3-5) will also infiltrate into various industrial fields, working with people. It will lead to a new era when the robot and human beings work cooperatively. Compared with fully automatic workstation of industrial robot, the cooperative industrial robot, which completes tasks along with people, has advantages of flexibility and cost.

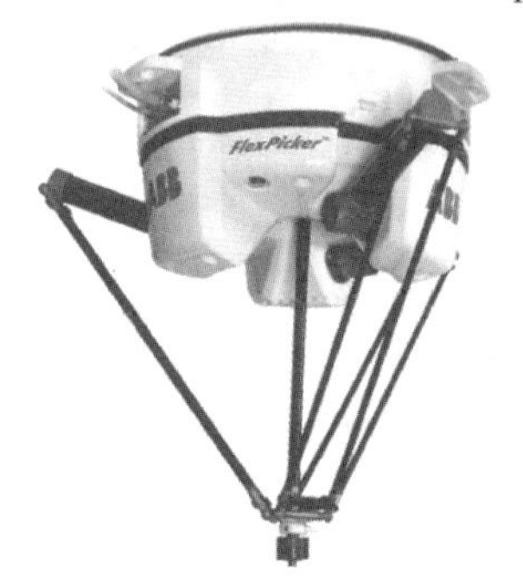


Figure 3-3　Parallel industrial robot

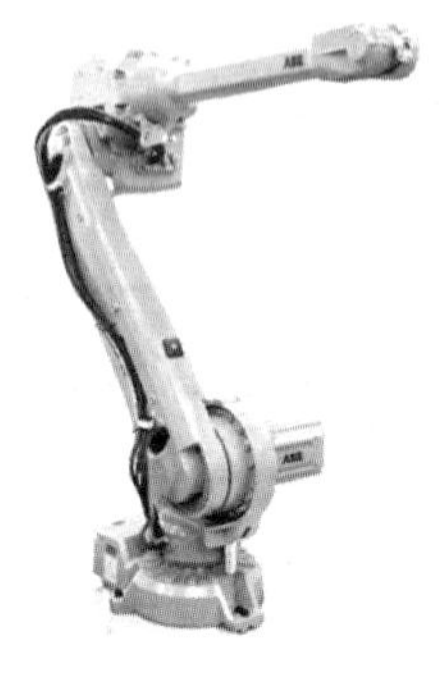

Figure 3-4　Serial industrial robot

Figure 3-5　Cooperative industrial robot

### 3.1.3.2 Safety Precautions for Industrial Robots

Safety precautions for industrial robots are as follows:

(1) Turn Off the Main Power! Do remember to turn off the main power while installing, repairing and maintaining the robot. Live working may lead to fatal consequences. Inadvertent high voltage shock may cause cardiac arrest, burns or other serious injuries.

(2) Keep Safe Distance from the Robot! When debugging or running the robot, it may move unexpectedly or irregularly. In addition, all the movements are powerful, which will seriously hurt people or damage any equipment within working space of the robot. So always be alert to keep enough safety distance from the robot.

(3) Electro-static Discharge Hazard! ESD (Electro-static Discharge) is the electrostatic conduction between two objects with different potential. It can be conducted through direct contact or induced electric field. While handling components or containers, ungrounded personnel may conduct a large amount of static charges. This discharge process may damage sensitive electronic equipment. Therefore, in the case of this mark, electrostatic discharge protection must be well done.

(4) Emergency Stop! The emergency stop takes precedence over any other robot control operation. It can disconnect driving power of the robot's motor, stop all moving parts, and cut off the power supply of the dangerous parts controlled by the robot system. Please press any emergency stop button immediately in case of the following situations:

1) While the robot is running, there are workers in the working area.

2) The robot injures the staff or damages the equipment.

(5) Put Out the Fire! In case of fire, please ensure that all personnel are evacuated safely before putting out the fire. The injured should be treated first. When electrical equipment (such as robots or controllers) is on fire, carbon dioxide fire extinguishers should be used instead of water or foam.

(6) Safety at Work! Robots may run slowly, but it is heavy and powerful. A pause or stop in movement can be dangerous. Even if the movement trajectory can be predicted, the external signal may change the operation, which will generate unexpected motion without any warning.

Therefore, when entering the protected space, all safety regulations must be followed:

1) If there are workers in the protected space, please operate the industrial robot system manually.

2) When entering the protected space, please prepare the FlexPendant to control the industrial robot at any time.

3) Pay attention to rotating or moving tools, such as cutting tools and saws. Make sure the tools stop moving before approaching the industrial robots.

4) Pay attention to the high temperature surfaces of workpieces and industrial robot systems. The temperature of robot's motor is very hot after long-term operation.

5) Pay attention to the clamp and make sure the workpiece is clamped. If the clamp is opened, the workpiece will fall off, which causes personal injury or equipment damage. The clamp is very

powerful and may cause personal injury if it is not operated correctly.

6) Pay attention to hydraulic and pneumatic systems and live parts. Even if the power is cut off, the residual power on these circuits is very dangerous.

(7) Safety of FlexPendant! FlexPendant is a high-quality hand-held terminal, which is equipped with first-class electronic equipment with high sensitivity. In order to avoid faults or damages caused by improper operation, please follow the instructions during operation:

1) Operate it carefully. Do not beat, throw or thump FlexPendant. When the device is not in use, it is hang on the special bracket to prevent it from falling to the ground accidentally.

2) The use and storage of FlexPendant should avoid being trampled on the cable.

3) Do not operate the touchscreen with sharp objects. Use fingers or touch pens to operate it.

4) Clean the touchscreen regularly.

5) Do not use solvents, detergents or scrubbing sponges to clean the FlexPendant. Clean it with a piece of soft cloth dipped in a little water or neutral detergent.

6) Always cover the USB port when not connecting the USB device.

(8) Safety in Manual Mode! In manual deceleration mode, the robot can only slow down (250mm/s or lower) to operate (move). As long as you work in a secure space, it is always operated at manual speed. In manual full speed mode, the robot moves at programmed speed. The manual full speed mode can be used only when all personnel are out of safe protection space, and the operator must be specially trained to know the potential danger.

(9) Safety in Automatic Mode! Automatic mode is used to run programs of robots in production. In the case of automatic mode, the general stop (GS), automatic stop (AS) and superior stop (SS) will be activated.

### 3.1.4 Evaluation

The table of task evaluation is shown in Table 3-1.

**Table 3-1 Task Evaluation**

| No. | Evaluation content | Achievement ratio | Selfevaluation | Teacher evaluation |
|---|---|---|---|---|
| 1 | Development and application of industrial robots | 15 | | |
| 2 | Characteristics of industrial robots | 15 | | |
| 3 | Typical structure of industrial robots | 40 | | |
| 4 | Safety precautions for industrial robots | 30 | | |

(1) According to the functions and uses, industrial robots can be roughly divided into four categories: machining robots, assembling robots, handling robots and packaging robots.

(2) The main enterprises include FANUC, YASKAWA, ABB, KUKA and so on, which are known as the 'Four Families' of industrial robots.

(3) The most remarkable characteristics of industrial robots include programmable, personification and versatility.

(4) Typical structures of industrial robots include cartesian-coordinate industrial robot, planar joint industrial robot, parallel industrial robot, serial industrial robot, and cooperative industrial robot.

(5) Safety precautions for industrial robots.

### 3.1.5 Development

ABB has installed more than 175000 robots around the world. The following is an introduction to the main models of ABB robots.

#### 3.1.5.1 IRB 120

IRB 120 (shown in Figure 3-6) is the smallest 6-axis robot manufactured by ABB so far, which is the newest member of ABB new forth-generation robot family. IRB 120 is agile, compact and lightweight, of which the control accuracy and path accuracy are both excellent. It is mainly used in logistics transportation, assembly and other work. IRB 120 only weighs 25kg, with a load of 3kg (4kg for the vertical wrist), and the working range is 580mm, as shown in Figure 3-7.

Figure 3-6 IRB 120

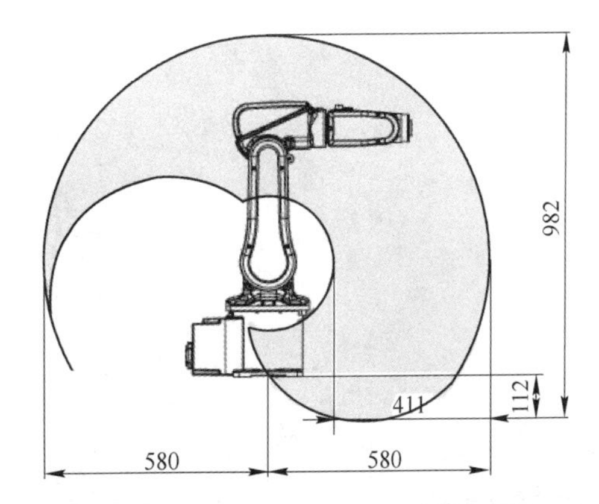


Figure 3-7 Working range of IRB 120

#### 3.1.5.2 IRB 260

IRB 260 (shown in Figure 3-8) robot is mainly designed and optimized for packaging applications. It has a small body and can be integrated into a compact packaging machine. The working range (shown in Figure 3-9) is closer to the base, minimizing the area covered at most. The robot adopts four-axis design, which is not only competent for all kinds of packaging operations, but also the guarantee of high productivity and flexibility.

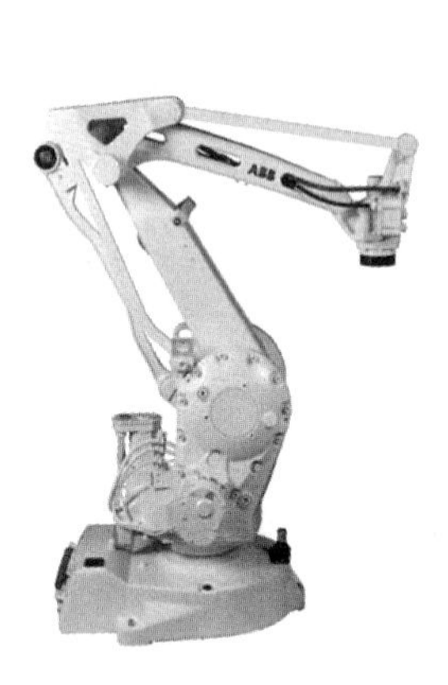

Figure 3-8 IRB 260

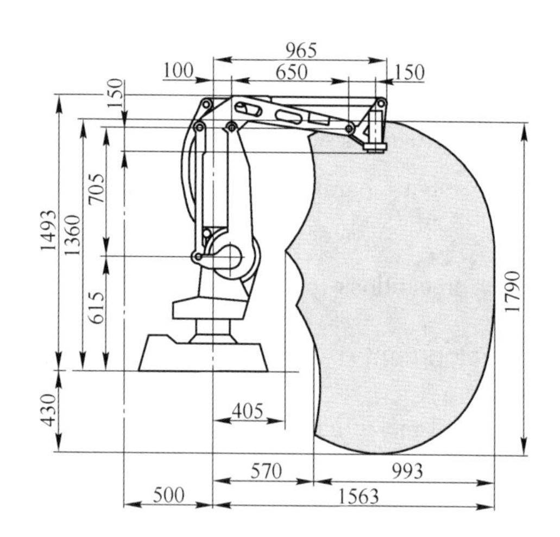


Figure 3-9 Working range of IRB 260

### 3.1.5.3 IRB 360

In the past 15 years, ABB IRB 360 (shown in Figure 3-10) has been leading the way in picking and packaging technology. IRB 360 covers a small area (shown in Figure 3-11), which is fast, flexible and heavy loaded (8kg). It adopts washable sanitary design, excellent tracking performance, integrated visual software, and synchronous integrated control of step conveyor. IRB 360 includes four series: compact type, standard type, heavy load type and long arm type.

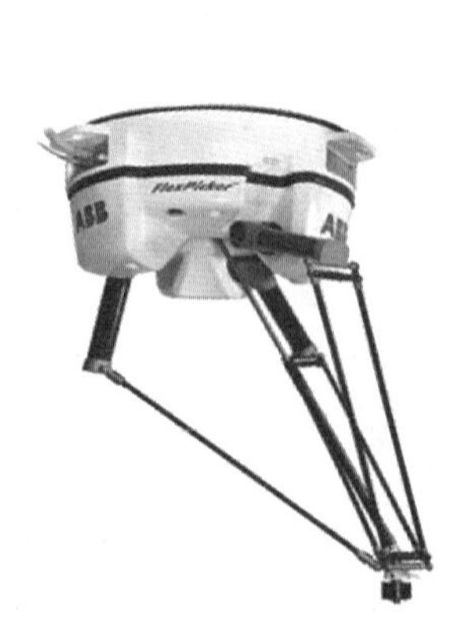

Figure 3-10 IRB 360

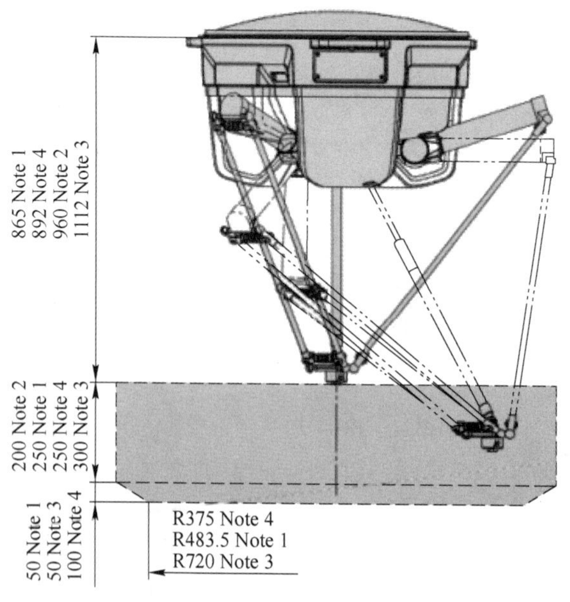


Note 1:IRB 360-1/1130 and IRB 360-3/1130
Note 2:IRB 360-1/800
Note 3:IRB 360-1/1600
Note 4:IRB 360-8/1130

Figure 3-11 Working range of IRB 360

### 3.1.5.4 IRB 910SC

IRB 910SC (known as SCARA), as shown in Figure 3-2, is a planar joint industrial robot produced by ABB. It adopts articulated robot arm, which is a robot of single arm that can operate in a narrow space. SCARA is fast and cost-effective, which is an ideal choice for small parts assembly, material handling and parts detection.

### 3.1.5.5 YuMi (IRB 14000)

Yumi, an industrial robot of double arms (shown in Figure 3-5), is the result of years of research and development, which increases the cooperation between human and robot. Yumi is designed for the new automation era, mainly used for the handling and assembly of small parts.

Yumi has a great prospect in the future and will change the way we think about assembly automation. Yumi means: you and me, exploring infinite possibilities together.

# Task 3.2 Recognition and Application of FlexPendant

## 3.2.1 Introduction

We should firstly contact with FlexPendant to operate ABB industrial robots. After learning this task, its components and main interface are know, the language and system time setted, the general information and event log of the robot viewed, and the data backup and recovery of industrial robots completed.

## 3.2.2 Related Knowledge

### 3.2.2.1 FlexPendant Components

FlexPendant is a kind of hand-held operation device, which is composed of hardware and software. It is a whole set of computer system, which is used to deal with many functions related to robot system operation, such as running programs, inching control operator, modifying robot programs, etc.

The main components of FlexPendant is shown in Figure 3-12. On FlexPendant, most of the operations are completed on the touch screen, while 12 special buttons (shown in Figure 3-13) are reserved. Table 3-2 shows the names and functions of each button.

**Table 3-2 The names and functions of each button**

| Code name | Button name | Function |
|---|---|---|
| A~D | Custom button | Users can set specific functions for these four keys as required; Programming these buttons can simplify program or test; They can also start menus on FlexPendant |

Continued Table 3-2

| Code name | Button name | Function |
|---|---|---|
| E | Selection button of mechanical units | Robot axis or external axis |
| F | Selection button between linear movement and reorient movement | Linear movement or reorient movement |
| G | Selection button of movement codes | 1-3 axis or 4-6 axis |
| H | Button of increment switch | Select the conesponding displacement and angle as required |
| J | Step back button | Move the program back to the previous command |
| K | START button | Start to execute programs |
| L | Step forward button | Move the program to the next command |
| M | STOP button | Stop program |

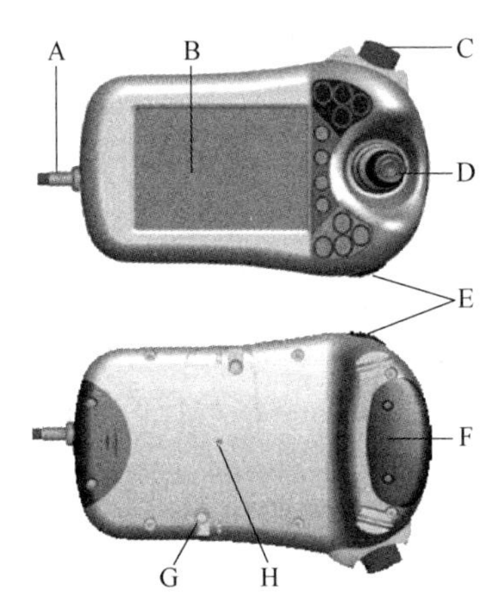


Figure 3-12 FlexPendant components

A—Connecting cable; B—Touch screen; C—Emergency stop switch; D—Operating lever; E—USB port; F—Enabler; G—Pen for touch screen; H—Reset button for FlexPendant

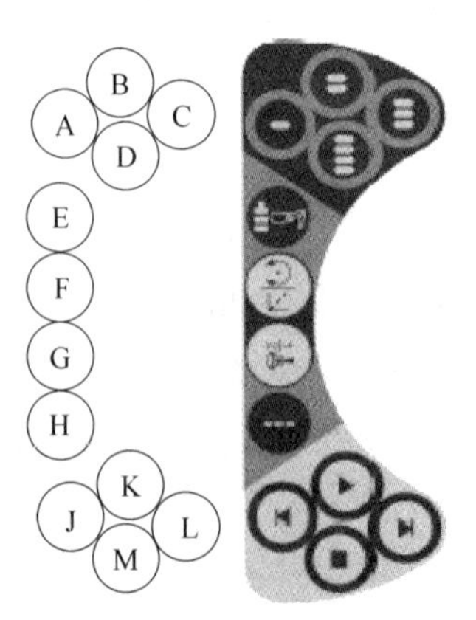


Figure 3-13 Buttons of FlexPendant

Generally the FlexPendant is operated with it held in hand. Right handed people usually hold the device with their left hand and operate the screen and buttons with their right hand (shown in Figure 3-14). On the contrary, the left-handed users can hold it by their right hand easily by rotating it 180 degrees (shown in Figure 3-15).

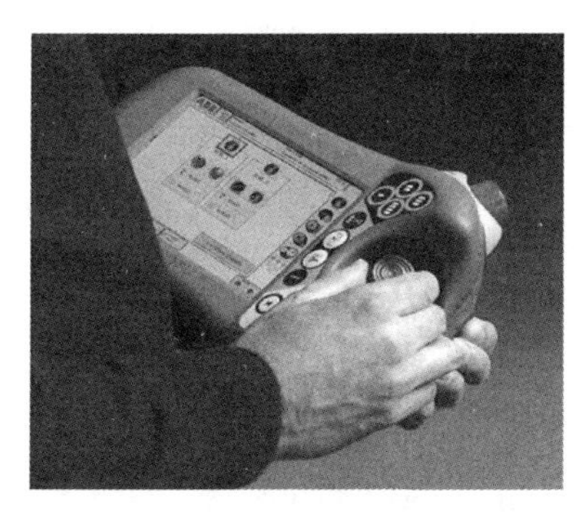

Figure 3-14 FlexPendant held in hand

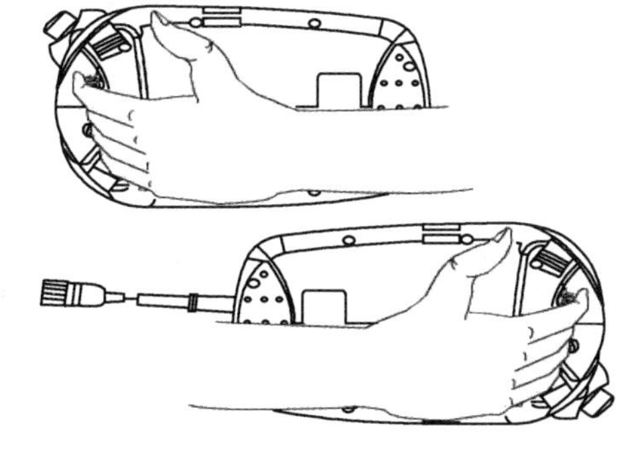

Figure 3-15 Holding FlexPendant with left/right hand

### 3.2.2.2 Recognize FlexPendant Main Interface

The initial interface of FlexPendant after startup is shown in Figure 3-16, and the names and functions of each part are shown in Table 3-3.

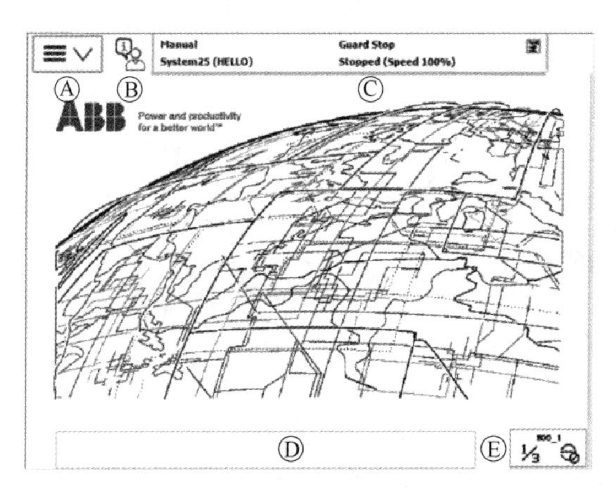


Figure 3-16 Initial interface of FlexPendant

**Table 3-3 Names and functions of each parts**

| Code name | Name | Function |
|---|---|---|
| A | Main menu | Options often used in the main menu interface include Inputs and Outputs, Jogging, Program Editor, Program Data, Calibration, Control Panel, etc. |
| B | Operator window | The operator window displays messages from the robot; This is often the case when the program requires some response from the operator to continue |
| C | Status bar | The status bar shows important information related to system status, such as operation mode, motor on/off, program status, etc. |

Continued Table 3-3

| Code name | Name | Function |
|---|---|---|
| D | Task bar | Through the main menu, the multiple views are opened, where only one can be operated once; The task bar displays all opened views which can be switched freely |
| E | Shortcut menu | The shortcut menu contains settings for inching control and program execution |

### 3.2.3 Implementation

#### 3.2.3.1 Setting the Language of FlexPendant

When FlexPendant is delivered from the factory, the default display language is English. If necessary, it can be set to another language through the following operations as follows:

(1) Click the main menu in the upper left corner and choose ‘Control Panel’ (shown in Figure 3-17).

(2) Choose ‘Language’.

(3) Choose the languse needed, such as ‘Chinese’ (shown in Figure 3-18), and then click ‘OK’.

(4) Click ‘Yes’, system restarted. After the restart, click the button in the upper left corner to see that the menu has been switched to Chinese interface.

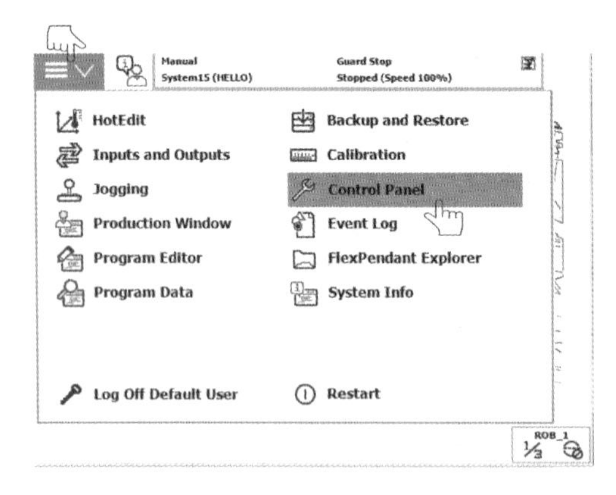


Figure 3-17 Click ‘Control Panel’

Figure 3-18 Choose ‘Chinese’

#### 3.2.3.2 Set System Time of Industrial Robot

In order to facilitate the management of documents and the consultation and management of faults, the robot system time should be set to local time before various operations. The operation processes are as follows:

(1) Click the main menu in the upper left corner and choose ‘Control Panel’.

(2) Choose ‘Date and Time’ (shown in Figure 3-19).

(3) In this screen, you can set the date and time. After that, click ‘OK’.

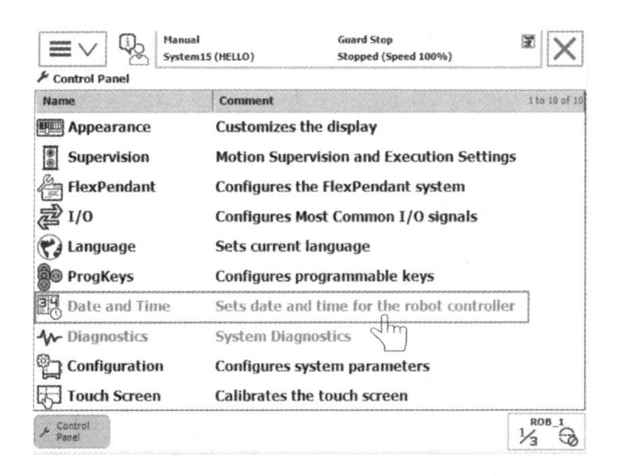


Figure 3-19 Click ‘Date and Time’

### 3.2.3.3 Check Common Information and Event Log of ABB Industrial Robot

The common information of ABB robot is shown on the status bar of FlexPendant, by which we can know the current status and some existing problems of the robot, as shown in Figure 3-20.

Click the status bar on FlexPendant to look overthe event log of the robot. The interface shows the event records of robot operation, including time and date, and provides accurate information for the analysis of related events and problems.

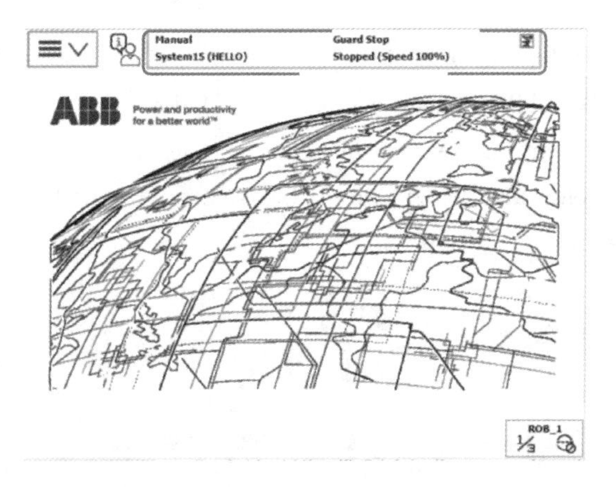


Figure 3-20 Common information of ABB robot

### 3.2.3.4 Data Backup and Restore of ABB Industrial Robot

It is a good habit to back up the data regularly to ensure ABB robots work well. The object of data backup is all RAPID programs and system parameters running in system memory. When robot system is disordered or new system is installed, the robot can be restored to the state of backup through backup operation quickly.

**Data Backup for ABB Industrial Robot**

The steps of data backup for ABB industrial robot are as follows:

(1) Click the main menu button in the upper left corner and select 'Backup and Restore' (shown in Figure 3-21).

(2) Click 'Backup Current System...'.

(3) Click 'ABC' button to set the folder name where the backup data is stored.

(4) Click '...' button to select the storage location (robot hard disk or USB storage device) and then click 'Backup' (shown in Figure 3-22).

(5) Wait for creating backup.

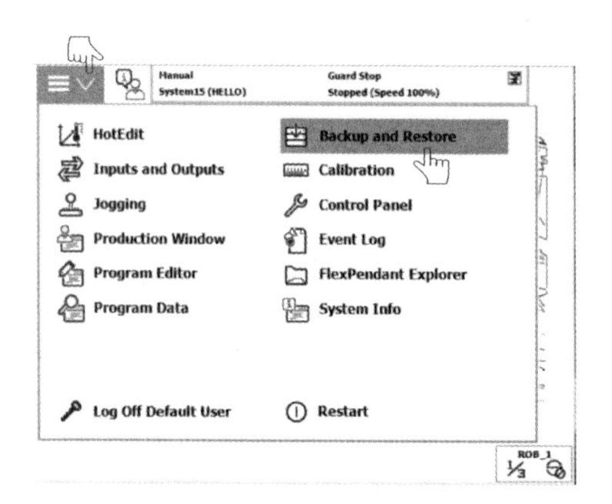


Figure 3-21 Click 'Backup and Restore'

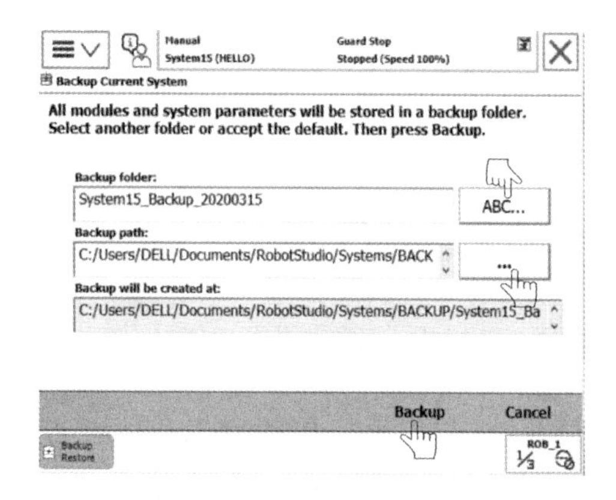


Figure 3-22 Set the name and storage location

## Data Restore for ABB Industrial Robot

The steps of data restore for ABB industrial robot are as follows:

(1) Click 'Restore System...' (shown in Figure 3-23).

(2) Click '...' to select the folder where the backup is stored and then click 'Restore' (shown in Figure 3-24).

(3) Click 'Yes'.

(4) Wait for restoring system.

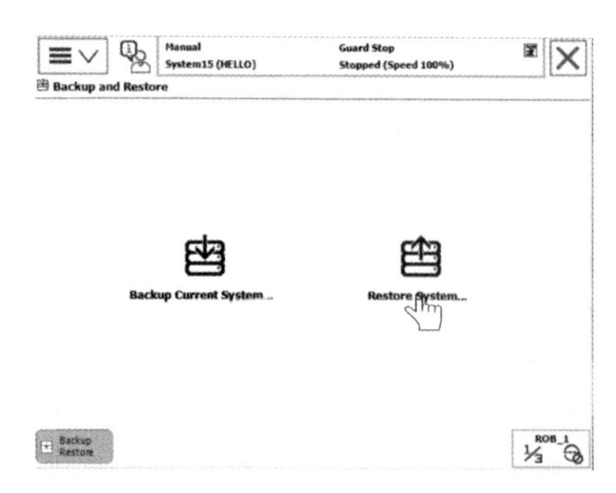


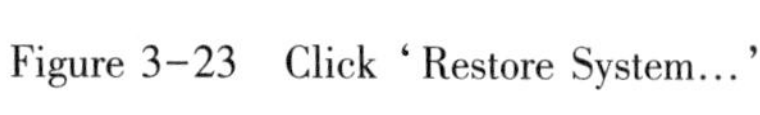

Figure 3-23 Click 'Restore System...'

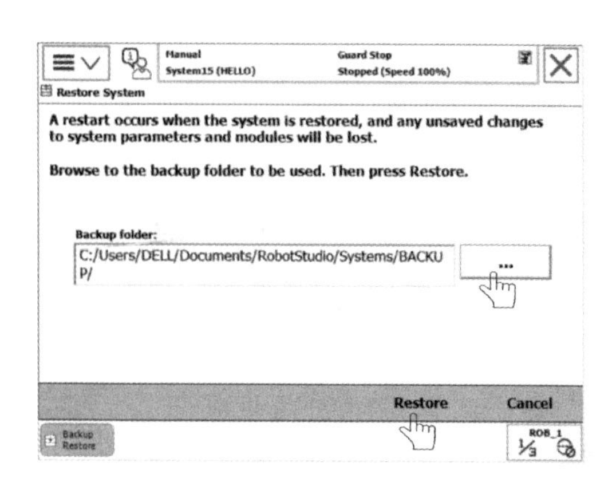


Figure 3-24 Choose the folder

### 3.2.4 Evaluation

The table of task evaluation is shown in Table 3-4.

**Table 3-4 Task evaluation**

| No. | Evaluation content | Achievement ratio | Self evaluation | Teacher evaluation |
|---|---|---|---|---|
| 1 | Components and hardware button of FlexPendant | 15 | | |
| 2 | Each part's names and functions of main interface of FlexPendant | 15 | | |
| 3 | Set the display language and system time of FlexPendant | 20 | | |
| 4 | Check common information and event log of the robot | 20 | | |
| 5 | Data backup and restore of ABB industrial robot | 30 | | |

(1) The components and functions of FlexPendant.

(2) The operation of setting the display language and system time of FlexPendant.

(3) The operation of checking common robot information and event log.

(4) Complete the backup and restoring of industrial robots.

### 3.2.5 Development

The backup data of a robot is unique, which can't be restored from one robot to another robot, otherwise system failure will be caused. However, the definition of program and I/O is often made commonly used, which is convenient for mass production. At this time, the program and EIO files are imported respectively to meet actual needs.

#### 3.2.5.1 Import Program Separately

The steps of import program separately are as follows:

(1) Click the main menu button in the upper left corner and choose 'Program Editor' (shown in Figure 3-25).

(2) Click the 'Modules' tab.

(3) Open 'File' menu and click 'Load Module...' to load the program module needed from 'Backup/RAPID' path (shown in Figure 3-26).

#### 3.2.5.2 Import EIO File Separately

The steps of import EIO file separately are as follows:

(1) Click the main menu in the upper left corner and choose 'Control Panel'.

(2) Choose 'Configuration' (shown in Figure 3-27).

(3) Open the 'File' menu and click 'Load Parameters...'.

(4) Choose ‘Delete exiting parameters before loading’.

(5) Find ‘EIO. cfg’ file from ‘Backup/SYSPAR’ path and click ‘OK’ (shown in Figure 3-28).

(6) Click ‘Yes’ to complete the import after restarting.

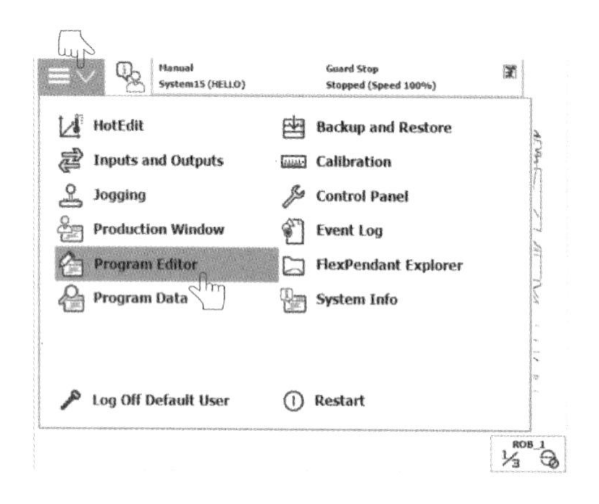


Figure 3-25 Click ‘Program Editor’

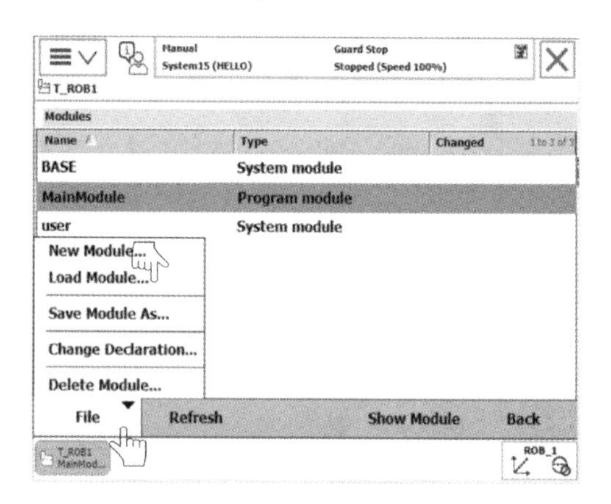


Figure 3-26 Click ‘Load Module…’

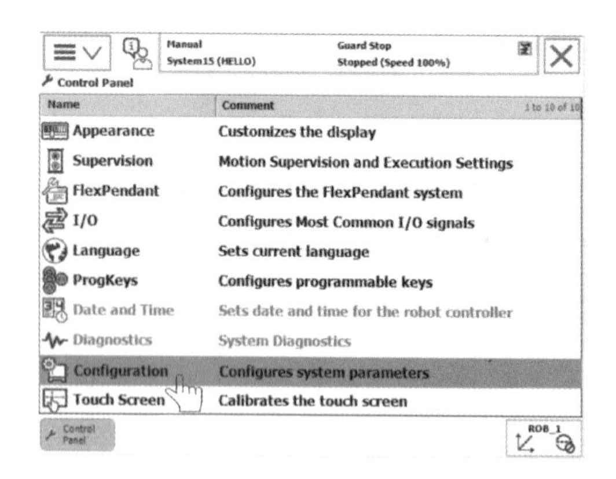


Figure 3-27 Click ‘Configuration’

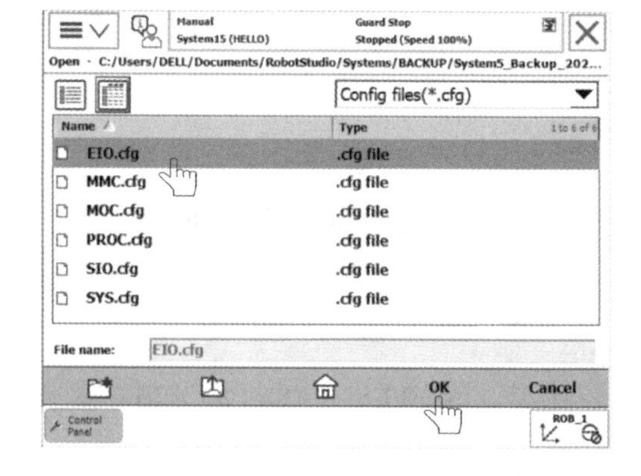


Figure 3-28 Choose ‘EIO. cfg’

# Task 3.3 Manual Operation of ABB Industrial Robots

## 3.3.1 Introduction

There are two operation modes of ABB industrial robot: manual mode and automatic mode. Manual mode is divided into deceleration speed mode and full speed mode. In manual deceleration speed mode, the maximum running speed can only reach 250mm/s. For beginners, operating a robot manually is the basis of learning industrial robots. This task mainly studies manual operation methods of single-axis motion, linear motion and reorient motion.

## 3.3.2 Related Knowledge

### 3.3.2.1 Enabler

The enabler button is on the right side of the manual operation lever of FlexPendant, as shown in Figure 3-29. The operator should operate it with four fingers of his hand, as shown in Figure 3-30. The enabler is divided into two gears: In the manual mode, pressing the first gear can make the industrial robot in 'Motor On' state; While pressing the second gear (press hard to the end) will make it in the 'Guard Stop' state.

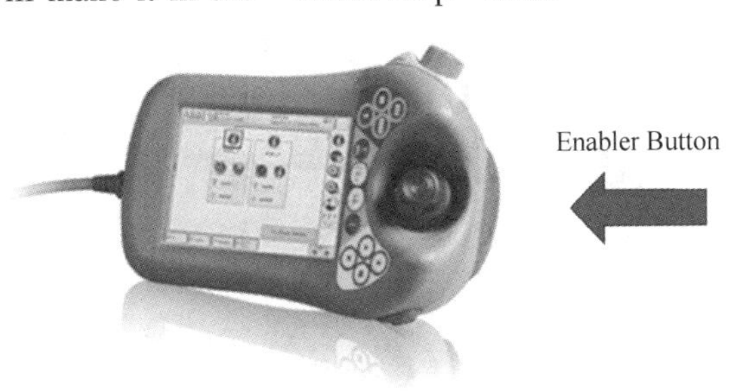


Figure 3-29 Enabler button

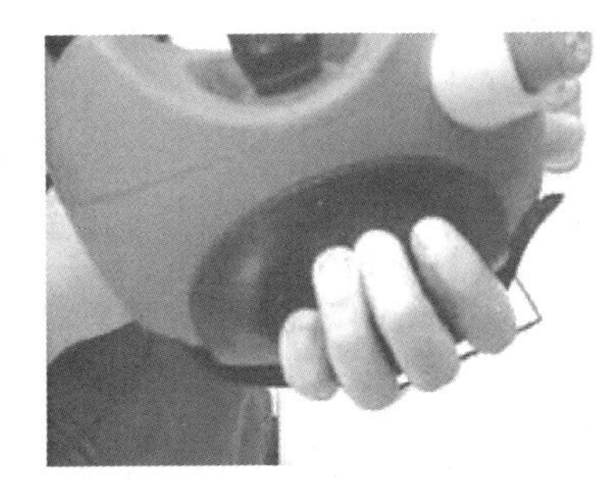

Figure 3-30 Operate the enable with four fingers

The enabler of the industrial robot are set up to ensure the personal safety of operators. Only when the enabler button is pressed and kept in the state of 'Motor On', the robot can be operated manually and the program can be debugged. In case of danger, people will instinctively release or press the button hard, which will make the robot will stop immediately to ensure safety.

### 3.3.2.2 Skills in Operating Lever

The operating lever of the robot can be compared with the throttle of the car. The operation range of the lever is related to movement speed of the robot. A small control range makes the robot move slowly while a large one make it move faster. So when we operate the lever, try a shorter range to make the robot move slowly to start learning manual operation.

### 3.3.2.3 Use of Increment

The robot running in manual mode contains two movement modes: default mode and incremental mode. The selection methods are as follows:

(1) Click the main menu in the upper left corner and choose 'Jogging'.

(2) Choose 'Increment' (shown in Figure 3-31).

(3) Select incremental mode as needed, and then click 'OK' (shown in Figure 3-32).

In the default mode 'None', the smaller the range of the lever, the slower the robot speed, and vice versa.

If you are not skillful in controlling the robot speed through the range of the operation lever, you can use the incremental mode to control it. In incremental mode, each time the lever swings, the robot will move one step (one increment); If the lever lasts one second or more, the robot will continue to move (at a rate of 10 steps/s). The incremental movement range can be selected between 'Small' 'Medium' and 'Large', or customized by users.

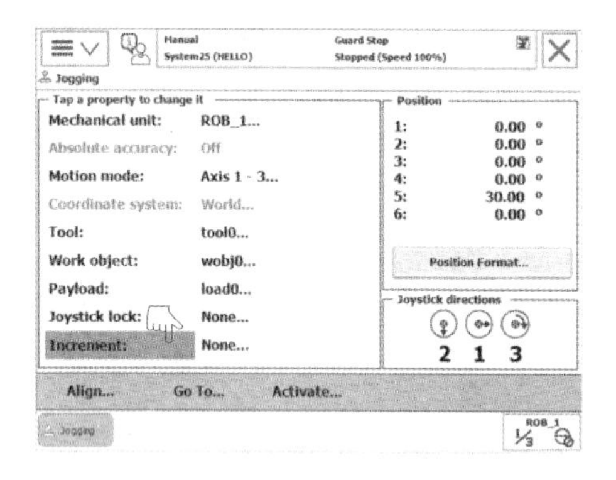


Figure 3-31 Click 'Increment'

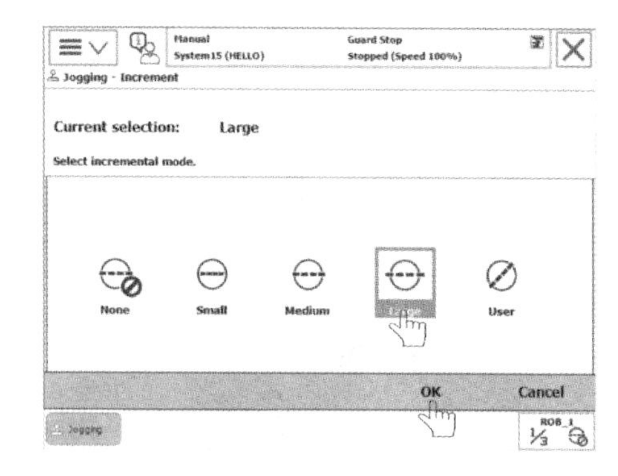


Figure 3-32 Select incremental mode

### 3.3.3 Implementation

There are three modes in industrial robot motion manually: single-axis motion, linear motion and reorient motion. The following describes how to operate the robot manually to complete the three kinds of motions.

#### 3.3.3.1 Single-axis Motion

Generally, six joint axes of ABB robot is driven by six servo motors separately (shown in Figure 3-33). Each time one joint is operated moving manually, which is called single-axis motion. The methods of manually controlling the single-axis motion are as follows:

(1) The robot mode key on is switched the control cabinet to manual deceleration speed mode (marked a small hand), as shown in Figure 3-34. In the status bar, it is confirmed that the status of the robot has been switched to manual.

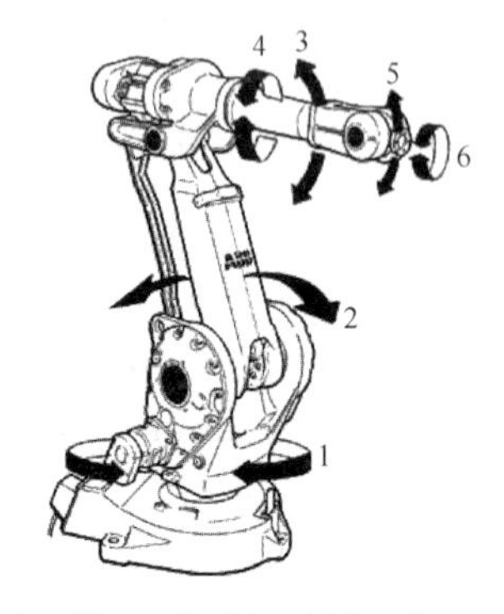


Figure 3-33 ABB robot

1~6—Joint axe

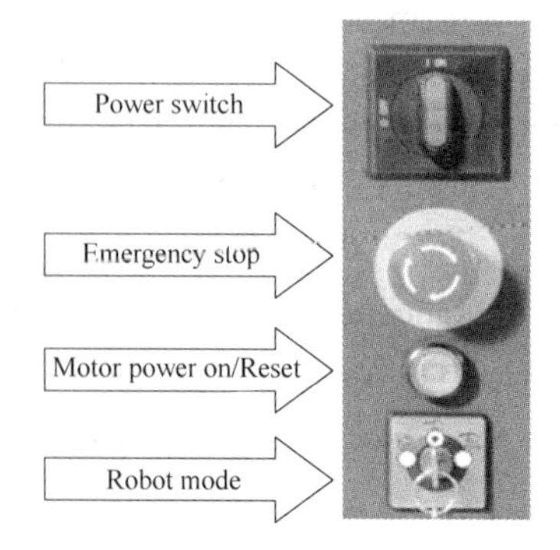


Figure 3-34 Buttons and switches on the control cabinet

(2) Click the main menu in the upper left corner, and select ‘Jogging’, and click ‘Motion Mode’. Then choose ‘Axis 1-3’ and click ‘OK’, as shown in Figure 3-35 (If choose ‘Axis 4-6’, axis 4-6 will move).

(3) Press the enabler by hand, and confirm that the robot has entered the ‘Motor On’ state in the status bar. The control lever direction of ‘Axis 1-3’ is shown in the lower right corner on the screen, where the arrow represents the positive direction (shown in Figure 3-36). At this time, the control lever is operated on FlexPendant to complete the single-axis motion.

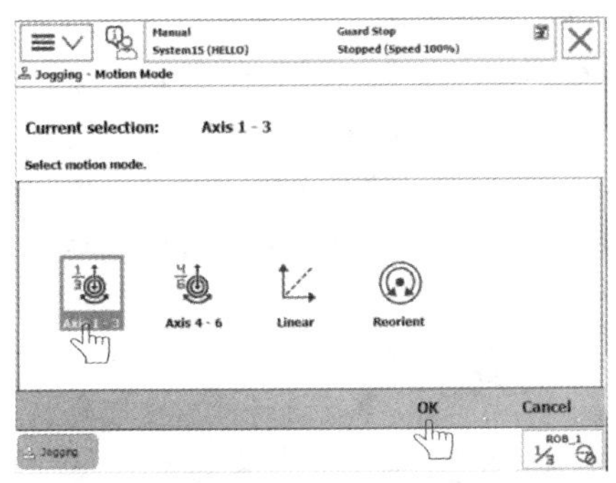


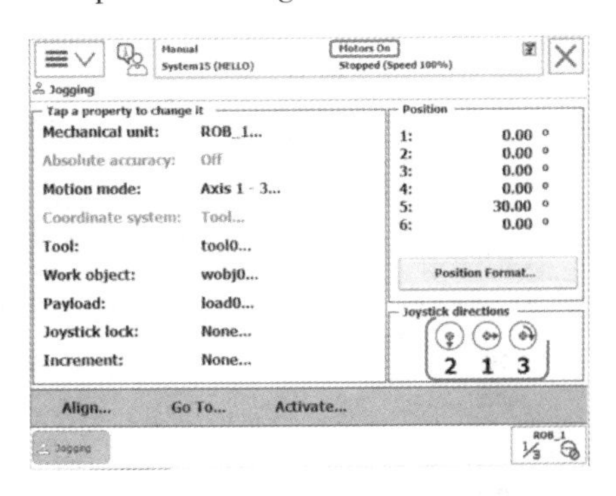

Figure 3-35 Choose ‘Axis 1-3’

Figure 3-36 Motor state and control lever direction

### 3. 3. 3. 2 Linear Motion

The linear motion of robot means that TCP of the tool, which is installed on the flange of the sixth axis, and makes linear motion in space. The methods of manually control linear motion are as follows:

(1) Select ‘Jogging’, and click ‘Motion Mode’. Select ‘Linear’, and click ‘OK’. Then click ‘Tool’ (Linear motion of the robot needs to specify the corresponding tool in the ‘Tool’), select the corresponding tool ‘Tool1’, and click ‘OK’, as shown in Figure 3-37 (Refering to Task 5. 2 for tool data setting).

(2) Press the enabler by hand, and confirm that the robot has entered the ‘Motor On’ state in the status bar. The control lever direction of ‘X, Y, Z’ is shown in the lower right corner on the screen, where the arrow represents the positive direction. At this time, the control lever is operated on FlexPendant and the TCP point moves linearly in space (shown in Figure 3-38).

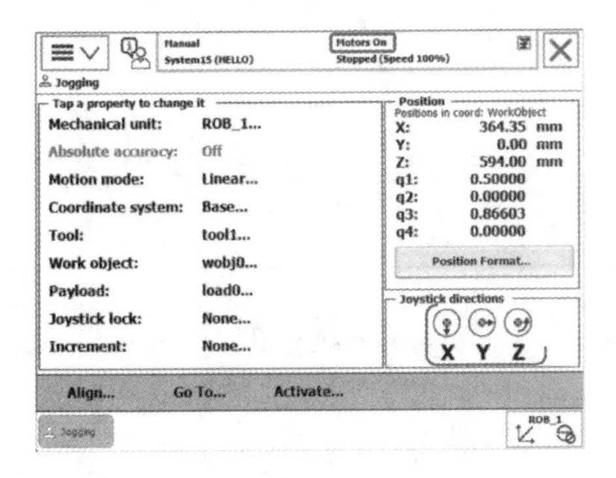

Figure 3-37 Motor star and control liver direction

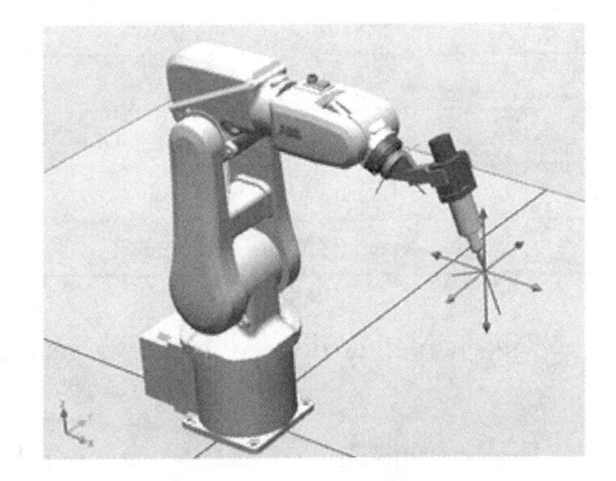
Figure 3-38 Linear motion

### 3. 3. 3. 3 Reorient Motion

The reorient motion of robot means that TCP of the tool, which is installed on the flange of the sixth axis, rotates around the coordinate axis in space. The methods of manually control reorient motion are as follows:

(1) Select 'Jogging', and click 'Motion mode'. Select 'Reorient', and click 'OK'. Click 'Coordinate system', choose 'Tool', and click 'OK' (shown in Figure 3-39). Then click 'Tool', select the corresponding tool 'tool1', and click 'OK'.

(2) Press the enabler by hand, and confirm that the robot has entered the 'Motor On' state in the status bar. The control lever direction of 'X, Y, Z' is shown in the lower right corner on the screen, where the arrow represents the positive direction. At this time, the control lever is operated on FlexPendant and the robot moves around TCP point for attitude adjustment (shown in Figure 3-40).

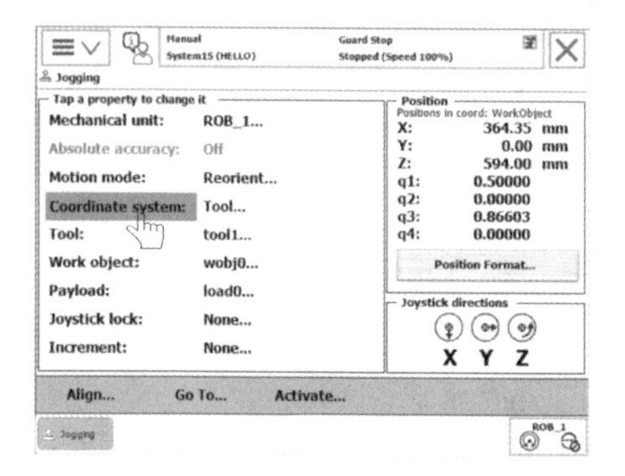


Figure 3-39 Click 'Coordinate system'

Figure 3-40 Reorient motion

### 3. 3. 4 Evaluation

The table of the task evaluation is shown in Table 3-5.

**Table 3-5 Task evaluation**

| No. | Evaluation content | Achievement ratio | Self evaluation | Teacher evaluation |
|---|---|---|---|---|
| 1 | Operate the enabler in the right way | 15 | | |
| 2 | Skills in Operating Lever | 15 | | |
| 3 | Use of Increment | 10 | | |
| 4 | Single-axis motion of the robot | 20 | | |
| 5 | Linear Motion of the robot | 20 | | |
| 6 | Reorient Motion of the robot | 20 | | |

(1) The enabler is divided into two gears, which is set up to ensure the personal safety of operators.

(2) The operation range of the lever is related to movement speed of the robot.

(3) The robot running in manual mode contains two movement modes: default mode and incremental mode.

(4) There are three modes in industrial robot motion manually: single-axis motion, linear motion and reorient motion.

### 3.3.5 Development

Manual operation of robots is widely used in the programming and debugging. If operated through the menu every time, it is too complicated. So FlexPendant provides shortcut buttons and shortcut menus to simplify the operation.

#### 3.3.5.1 Shortcut Buttons of Manual Operation

As required, four shortcut buttons of manual operation are arranged on FlexPendant of ABB industrial robot, as shown in Figure 3-41.

#### 3.3.5.2 Shortcut Menu of Manual Operation

Click the shortcut menu button in the lower right corner, choose 'Jogging', and click 'Show Details' button (shown in Figure 3-42). The functions of each part in the interface are shown in Figure 3-43. Then click 'Increments' on the shortcut menu to choose the increment needed, as shown in Figure 3-44.

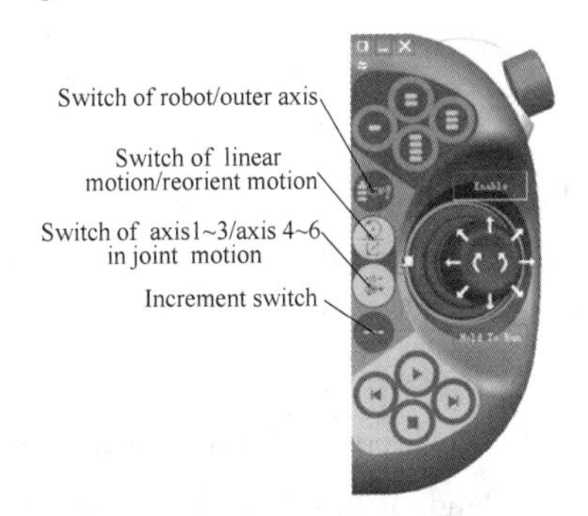


Figure 3-41 Four shortcut buttons of Manual operation

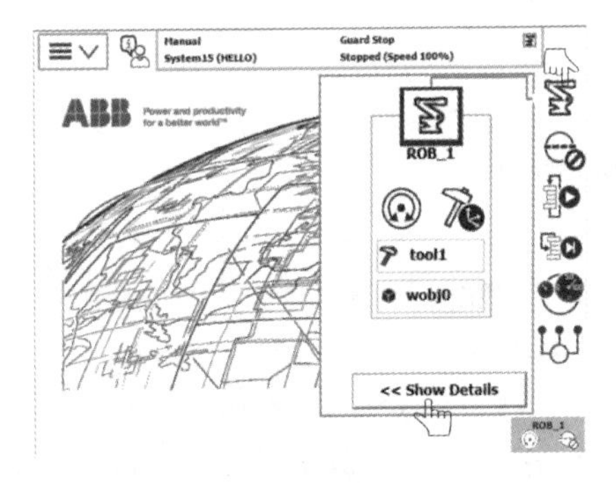


Figure 3-42 Click 'Jogging'

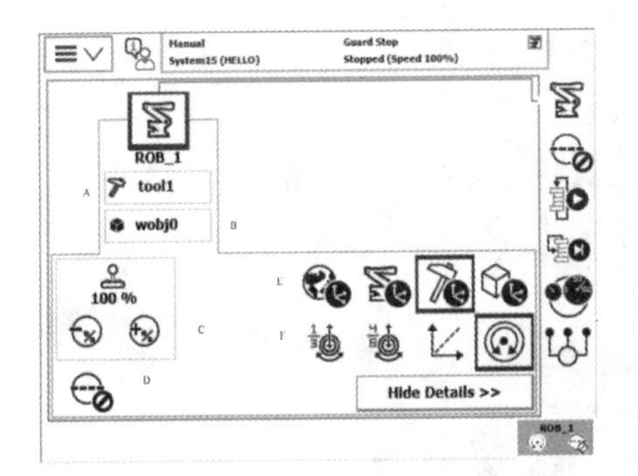


Figure 3-43 Interface of 'Show Details'

A—Select tool data currently used;

B—Select workpiece coordinate currently used;

C—Operating lever speed; D—Increment on/off;

E—Select coordinate system; F—Select motion mode

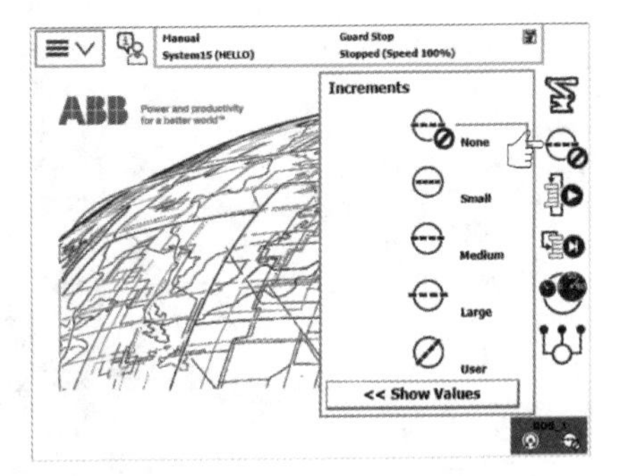


Figure 3-44 Click 'Increments'

# Task 3.4 Update Revolution Counters of ABB Robots

## 3.4.1 Introduction

The update of revolution counters is to stop each axis of the robot at the mechanical origin, align the synchronous marks on each joint axis, and then calibrate and update on FlexPendant. After learning this task, the mechanical origin position of the industrial robot is known, and each joint axis is moved to the mechanical origin by manual operation to update the revolution counter.

## 3.4.2 Related Knowledge

There is a mechanical origin position on all six joint axes of ABB industrial robot. In the following cases, it is necessary to update the revolution counters for the position of the mechanical origin: (1) after replacing the revolution counter battery of the servo motor; (2) when the revolution counter breaking down and repaired; (3) after disconnection between revolution counter and measuring board; (4) after power off, the joint axis of the robot displaced; and (5) when the system alarms and prompts '10036, Revolution counter not updated'.

## 3.4.3 Implementation

The operations of the revolution counter update of IRB1200 are as follows:

(1) By the method of 'Single-axis Motion' in last task, each joint axis is returned to the mechanical origin position one by one in the order of axis 4-5-6-1-4-3. There is an obvious mark of mechanical origin on each axis of the robot, which is easy to identify.

(2) After all six axes return to the mechanical origin, the robot posture, as shown in Figure 3-45. Next, continues the following operations with FlexPendant to update the counter:

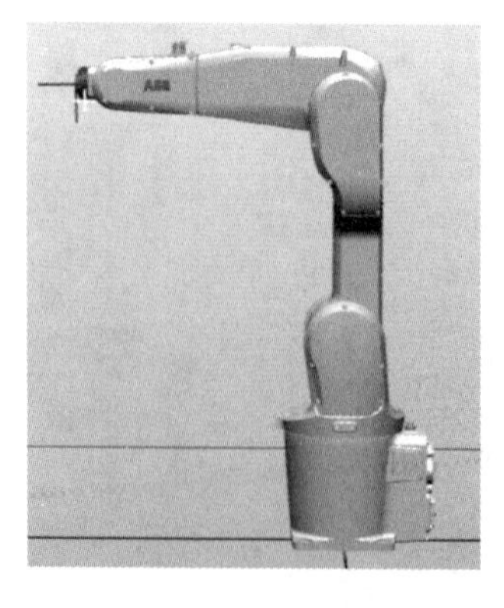

Figure 3-45 Six axes returning to mechanical origin

1) Click the main menu in the upper left corner and select 'Calibration' (shown in Figure 3-46)

2) Click 'ROB_1'.

3) Select ‘Calib. Parameters’, and then click ‘Edit Motor Calibration Offset...’, as shown in Figure 3-47.

4) Click ‘Yes’.

5) Record the offset value of motor calibration on the robot body, as shown in Figure 3-48, input it into the edit motor calibration offset interface, and then click ‘OK’, as shown in Figure 3-49. If the value displayed on FlexPendant is the same as the label value on the robot body, there is no need to modify it, and then click ‘Cancel’.

6) To activate these parameters, the controller needs to be restarted, and click ‘Yes’.

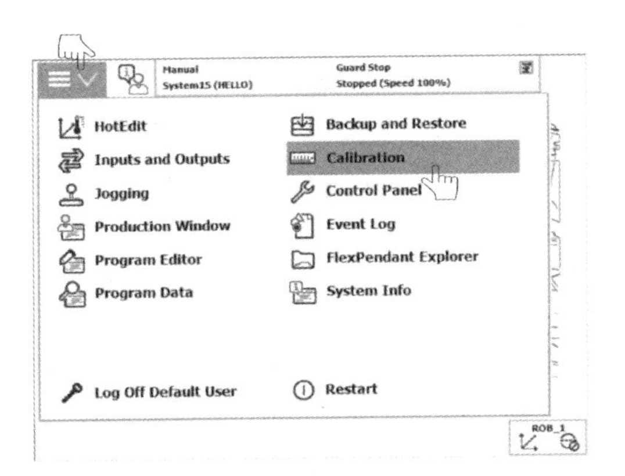


Figure 3-46 Click ‘Calibration’

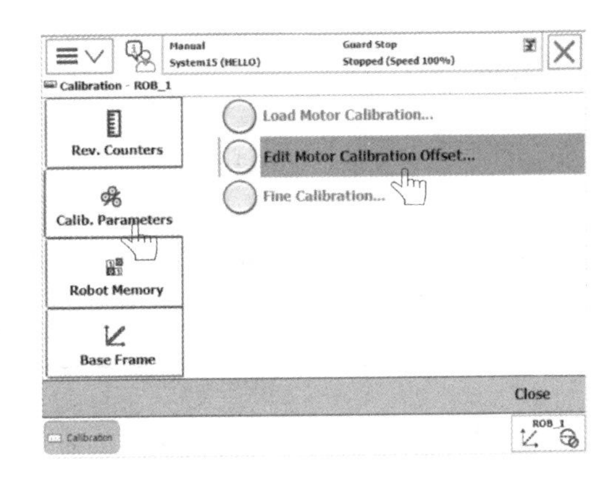


Figure 3-47 Click ‘Edit Moto Calibration Offset...’

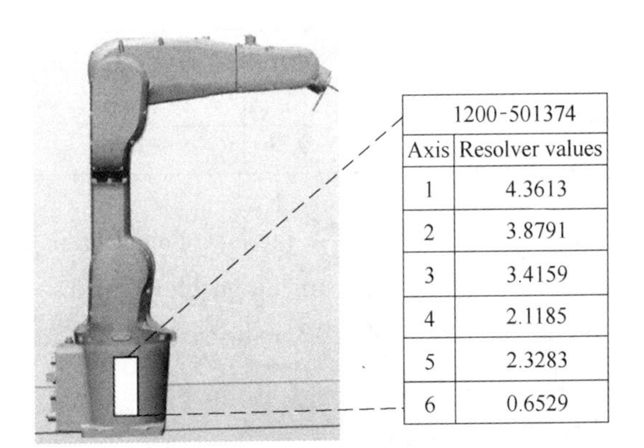

| 1200-501374 | |
|---|---|
| Axis | Resolver values |
| 1 | 4.3613 |
| 2 | 3.8791 |
| 3 | 3.4159 |
| 4 | 2.1185 |
| 5 | 2.3283 |
| 6 | 0.6529 |

Figure 3-48 Offset value of motor calibration on the robot body

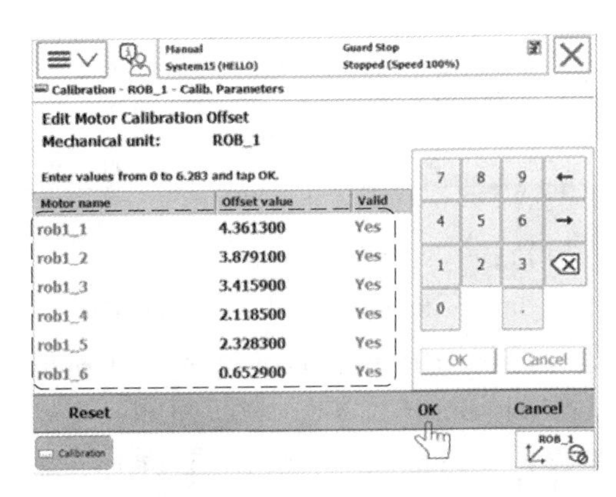


Figure 3-49 Interface of edit motor calibration offset

(3) After restarted, select ‘Calibration’ again, click ‘ROB_1’, select ‘Rev. Counters’, and then click ‘Update Revolution Counters...’, as shown in Figure 3-50. Then click ‘Yes’, click ‘OK’, click ‘Select All’ (shown in Figure 3-51), and then click ‘Update’. If six axes cannot reach their mechanical origin at the same time due to the installation position of the robot, the revolution counters of joint axes can be updated one by one. Then click ‘Update’. After these operation, the revolution counters update is complete.

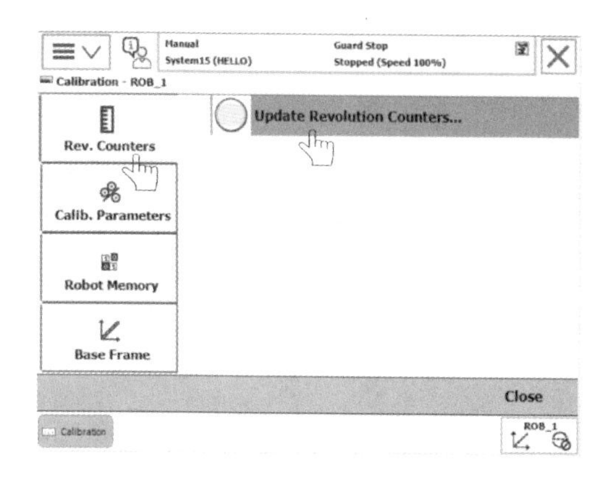


Figure 3-50 Click ‘Update Revoluation Counters...’

Figure 3-51 Click ‘Select All’

### 3. 4. 4 Evaluation

The table of the task evaluation is shown in Table 3-6.

**Table 3-6 Task evaluation**

| No. | Evaluation content | Achievement ratio | Self evaluation | Teacher evaluation |
|---|---|---|---|---|
| 1 | Mechanical origin position on each axis of the industrial robot | 10 | | |
| 2 | The cases that revolution counters need to be updated | 20 | | |
| 3 | Move each joint axis to the mechanical origin by manual operation | 30 | | |
| 4 | Update the revolution counters on FlexPendant | 40 | | |

(1) The update of revolution counters is to stop each axis of the robot at the mechanical origin, align the synchronous marks on each joint axis, and then calibrate and update on FlexPendant.

(2) There is a mechanical origin position on all six joint axes of ABB industrial robot. Each joint axis is moved to the mechanical origin by manual operation.

(3) The revolution counters is updated on FlexPendant.

### 3. 4. 5 Development

#### 3. 4. 5. 1 Function of the Battery in Industrial Robots

In this book, zero information data of the robot is stored on the serial measurement board of the body, while the serial measurement board is powered by the main power when the robot system connects with external main power. When the system disconnects from the main power, the battery of the serial measurement board (the battery of the body) is required to power it.

If the serial measurement board is powered off, the zero information will be lost, which will cause each joint axis of the robot cannot move according to the correct benchmark. In order to retain the storage of mechanical zero position data of the robot, serial measurement board must be powered constantly. When the battery reserve of the serial measurement board can lasts less than 2 months (the power of industrial robot is off), the warning of battery low (38213, battery low) will appear, and the battery needs to be replaced. Otherwise, the battery power is exhausted, every time the main power is on again after power off, and the revolution counters update is always required.

#### 3. 4. 5. 2 Battery Life of Industrial Robots

Generally, if the power of industrial robot is turned off 2 days a week, the service life of new battery is 36 months. While the power is turned off 16 hours a day, the service life of new battery is 18 months. For long production interruption, the battery life can be extended (approximately three times longer) by closing service routine through the battery.

# Project 4 I/O Communication of ABB Industrial Robots

In the actual production process, in order to complete the work task better, the industrial robot often needs to carry on the communication with the peripheral equipment, and obtains the environment and the equipment information. Industrial robot communication can be realized by different communication interfaces such as RS485, optical fiber, Modbus protocol, TCP/Ip protocol etc. This project introduces the communication types of ABB industrial robots, standard I/O communication board, DSQC651 board configuration, communication bus connection and so on, and the ABB industrial robots I/O communication and use logic methods are understood.

## Task 4.1 ABB Industrial Robot I/O Communication Types

### 4.1.1 Introduction

The default configuration of ABB industrial robot is I/O communication board based on DeviceNet protocol, which is used to communicate between host control system and industrial equipment. DeviceNet is a kind of CAN Controller Area Net. This kind of sensor/actuator bus system was originated from America, but it is more and more widely used in Europe and Asia.

### 4.1.2 Related Knowledge

I/O is an abbreviation for Input/Output, or Input/Output port. Robots can interact with peripheral equipment via I/O. For example:

(1) Digital input, which includes: various switch signal feedback (such as button switch, transfer switch, proximity switch, etc.), sensor signal feedback (such as photoelectric sensor and optical fiber sensor), and contact signal feedback of contactor and relay. There is also the switch signal feedback in the touch screen.

(2) Digital output, which includes: controling various relay coils (such as contactors, relays and solenoid valves), and control various indicator signals (such as indicator lights and buzzers).

Input and output of standard I/O boards for ABB robots are of the PNP type.

#### 4.1.2.1 ABB Industrial Robot I/O Communication

ABB Robot provides a rich I/O communication interface, such as ABB standard communication, PLC fieldbus communication, and data communication with PC (shown in Figure 4-1). I/O communication can easily achieve communication with peripheral equipment.

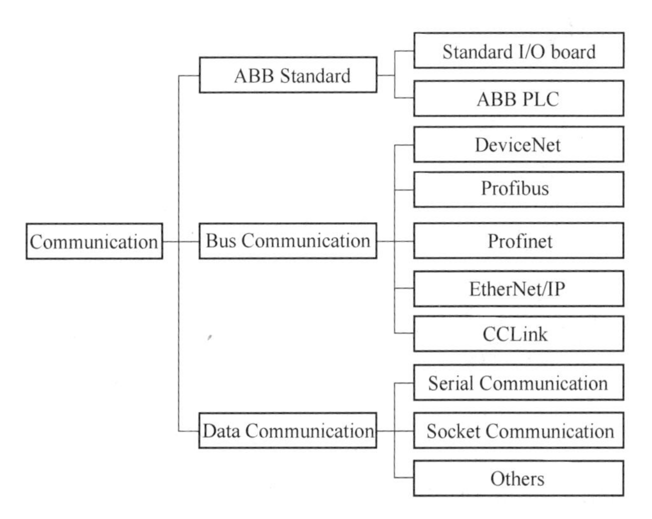


Figure 4-1 ABB robot I/O interface

ABB standard I/O board provides common signal processing including digital input, digital output, group input, group output, analog input, and analog output.

ABB robot can choose standard ABB PLC, which saves the trouble of communication with External PlC, and it can realize the related operation with PLC on the robot's demonstrator.

Here, the most commonly used ABB standard I/O board DSQC651 and Profibus-DP are taken as examples to explain in detail.

#### 4.1.2.2 Introduction of ABB Robot Communication

Main computer unit, ABB standard I/O board general installation location is shown in Figure 4-2, the robot bus and serlal port is shown in Figure 4-3, and Profibus and location of memory card is shown in Figure 4-4.

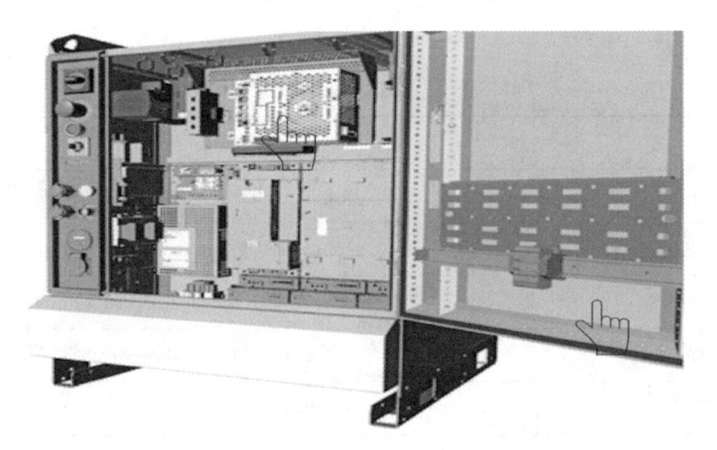

Figure 4-2 ABB standard I/O board installation location

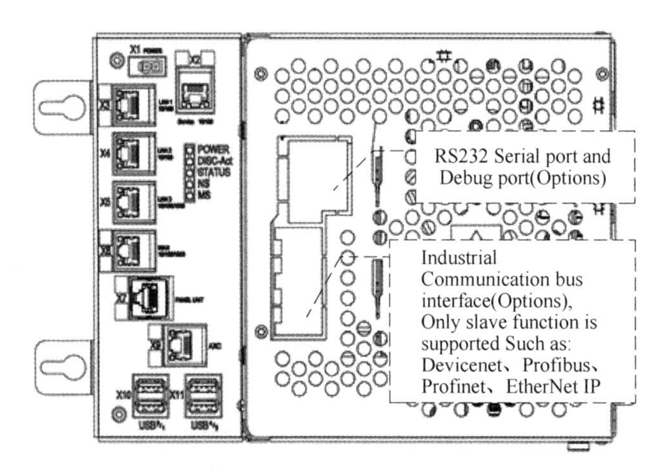


Figure 4-3  Robot bus and serlal port

Note: The WAN interface requires the option 'PCINTERFACE' to be used. And it is optional depending on your needs to choose fieldbus.

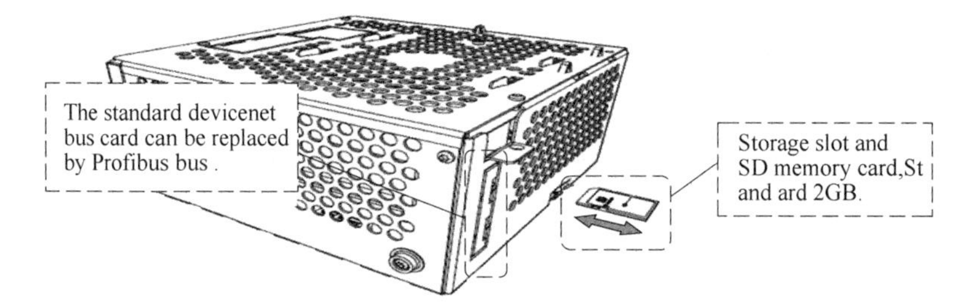


Figure 4-4  Profibus bus and location of memory card

Note: If you use ABB standard I/O board, you must have the bus of DeviceNet.

#### 4.1.2.3 ABB Standard I/O Card

Common Abb Standard I/O Board is shown in Table 4-1. The specific specifications to ABB official latest release shall prevail.

**Table 4-1  ABB standard I/O card**

| Type | Introduction |
|---|---|
| DSQC651 | Distributed I/O module di8 \ do8 \ ao2 |
| DSQC652 | Distributed I/O module di16 \ do16 |
| DSQC653 | Distributed I/O module di8 \ do8 with relay |
| DSQC355A | Distributed I/O module ai4 \ ao4 |
| DSQC377A | Conveyor chain tracking unit |

DSQC651 board mainly provides the processing of 8 digital input signals, 8 digital output signals and 2 analog output signals.

**Module Interface Description**

As shown in Figure 4-5, A represents digital output signal indicator, B represents X1 digital output interface, C represents X6 analog output interface, D represents X5 DeviceNet interface, E represents module status indicator, F represents X3 digital input interface, and G represents digital input signal indicator.

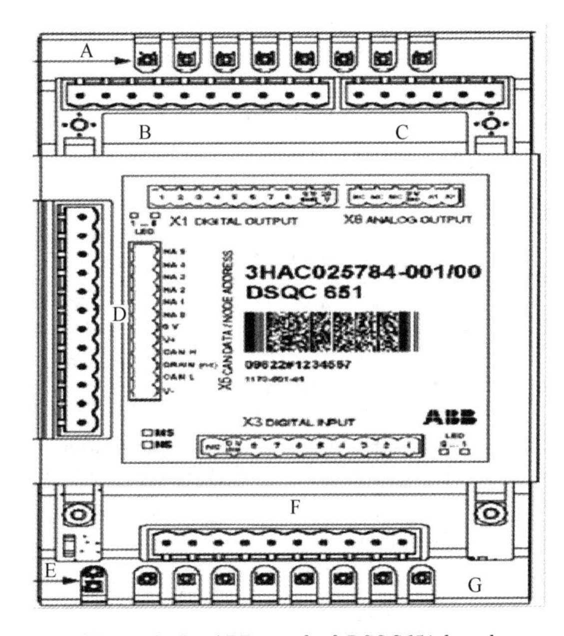


Figure 4-5 ABB standard DSQC651 board

**Module Interface Connection Description**

The direction of X1 terminal is shown in Table 4-2.

**Table 4-2 X1 terminal interface specification**

| X1 terminal number | Usage definition | Address assignment |
|---|---|---|
| 1 | OUTPUT CH1 | 32 |
| 2 | OUTPUT CH2 | 33 |
| 3 | OUTPUT CH3 | 34 |
| 4 | OUTPUT CH4 | 35 |
| 5 | OUTPUT CH5 | 36 |
| 6 | OUTPUT CH6 | 37 |
| 7 | OUTPUT CH7 | 38 |
| 8 | OUTPUT CH8 | 39 |
| 9 | 0V | — |
| 10 | 24V | — |

The direction of X3 terminal is shown in Table 4-3.

**Table 4-3  X3 terminal interface specification**

| X3 terminal number | Usage definition | Address assignment |
|---|---|---|
| 1 | INPUT CH1 | 0 |
| 2 | INPUT CH2 | 1 |
| 3 | INPUT CH3 | 2 |
| 4 | INPUT CH4 | 3 |
| 5 | INPUT CH5 | 4 |
| 6 | INPUT CH6 | 5 |
| 7 | INPUT CH7 | 6 |
| 8 | INPUT CH8 | 7 |
| 9 | 0V | — |
| 10 | Unused | — |

The direction of X5 terminal is shown in Table 4-4.

**Table 4-4  X5 terminal interface specification**

| X5 terminal number | Usage definition |
|---|---|
| 1 | 0V BLACK (BLACK) |
| 2 | CAN The signal line low BLUE (BLUE) |
| 3 | Shielding wire |
| 4 | CAN The signal line high WHITE (WHITE) |
| 5 | 24V RED (RED) |
| 6 | GND Address selection common |
| 7 | Modular ID bit 0 (LSB) |
| 8 | Modular ID bit 1 (LSB) |
| 9 | Modular ID bit 2 (LSB) |
| 10 | Modular ID bit 3 (LSB) |
| 11 | Modular ID bit 4 (LSB) |
| 12 | Modular ID bit 5 (LSB) |

Note: the ABB standard I/O board is hung on the DeviceNet network, so the address of the module in the network should be set. The 6~12 jumper of terminal X5 is used to determine the address of the module, and the available address range is 10~63.

As shown in Figure 4-6, the jumpers of the 8th and 10th pins are cut off, and '2+8=10' will get the address of 10.

The direction of X6 terminal is shown in Table 4-5.

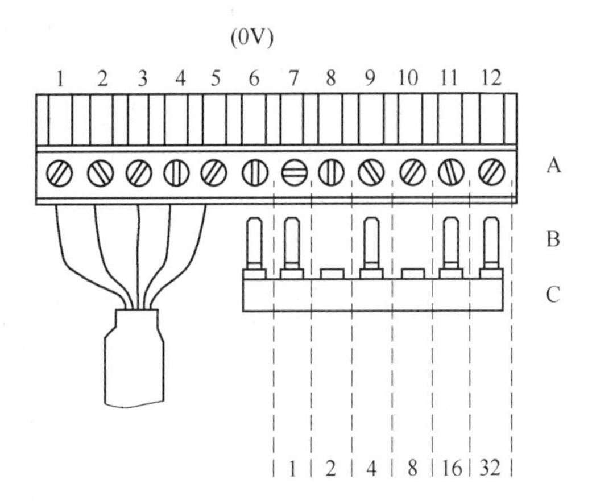

Figure 4-6 Jumper of ABB standard I/O board address

**Table 4-5 X6 terminal interface specification**

| X6 terminal number | Usage definition | Address assignment |
|---|---|---|
| 1 | not used | — |
| 2 | not used | — |
| 3 | not used | — |
| 4 | 0V | — |
| 5 | Analog outputAO1 | 0-15 |
| 6 | Analog outputAO2 | 16-31 |

Note: Analog output range is from 0V to +10V.

### 4. 1. 3 Implementation

ABB standard I/O board DSQC651 is the most commonly used module. The following takes the creation of digital input signal di1, digital output signal do1, group input signal gi1, and group output signal go1 as an example to implement.

#### 4. 1. 3. 1 Define the Bus Connection of DSQC651 Board

ABB standard I/O boards are all devices under DeviceNet fieldbus, and communicate with DeviceNet fieldbus through X5 port.

Table 4-6 shows the description of the parameters defining the bus connection of DSQC651 board.

**Table 4-6 Parameters of the bus connection of the DSQC651 board**

| Parameter name | Set value | Explain |
|---|---|---|
| Name | boardI/O | Set up I/O the name of the board in the system |
| Network | DeviceNet | I/O board connected bus |
| Address | I/O | Set up I/O address of the board in the bus |

To define the DSQC651 Card in the system, the following steps are used:

(1) Click the main menu button in the upper-left corner and select control panel, select 'Configuration', and double click 'DeviceNet Device'. Then click 'Add to', click 'Use values from template' corresponding drop-down arrow, and choose 'DSQC 651 Combi I/O Device' (shown in Figure 4-7).

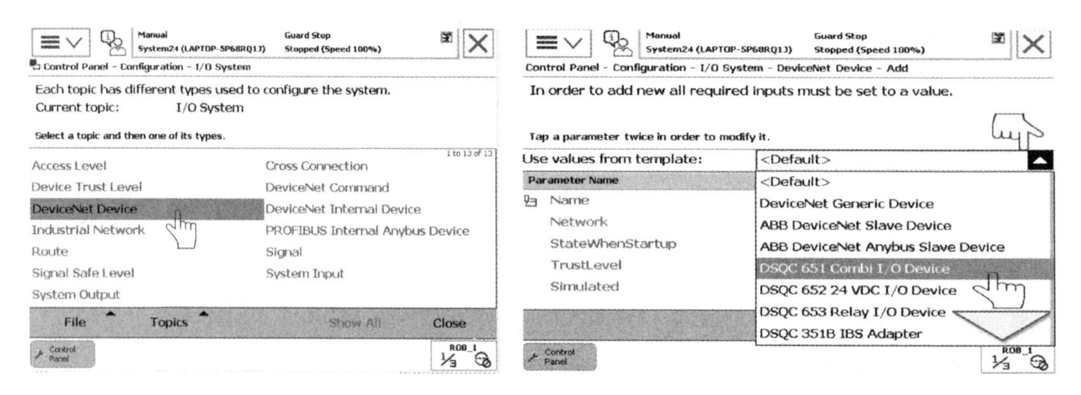


Figure 4-7 Define standard DSQC651 board

(2) Double click 'Name' conduct DSQC651 setting the name of board in the system. If not modified, the name is the default 'd651'. In the system DSQC651, the name of the board is set to 'board10' (10 represents the address of the module in the DeviceNet Bus for easy identification). Take 'Address' set to 10, and click 'Determination'. Then click 'Yes', and the DSQC651 board definition complete.

### 4.1.3.2 Define Digital Input Signal di1

Table 4-7 shows the relevant parameters of digital input signal di1.

**Table 4-7 Parameters of the digital input signal di1**

| Parameter name | Set value | Explain |
| --- | --- | --- |
| Name | di1 | Set the name of the digital input signal |
| Type of Signal | Digital Input | Set the type of signal |
| Assigned to Device | board10 | Set the I/O module of the signal |
| Device Mapping | 0 | Set the address occupied by the signal |

The steps of operation are as follows:

(1) Click the main menu button in the upper left corner, choose 'Control Panel', choose 'Configuration'. Double click 'Signal', and click 'Add to'. Double click 'Name', input 'di1', and click 'Determine'. Then double click 'Type of Signal', and choose 'Digital Input' (shown in Figure 4-8).

(2) Double click 'Assigned to Device', choose 'board10'. Double click 'Device

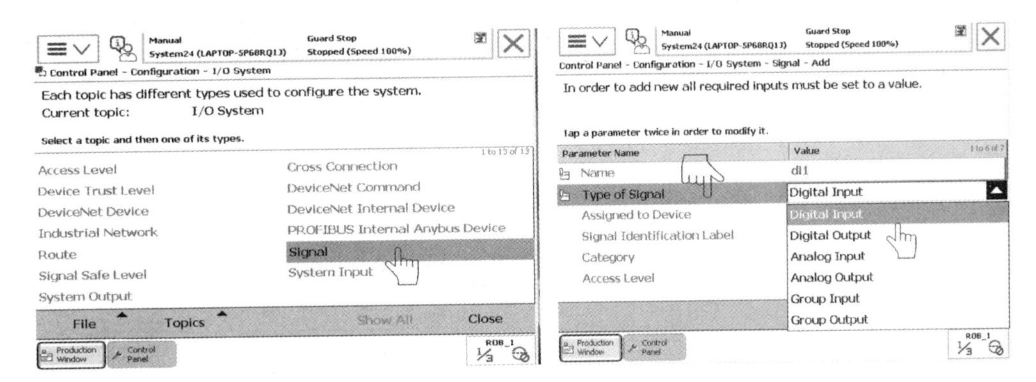


Figure 4-8 Define digital input signal di1

Mapping', and Input '0' . Then click 'Determine', and click 'Yes'.

#### 4.1.3.3 Define Digital Output Signal do1

Table 4-8 shows the relevant parameters of digital output signal do1.

**Table 4-8 Parameters of the digital output signal do1**

| Parameter name | Set value | Explain |
|---|---|---|
| Name | do1 | Set the name of the digital output signal |
| Type of Signal | Digital Output | Set the type of signal |
| Assigned to Device | board10 | Set the I/O module of thesignal |
| Device Mapping | 32 | Set the address occupied by the signal |

The steps of operation are as follows:

(1) Click the main menu button in the upper left corner, choose 'Control Panel', and choose 'Configuration'. Double click 'Signal', click 'Add to'. Double click 'Name', Input 'do1', and click 'Determine'. Then double click 'Type of Signal', and choose 'Digital Output' (shown in Figure 4-9).

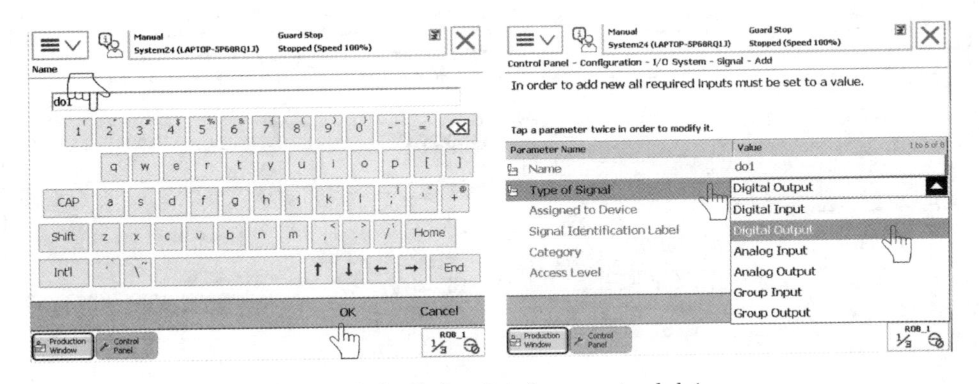


Figure 4-9 Define digital output signal do1

(2) Double click 'Assigned to Device', and choose 'board10'. Double click 'Device Mapping', and Input '32'. Then click 'Determine', and click 'Yes'.

#### 4.1.3.4 Define Group Input Signal gi1

Tables 4-9 and 4-10 show the parameters and status of group input signal gi1.

**Table 4-9 Parameters associated with group input signal gi1**

| Parameter name | Set value | Explain |
|---|---|---|
| Name | gi1 | Set the name of the group input signal |
| Type of Signal | Group Input | Set the type of signal |
| Assigned to Device | board10 | Set the I/O module of the signal |
| Device Mapping | 1-4 | Set the address occupied by the signal |

**Table 4-10 Associated State of group input signal gi1**

| State | Address1 | Address2 | Address3 | Address4 | Decimal number |
|---|---|---|---|---|---|
| | 1 | 2 | 4 | 8 | |
| State 1 | 0 | 1 | 0 | 1 | 2+8=10 |
| State 2 | 1 | 0 | 1 | 1 | 1+4+8=13 |

Note: Group input signal is to use several digital input signals together to receive the decimal number of BCD code input from peripheral devices.

In this example, gi1 takes up 4 bits of address of 1~4, which can represent 0~15 decimal number. By analogy, if the address occupies 5 bits, it can represent the decimal number of 0~31.

The steps of operation are as follows:

(1) Click the main menu button in the upper left corner, and choose 'Control Panel', and choose 'Configuration'. Double click 'Signal', click 'Add to'. Double click 'Name', and input 'gi1'. Then click 'Determine', double click 'Type of Signal', and choose 'Group Input' (shown in Figure 4-10).

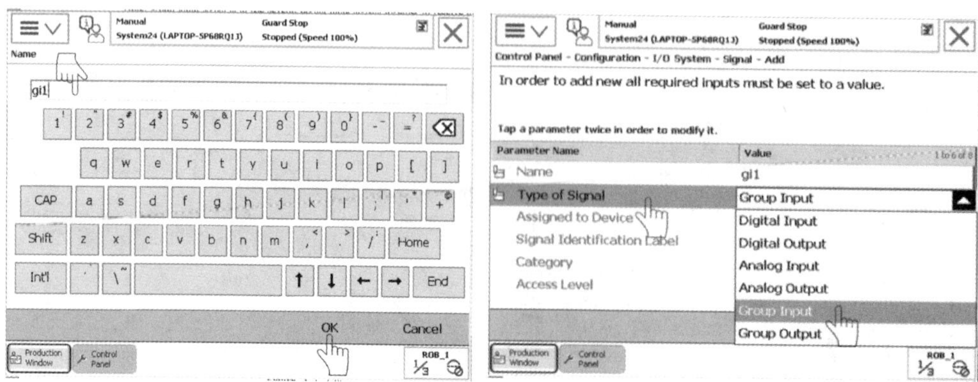


Figure 4-10 Define group input signal gi1

(2) Double click 'Assigned to Device', and choose 'board10'. Double click 'Device Mapping', and input '1~4'. Then click 'Determine', and click 'Yes'.

### 4.1.3.5 Define Group Output Signal go1

Tables 4-11 and 4-12 show the parameters and status of group output signal go1.

**Table 4-11 Parameters associated with group output signal go1**

| Parameter name | Set value | Explain |
|---|---|---|
| Name | go1 | Set the name of the group input signal |
| Type of Signal | Group Output | Set the type of signal |
| Assigned to Device | board10 | Set the I/O module of the signal |
| Device Mapping | 33-36 | Set the address occupied by the signal |

**Table 4-12 Defines the Associated State of group output signal go1**

| State | Address33 | Address34 | Address35 | Address36 | Decimal number |
|---|---|---|---|---|---|
| | 1 | 2 | 4 | 8 | |
| State 1 | 0 | 1 | 0 | 1 | 2+8=10 |
| State 2 | 1 | 0 | 1 | 1 | 1+4+8=13 |

Note: Group output signal is a combination of several digital output signals for output of BCD coded decimal number.

In this example, go1 occupies 4 bits of address of 33-36, which can represent 0~15 decimal number. By analogy, if the address occupies 5 bits, it can represent the decimal number 0~31.

The steps of operation are as follows:

(1) Click the main menu button in the upper left corner, choose 'Control Panel', and choose 'Configure'. Double click 'Signal', and click 'Add to'. Double click 'Name', and input 'go1'. Then click 'Determine', double click 'Type of Signal', and choose 'Group Output'.

(2) Double click 'Assigned to Device', and choose 'board10'. Double click 'Device Mapping', and input '33-36'. Then click 'Determine', and click 'Yes' (shown in Figure 4-11).

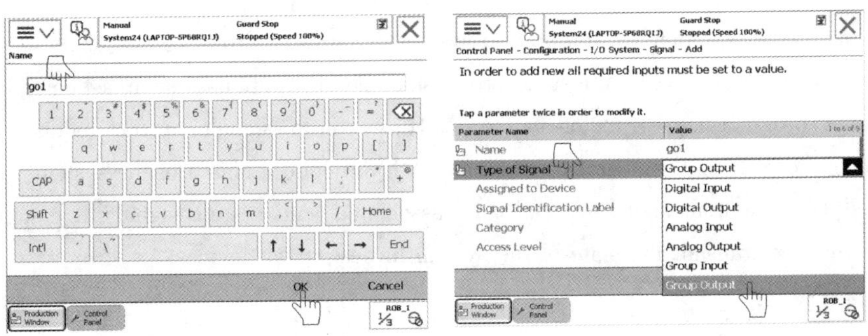


Figure 4-11 Define group output signal go1

### 4. 1. 4 Evaluation

The table of the task evaluation is shown in Table 4-13.

**Table 4-13 Task evaluation**

| No. | Evaluation content | Achievement ratio | Self evaluation | Teacher evaluation |
|---|---|---|---|---|
| 1 | ABB industrial robot I/O communication type | 15 | | |
| 2 | Board and card configuration of ABB industrial robot DSQC652 | 35 | | |
| 3 | Configuration of ProfinetAnybus board for ABB industrial robot | 15 | | |
| 4 | Configuration of ProfibusAnybus board of ABB industrial robot | 15 | | |
| 5 | Safety precautions for operating industrial robots | 20 | | |

(1) The default configuration of ABB industrial robot is I/O communication board based on DeviceNet communication bus protocol, through which the communication between host control system and industrial equipment can be carried out. DeviceNet protocol is a kind of CAN (Controller Area Net, or Controller Area Network) bus protocol, This sensor/actuator bus system originated in the United States, But it is more and more widely used in Europe and Asia.

(2) DSQS651 board can provide 8-bit digital input signal, 8-bit digital output signal and 2-bit analog output signal processing.

(3) The standard configuration of ABB industrial robot I/O communication board is the equipment hanging under the DeviceNet field communication bus module. Its X5 port can be connected to the DeviceNet field communication bus through a special DeviceNet cable.

### 4. 1. 5 Development

(1) Define an I/O board of DSQC652 in the teaching pendant.

(2) Please define di1, do2, gi3, go4 signals for DSQC652 board.

## Task 4. 2 Correlation of System I/O and I/O Signals

### 4. 2. 1 Introduction

There are many control signals in the control system of industrial robot. Some of these signals are associated with physical keys, and some are system control actions, such as emergency stop, program start, program pause, motor start, etc. These control signals can be divided into input/output signals. Correlating the digital input signal with the control signal of the system, the system is controlled. Digital output signals can also be associated with system signals. It can be used to control the robot or transmit the state of the system to peripheral devices.

### 4. 2. 2 Relevant Knowledge

Through the connection of I/O signal of external controller and I/O signal of industrial robot controller, the robot receives the external input signal, The output signal of the robot is sent to the external controller, so that the robot and other devices can work together.

Associating the digital input signal with the control signal of the system, the system is controlled (such as motor start, program start, etc.).

The state signal of the system can also be associated with the digital output signal, and output the state of the system to peripheral devices for control.

## 4.2.3 Implementation

### 4.2.3.1 Correlation between System I/O and I/O Signal

The steps of establishing the relationship between the system input 'Motor On' and the digital input signal di1 are as follows:

(1) Click the main menu button in the upper left corner, choose 'Controlpanel', and choose 'Configuration'. Double click 'System Input', and double click 'Add'. Double click 'Signal Name', choose 'di1', and click 'Sure'. Then double click 'Action', choose 'Motors On', and click 'Sure' (shown in Figure 4-12).

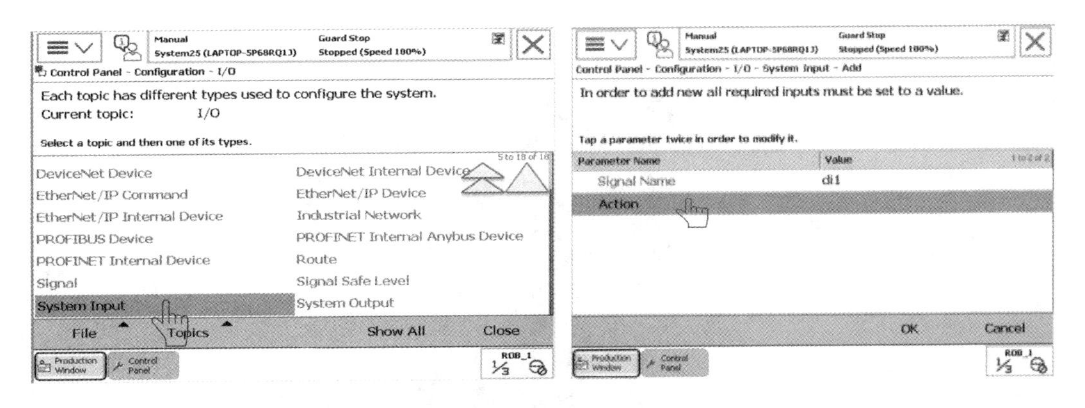


Figure 4-12 Establish system input and digital input signal di1

(2) Click 'Sure', and then click 'Yes'.

The steps of establishing the relationship between the system output 'Motor On' status and the digital output signal do1 are as follows:

(1) Click the main menu button in the upper left corner, choose 'Control Panel', and choose 'Configuration'. Double click 'System Output', and double click 'Add'. Double click 'Signal Name', choose 'do1', and click 'Sure'. Then double click 'Status', choose 'Motors On State', and then click 'Sure' (shown in Figure 4-13).

(2) Click 'Sure', and then click 'Yes'.

Note: For the definition details of system input/system output, please refer to the manual of ABB Robot CD-ROM. At this time, the change of relevant signals is checked in the 'I/O' menu on the teaching pendant.

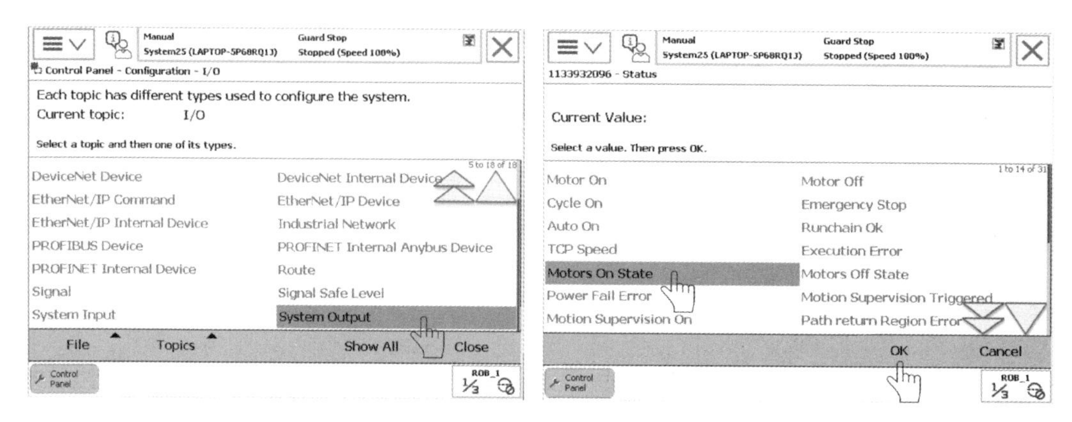


Figure 4-13 Establish system output and digital output signal do1 association

### 4.2.3.2 The Use of Programmable Key in Teaching Device

Programmable keys can be assigned I/O signals to control quickly, so as to facilitate the forced and simulation operation of I/O signals. The programmable button on the demonstrator is shown in Figure 4-14.

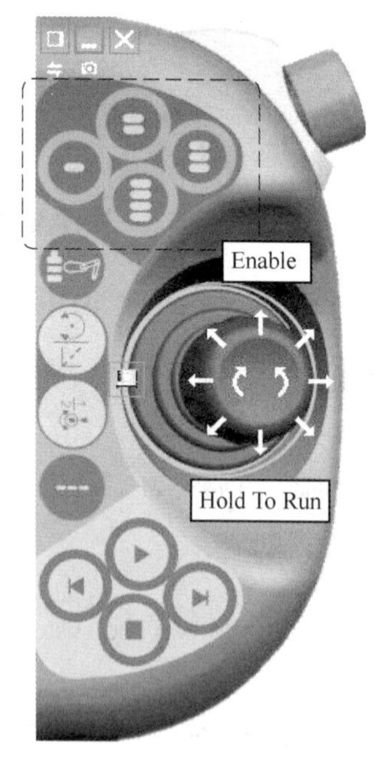


Figure 4-14 Programmable button on the demonstrator

For example, the steps of operation of configuring the digital output signal do1 for the programmable key 1 are as follows:

(1) Click the main menu button in the upper left corner, choose 'Control Panel', and choose 'Configure Programmable Keys'. In type, chosse 'Output' (shown in Figure 4-15).

(2) Select 'do1', and select 'Press/Release' in 'Press Button'. You can also select the

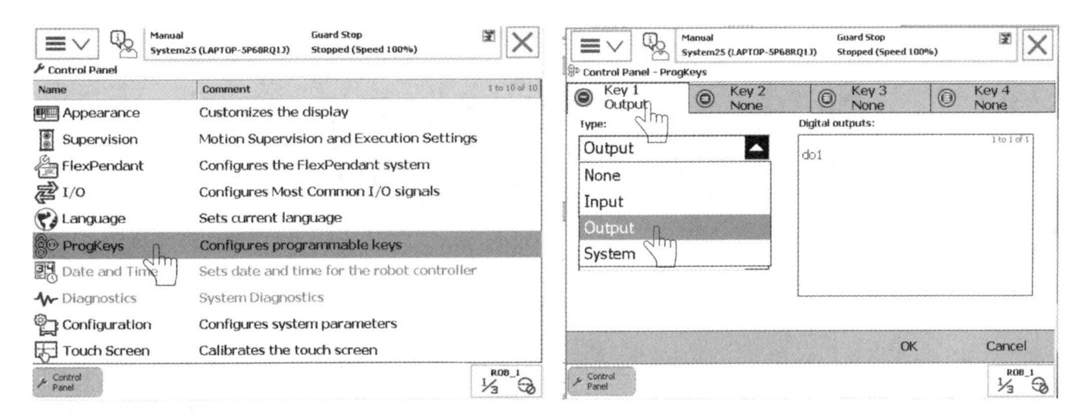


Figure 4-15 Programmable button configuration

action characteristics of the key according to the actual needs, and then click 'Sure'. Now the programmable key ⟨1⟩ is used to force the operation of do1 in the manual state.

### 4.2.4 Evaluation

The table of the task evaluation is shown in Table 4-14.

**Table 4-14 Task evaluation**

| No. | Evaluation content | Achievement ratio | Self evaluation | Teacher evaluation |
|---|---|---|---|---|
| 1 | Input and output setting of ABB industrial robot system | 15 | | |
| 2 | The use of programmable shortcut key for ABB industrial robot | 15 | | |
| 3 | ABB industrial robot system control motor power on operation | 25 | | |
| 4 | Association of control lights in ABB industrial robot system | 35 | | |
| 5 | Safety precautions for operating industrial robots | 15 | | |

There are many control signals in ABB industrial robot control system. Some of these signals are associated with physical keys, and some are system control actions, such as emergency stop, program start, program pause, etc. By associating the digital input signal with the control signal of the system, the system can be controlled, such as starting the motor, starting the program, etc.

### 4.2.5 Development

(1) Please configure an associated system input signal to power on the motor.

(2) Please configure an associated system output signal of lamp and motor status.

(3) Please configure the input and output signals of the two lights and the system associated with the power on and power off of the motor.

(4) Please configure an industrial robot to grasp the open and close programmable shortcut keys.

(5) Please configure an industrial robot suction cup to hold the programmable shortcut key of suction and discharge.

# Project 5 Program Data of ABB Industrial Robots

Industrial robot data is a general term for numbers, letters, symbols and analog quantities with specific meanings that can be processed by robot programs. Industrial robot data can be either environment data or pure values. Programming ABB industrial robots requires the use of the specific language RAPID and ABB robot-specific programming environment. ABB industrial robot data is created in the program module or system module of the programming environment and it can be referenced by instructions in the same module or other modules. This project introduces ABB industrial robot understanding and building program data. So the ABB industrial robot programming data can be understood quickly, used in the type and classification, how to create program data, and master the most important three key program data settings.

## Task 5. 1 Understand and Establish Program Data

### 5. 1. 1 Introduction

Industrial robot is a typical mechatronics equipment. Before operating ABB industrial robot equipment, it is necessary to learn the relevant program data of industrial robot. Program data is used to edit the instructions, symbols, operations and other functions of the robot program, so that users can have simple human-computer interaction with the robot system.

### 5. 1. 2 Related Knowledge

Data declared within a program is called program data. Data is the carrier of information, which can be recognized, stored and processed by computer. It is the raw material of computer program processing, and the application program processes all kinds of data. In computer science, the so-called data is the object of computer processing. It can be numerical data or non numerical data. Numeric data are integers, real numbers, or complex numbers. It is mainly used in engineering calculation, scientific calculation, business processing and so on; Non numerical data includes characters, text, graphics, images, voice, etc. The types and classifications of program data can be understood used in ABB robot programming, how to create program data, and the three most important key program data (tooldata, wobjdata and loaddata).

There are two topics in the study of program data type classification and storage type, so that we can have an understanding of the program data, choose the program data according to the actual needs and different data purposes, and define different program data.

### 5.1.3 Implementation

#### 5.1.3.1 Cognition of Program Data

Program data refers to setting values and defining some environment data in program module or system module. The created program data is referenced by instructions in the same module or other modules. As shown in Figure 5-1, dashed box is a common command of robot joint motion (MoveJ), invoking 4 program data.

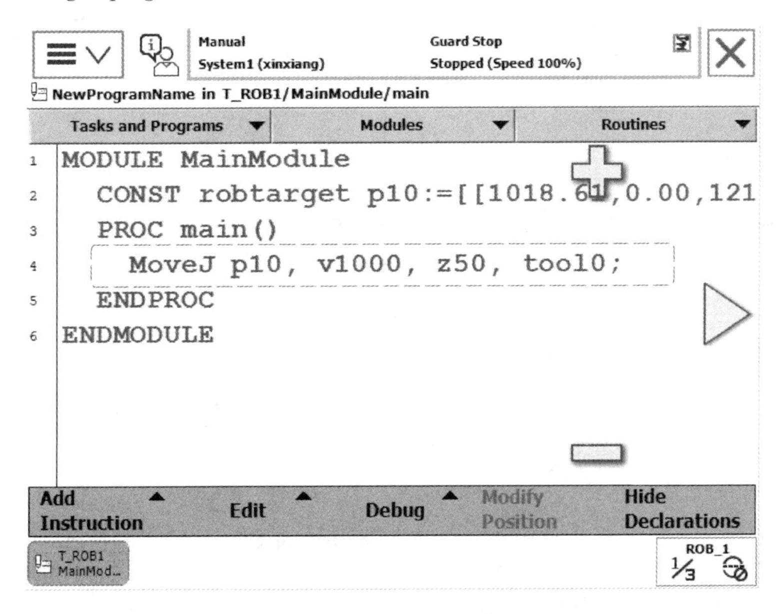


Figure 5-1 Program data

The program data is shown in Table 5-1.

**Table 5-1 Program data**

| Program data | Data type | Explain |
|---|---|---|
| p10 | Robtarget | Robot moving target position data |
| v1000 | Speeddata | Robot speed data |
| z50 | Zonedata | Robot turning data |
| tool0 | Tooldata | Robot tool data TCP |

#### 5.1.3.2 Establishment of Program Data

The establishment of program data can be generally divided into two forms: One is to directly create program data in the program data screen of the teaching device; The other is to automatically generate the corresponding program data when establishing the program instructions.

In the task, the methods of creating program data directly in the program data screen of the teaching device will be completed. This paper takes boolean data (bool) and numerical data (num) as examples.

**Operation of Building Bool Type Program Data**

The steps of operation of building bool type program data are as follows:

(1) Click the main menu button in the upper left corner, choose 'Program Data', select data type 'Bool', and then click 'Display Data' (shown in Figure 5-2).

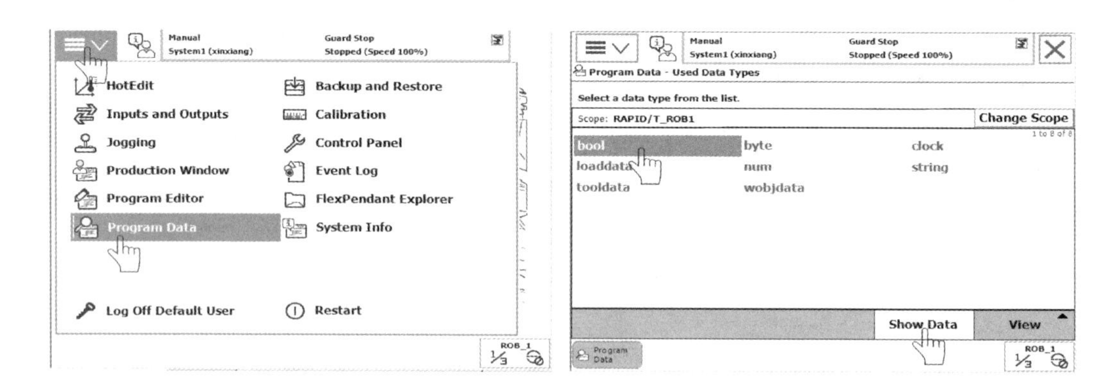


Figure 5-2 Create bool type program data

(2) Click 'Newly Build...', click the button to set the name, click the drop-down menu to select the corresponding parameters, and then click 'Determine'.

The datas of setting parameters and descriptions are shown in Table 5-2.

**Table 5-2 Data set parameter**

| Data setting parameters | Explain |
|---|---|
| Name | Set the name of the data |
| Range | Set the range of data available |
| Storage type | Set the storage type of data |
| Task | Set the task of data |
| Modular | Set the module of data |
| Routine | Set routine of data |
| Dimension | Set the dimension of data |
| Initial value | Set the initial value of the data |

**Operation of Building Program Data of Num Type**

The steps of operation of building program data of num type are as follows:

Click the main menu button in the upper left corner, choose 'Program Data', and select data type 'num'. Click 'Display Data', click 'Newly build...', and click the button to set the name. Click the drop-down menu to select the corresponding parameters, and then click 'Determine' (shown in Figure 5-3).

So far, You have mastered the basic method of establishing program data, and the definition methods of related parameters.

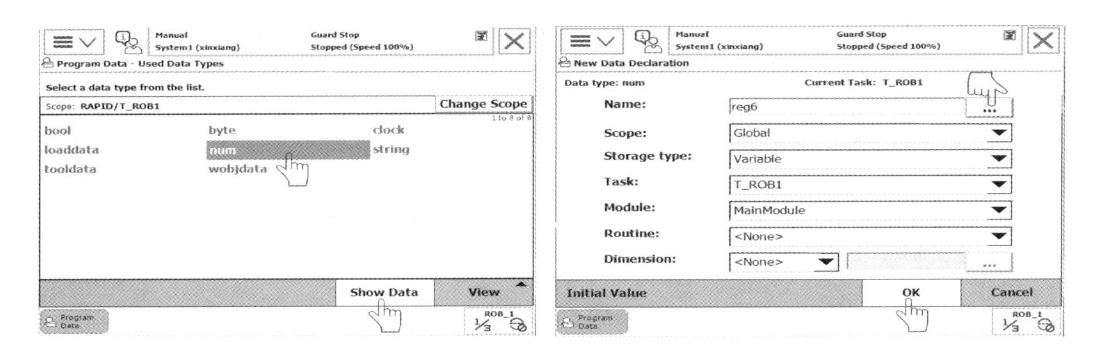


Figure 5-3 Establish 'num' type program data

## Type Classification of Program Data

There are about 100 program data of ABB Robot. The program data can be created according to the actual situation, and it brings infinite possibility for the program design of ABB Robot.

In the teaching device 'Program Data' window, you can view and create the program data you need (shown in Figure 5-4).

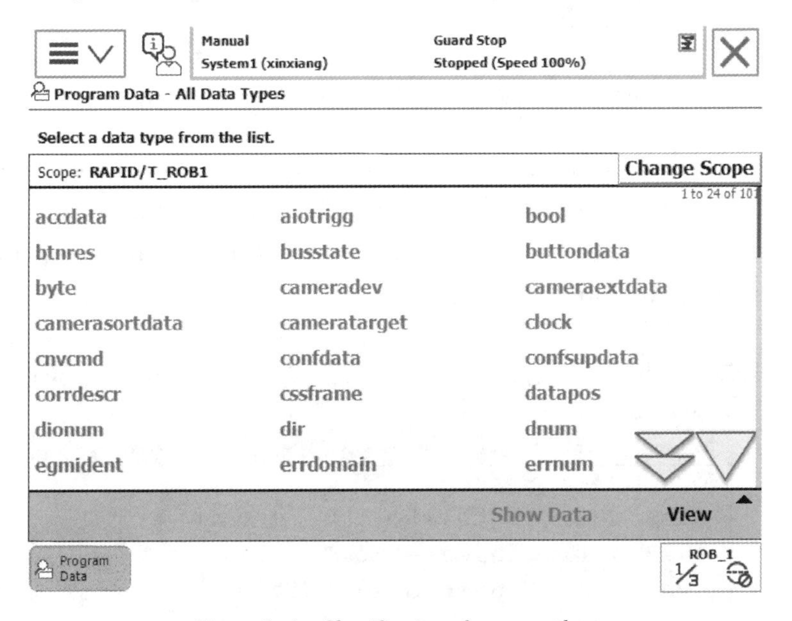


Figure 5-4 Classification of program data

## Storage Type of Program Data

The storage types of program data are as follows:

(1) Variable VAR: Variable data will maintain the current value during program execution when it is stopped. However, if the program pointer is reset or the robot controller is restarted, the value will return to the initial value assigned when the variable was declared.

Note: 'VAR' indicates that the storage type is variable; 'num' indicates that the declared data is numeric (Stored content is digital).

As shown in Figure 5-5, 【VAR num length: =0;】 means variable type numeric data named 'length'. 【VAR string name: ="Tom";】 means variable character data named 'name'. 【VAR bool finished: =FALSE;】 means variable type of boolean data named 'finished'.

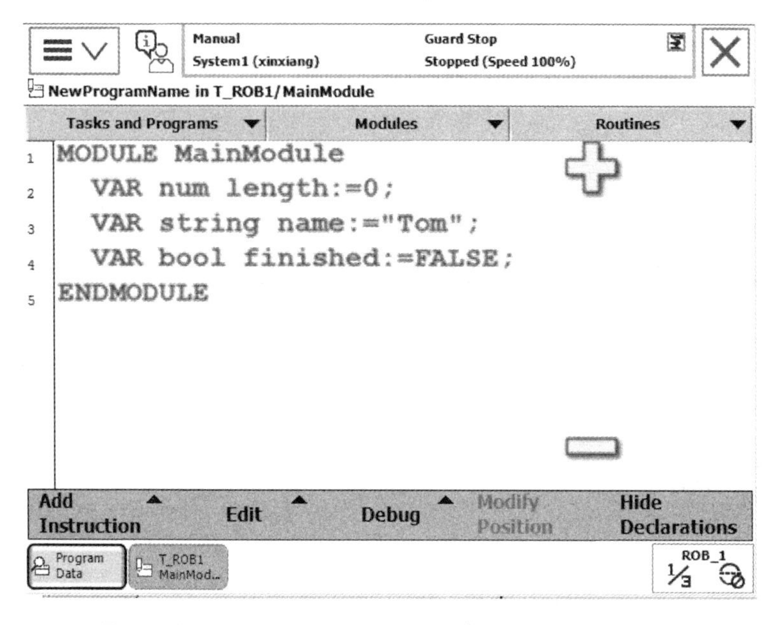


Figure 5-5 Program editing window (variable storage type)

Note: When declaring data, the initial value of variable data can be defined. For example, the initial value of length is 0, the initial value of name is Tom, and finished initial value is FALSE.

In the rapid program executed by the robot, the data of variable storage type program can also be assigned, as shown in Figure 5-6.

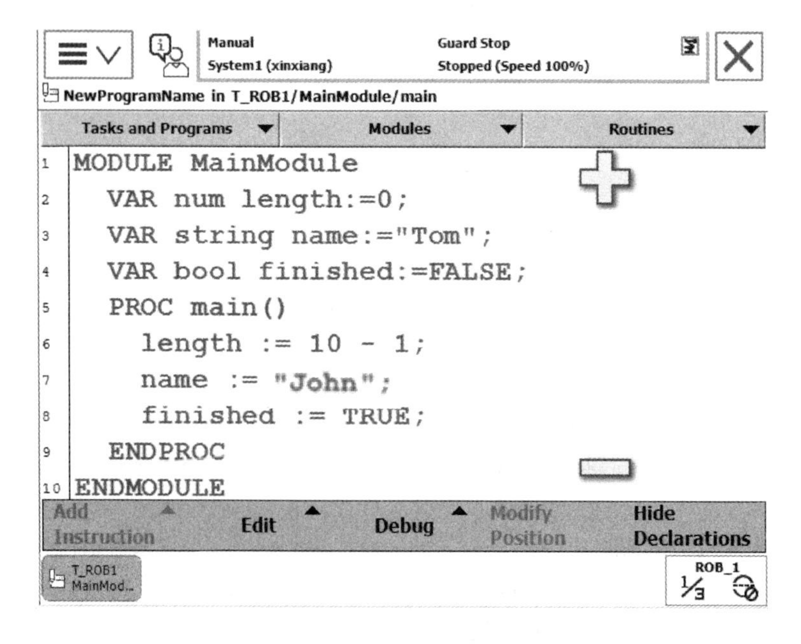


Figure 5-6 Program editing window (variable assignment)

Note: The assignment of variable program data is performed in the program. After the pointer resets or the robot controller restarts, it will return to the initial value.

(2) Variable PERS: No matter how the program pointer changes, whether or not the robot controller is restarted, the data of variable type will keep the last given value.

The display in the program editing window is as shown in Figure 5-7. 【PERS num numb: = 1;】 means numeric data named ‘num’. 【PERS string text: =" Hello" ;】 means Character data named ‘text’.

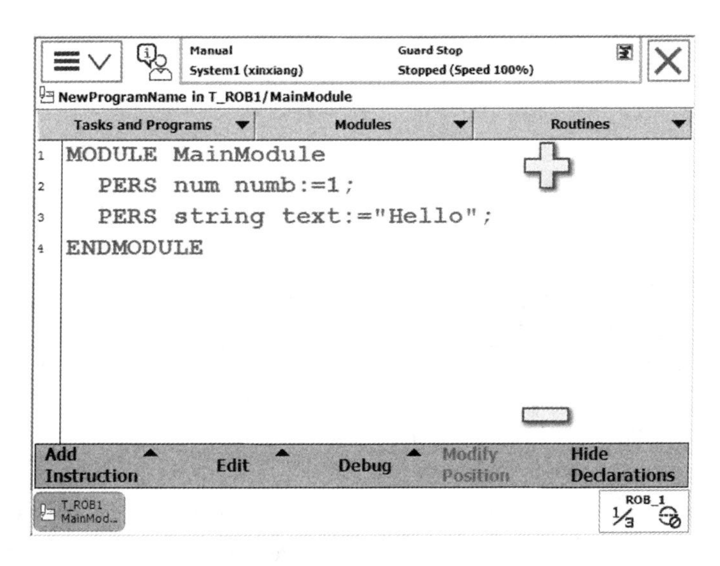


Figure 5-7 Program editing window (variable storage type)

Note: PERS indicates that the storage type is variable.

In the RAPID program executed by the robot, you can also assign values to the variable storage type program data, as shown in Figure 5-8.

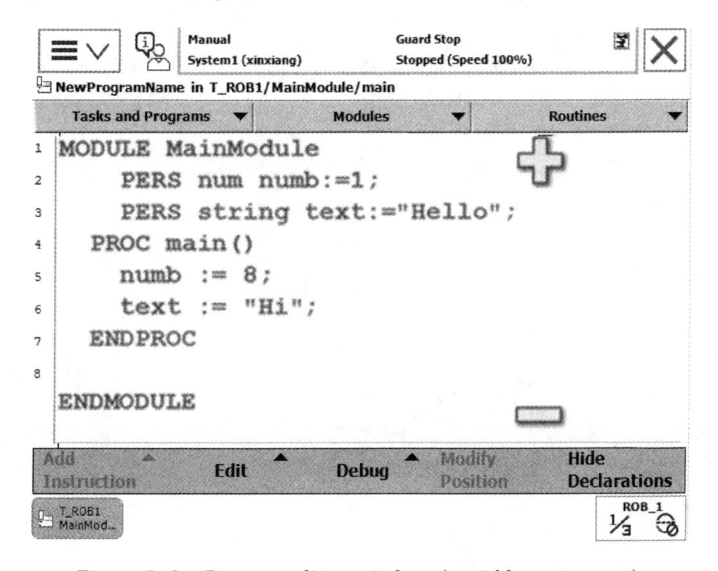


Figure 5-8 Program editing window (variable assignment)

After the program is executed, the result of the assignment will remain until the next assignment, as shown in Figure 5-9.

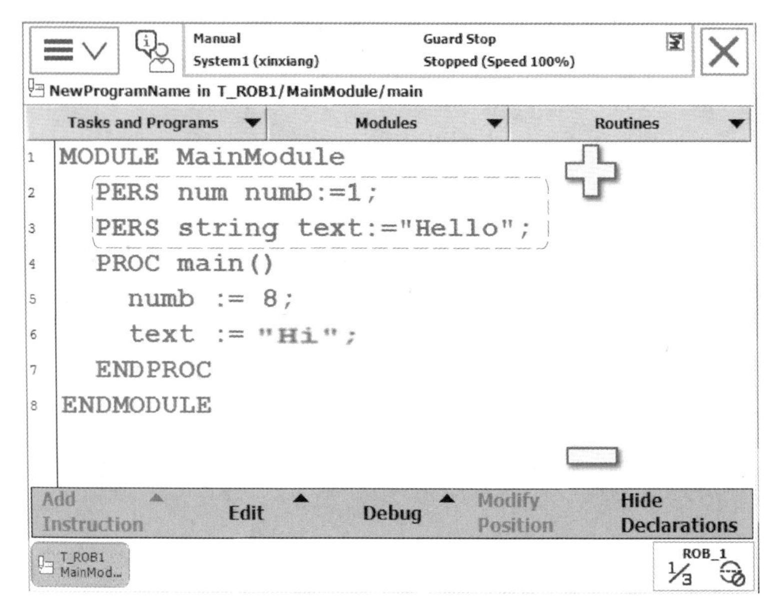


Figure 5-9 Program editing window (variable assignment)

(3) CONST constant: CONST constant is characterized by a value given at the time of definition, which cannot be modified in the program, and it can only be modified manually. For example, 【CONST num gravity: =9.81;】 means numeric data named 'gravity'. 【CONST string greating: =" Hello";】 means character data named 'greeting'.

The display in the program editing window is shown in Figure 5-10.

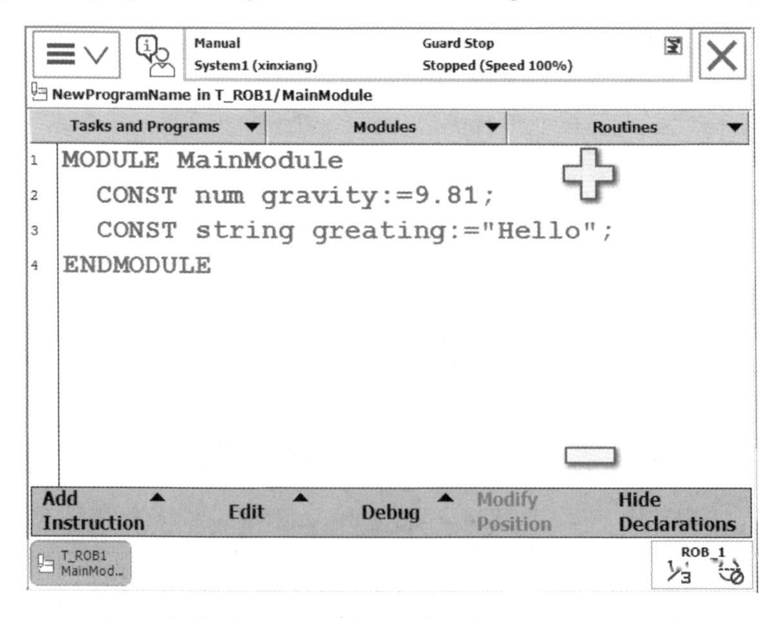


Figure 5-10 Program editing window (constant storage type)

Note: Storing the program data of constant type, the assignment operation in the program is not allowed.

**Common Program Data Description**

The common program data are as follows:

(1) Numerical data num: Num is used to store numerical data, such as a counter. The value of num data includes integer (−5), decimal fraction (3.45), and written in the form of index, such as 2E3 (=2 * 10^3=2000), 2.5E−2 (=0.025) etc.

Integer value always store integers between −8388607 and +8388608 as accurate integers. Decimal values are approximate numbers. Therefore, it shall not be used for equal or unequal comparison. If decimal division and operation are used, the result will also be decimal.

Numerical data num example: Assign integer 3 to numerical data named count1 (shown in Figure 5−11).

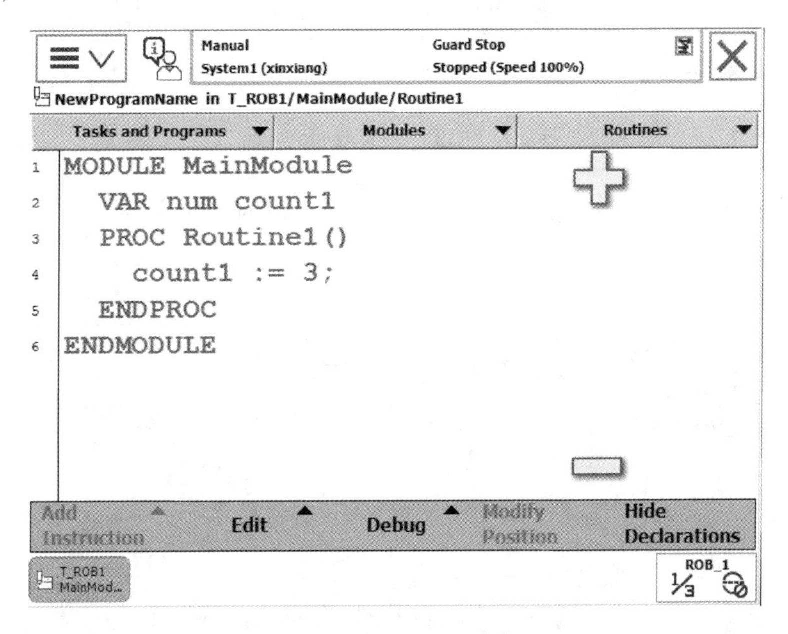


Figure 5−11 Numerical data

(2) Logical value data bool: Bool is used to store logical value (true/false) data, and the data value of bool type can be TRUE or FALSE.

Example of logical value data bool is: Firstly, judge whether the value in count1 is greater than 100; If it is greater than 100, assign TRUE to high value, otherwise assign FALSE (shown in Figure 5−12).

(3) String datastring string: String is used to store string data. A string is a string with quotation marks around it ("") character (up to 80) form, such as," This is a character string".

If a backslash ( \ ) is included in the string, two backslashes must be written, such as, " This string contains a \\ character".

String data string example: assign start welding pipe 1 to text. After running the program, the

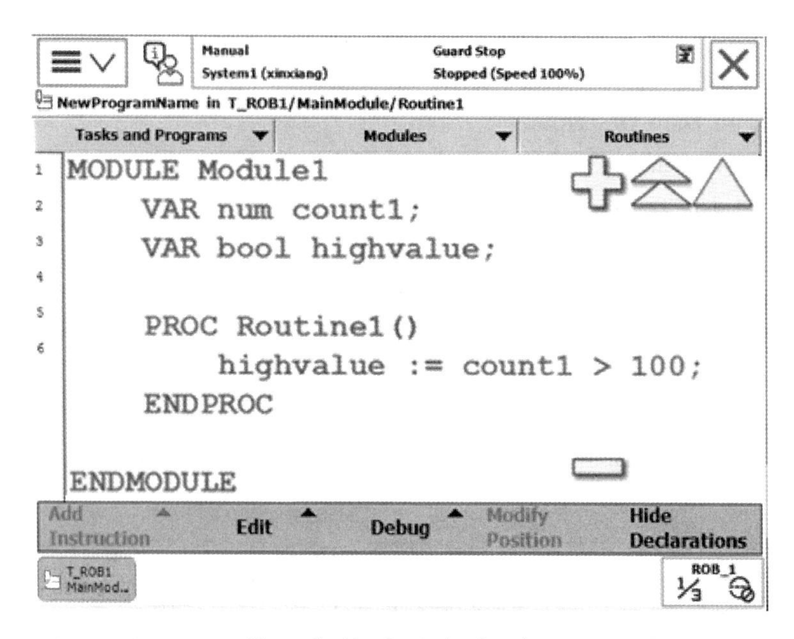


Figure 5-12 Logical value data

operator window in the teaching pendant will be displayed start welding pipe 1, as shown in Figure 5-13.

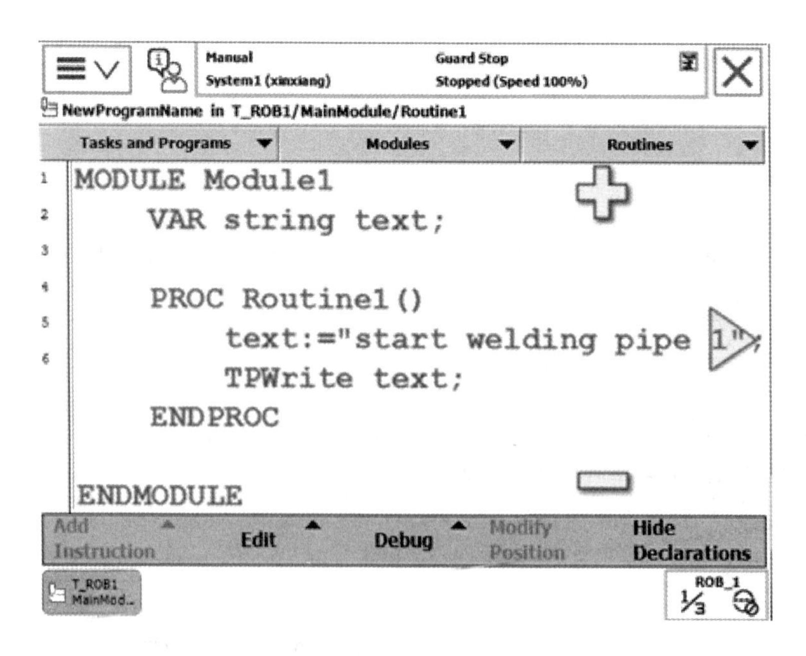


Figure 5-13 String data

### 5.1.4 Evaluation

The table of the task evaluation is shown in Table 5-3.

Table 5-3 **Task evaluation**

| No. | Evaluation content | Achievement ratio | Self evaluation | Teacher evaluation |
|---|---|---|---|---|
| 1 | Program data of industrial robot | 10 | | |
| 2 | Program data structure of industrial robot | 20 | | |
| 3 | Program data establishment of industrial robot | 25 | | |
| 4 | Type and description of industrial robot program data | 25 | | |
| 5 | Safety precautions for operating industrial robots | 20 | | |

(1) In RAPID language, different information such as tool, location and load are saved in the form of data. Data is created and declared by the user and can be named arbitrarily.

(2) When establishing data, the storage mode of data is defined. Common data storage types include variable, variable and constant, which are represented by VAR、PERS and CONST respectively.

## 5. 1. 5 Development

(1) Please indicate the program data in Figure 5-14.

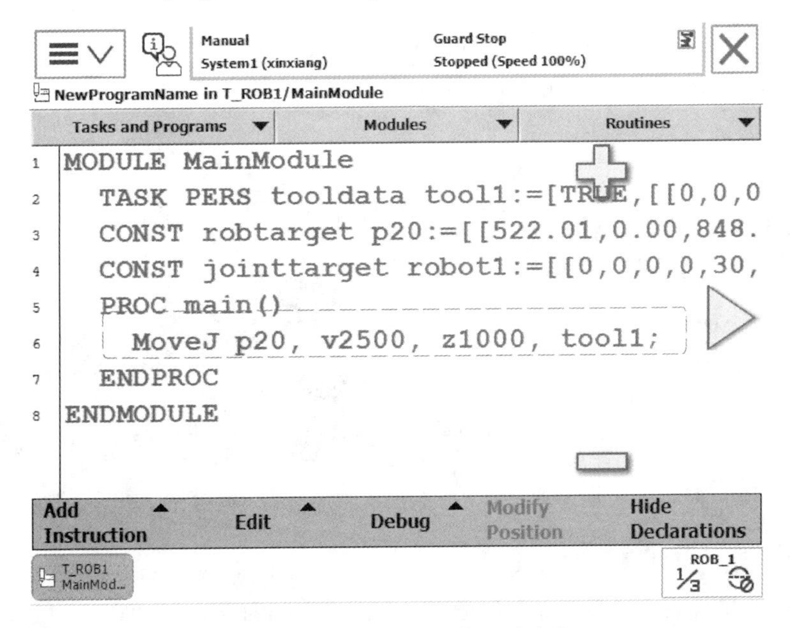


Figure 5-14 Programe data of (1)

(2) Create a bool type program data named robot 1, and set the available range of data to be local.

(3) Create a num type program data named robot 2, and set the range of data available as task.

(4) Please briefly describe the functions and differences between VaR and pers and const.

(5) Create a string program data that assigns 'Hello my friend' to text.

# Task 5.2 Setting of Three Key Program Date

## 5.2.1 Introduction

Before working, industrial robot must program according to the work content and edit the application program of industria. The first step is to establish three necessary basic datas including tooldata, wobjdata and loaddata.

## 5.2.2 Related Knowledge

### 5.2.2.1 Tooldata

Tooldata is used to describe the TCP, weight, center of gravity and other parameter data of the tool installed on the sixth axis of the robot.

Different robot applications may be equipped with different tools. For example, the arc welding robot uses the arc welding gun as a tool, while the robot used to carry plates uses the sucker type fixture as a tool (shown in Figure 5-15).

The tool center point of default tool (tool0) is located in the center of the robot mounting flange. The original TCP point is shown in Figure 5-16.

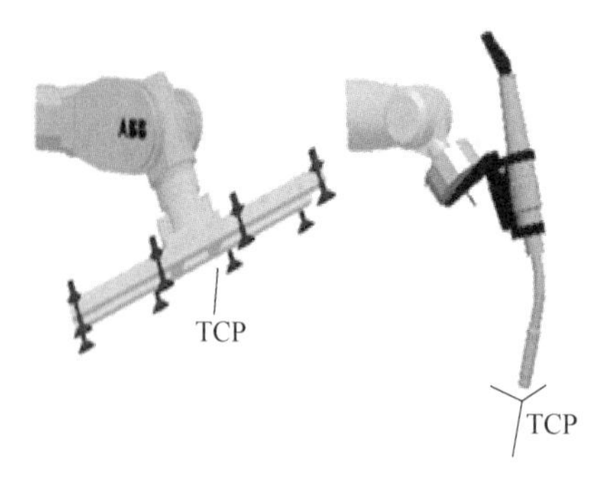


Figure 5-15 Robot tools

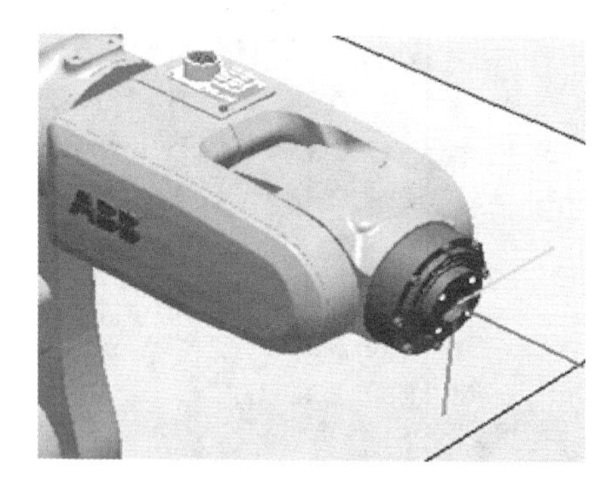

Figure 5-16 Robot TCP

Tooldata is used to describe tools characteristics, such as welding guns and clamps. The features include the location and orientation of the tool center point (TCP), and the physical characteristics of the tool load.

Note: If you are using a fixed tool, the defined tool coordinate system is relative to the world coordinate system.

An example of tool data is as follows:

```
PERS tooldata gripper:=[TRUE,[[97.4,0,223.1],[0.924,0,0.383,0]],[5,[23,0,75],[1,0,0,0],0,
0,0]];
```

The definitions of gripper are as follows:

(1) The arm is holding the tool.

(2) The point where TCP is located is offset 97.4mm along the X direction of the tool coordinate system, and 223.1mm along the Z direction of the tool coordinate system.

(3) The X and Z directions of the tool rotate 45° relative to the Y direction of the wristcoordinate system.

(4) The weight of the tool is 5kg.

(5) The center of gravity is offset 23mm along the X direction of the wrist coordinate system and 75mm along the Z direction of the wrist coordinate system. The load can be considered as a point mass, i. e. , without torque inertia.

#### 5.2.2.2 Wobjdata

Workpiece coordinate system corresponds to workpiece defines the position of workpiece relative to geodetic coordinate system (or other coordinate system). Robots can have several workpiece coordinate systems, or represent different workpieces, or represent several copies of the same workpiece in different positions.

When programming the robot, the target and path are created in the workpiece coordinate system, which brings many advantages as follows:

(1) When repositioning the workpiece in the workstation, you only need to change the position of the workpiece coordinate system, and all paths will be updated immediately.

(2) It is allowed to operate the workpiece moved by the external axis or transfer guide rail. Because the whole workpiece can be moved together with its path.

Note: A is the geodetic coordinate of the robot. In order to facilitate the programming, a workpiece coordinate B is established for the first workpiece, and the trajectory is programmed in this workpiece coordinate B.

If there is a same workpiece on the table that needs to follow the same track, you only need to create a workpiece coordinate C, copy the track in the workpiece coordinate B, and then update the workpiece coordinate from B to C. Then there is no need to program the same workpiece repeated track (shown in Figure 5-17).

Note: In workpiece coordinate B, the track of object A is programmed. If the position of the workpiece coordinate changes to the workpiece coordinate D, only the workpiece coordinate D needs to be redefined in the robot system. Then the trajectory of the robot will be automatically updated to C, and there is no need to program the trajectory again. Because the relationship between A and B, and C and D are the same, they does not change due to the overall offset (shown in Figure 5-18).

If the workpiece is specified in the motion instruction, the target point position will be based on the workpiece coordinate system. The advantages are as follows:

(1) It is convenient manual input of location data. For example, offline programming can obtain location values from drawings.

(2) The track program can be reused quickly according to the change. For example, if the

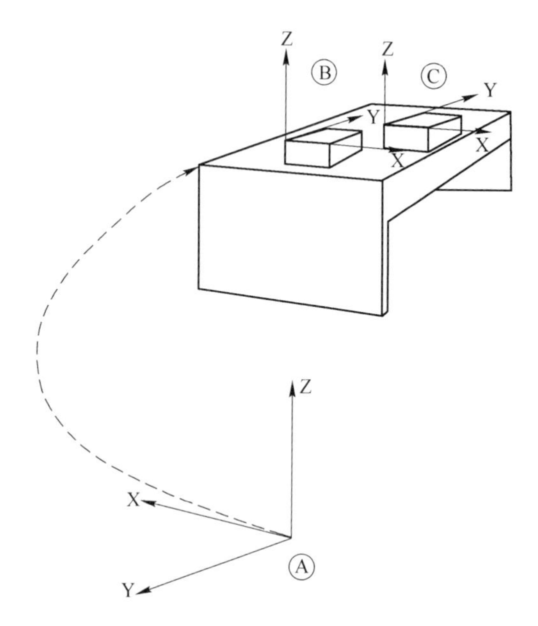


Figure 5-17　Workpiece coordinate system 1

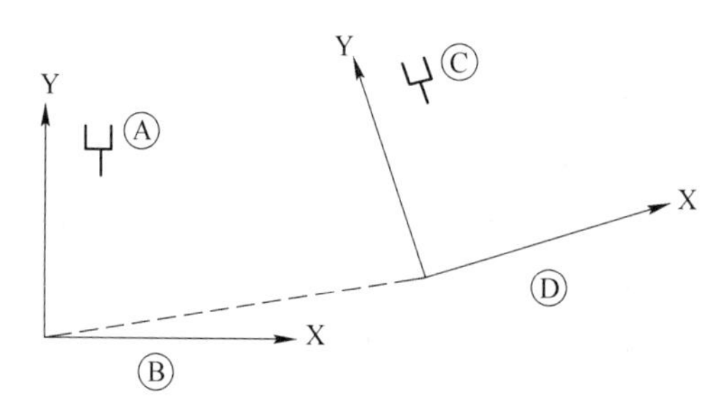


Figure 5-18　Workpiece coordinate system 2

worktable is moved, you only need to redefine the worktable workpiece coordinate system.

(3) The workpiece coordinate system can be compensated according to the change. For example, the sensor to get the deviation data to locate the workpiece is used.

An example of wobjdata is as follows:

```
PERSwobjdatawobj1:=[FALSE,TRUE,"",[[300,600,200],[1,0,0,0]],[[0,200,30],[1,0,0,0]]];
```

Wobj1 is defined as follows:

(1) The manipulator does not hold the workpiece.

(2) It uses a fixed user coordinate system.

(3) The user coordinate system does not rotate, and the origin of the user coordinate system in the geodetic coordinate system is $x=300$mm, $y=600$mm, and $z=200$mm.

(4) The target coordinate system does not rotate, and the origin of the target coordinate system in the user coordinate system is $x=0$mm, $y=200$mm, and $z=30$mm.

#### 5. 2. 2. 3 Loaddata

For the robot of handling application, the weight of the fixture, the center of gravity tooldata, and the weight and center of gravity data loaddata of the object to be handled should be set correctly (shown in Figure 5-19).

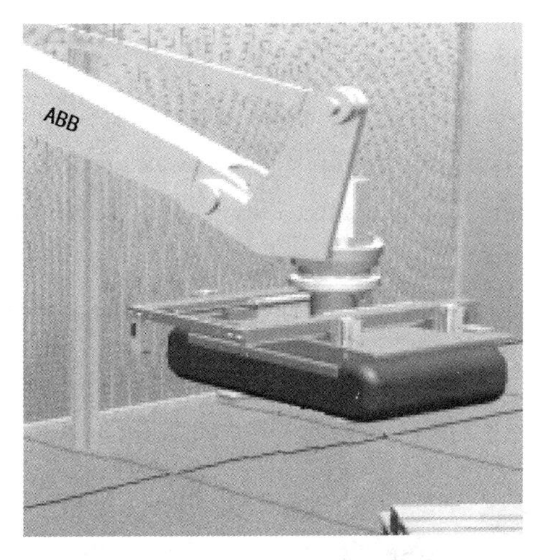


Figure 5-19 Payload

Loaddata is used to set robot axis 6 up Load data for flange mounting. Loaddata often defines the payload or grab of a robot load (setting by instruction gripload or mechunitload), i. e., the load held by the fixture. At the same time, loaddata is used as the component to describe the tool load.

An example of loaddata is as follows:

```
PERS loaddata piece1:=[5,[50,0,50],[1,0,0,0],0,0,0];
```

Piece1 is defined as follows:

(1) The weight is 5kg.

(2) The center of gravity is $x = 50$mm, $y = 0$mm, and $z = 50$mm, relative to the tool coordinate system.

(3) Payload is a point mass.

### 5. 2. 3 Implementation

#### 5. 2. 3. 1 Setting of Tool Coordinate Tooldata

The setting principles of tool center point are as follows:

(1) Firstly, a very precise fixed point is found as the reference point in the working range of the robot.

(2) It determines a reference point (preferably the center point of the tool) on the tool.

(3) Through the method of manual control of the robot that we learned before, the reference point on the moving tool can meet the fixed point as much as possible with at least four different ro-

bot postures. In order to obtain more accurate TCP, the six point method is used to operate in the following examples. The fourth point is that the reference point of the tool is perpendicular to the fixed point. The fifth point is that the reference point of the tool moves from the fixed point to the X direction to be set as TCP. The sixth point is that the reference point of the tool moves from the fixed point to the Z direction to be set as TCP.

(4) The robot can calculate the data of TCP through the location data of these four points, and then the data of TCP is saved in the program data of tooldata and called by the program.

The operation steps of setting of tool coordinate tooldata are as follows:

(1) Click the main menu button in the upper left corner and choose manual operation, select tool coordinates, and then click 'New'. After setting the tool data attribute, click 'OK'. After selecting tool1, click 'Definition' in 'Edit' menu, and select 'TCP and Z, X' method to set TCP. Select the appropriate manual operation mode, press the enable key, and use the rocker to make the tool reference point lean against the fixed point as the first point (shown in Figure 5-20).

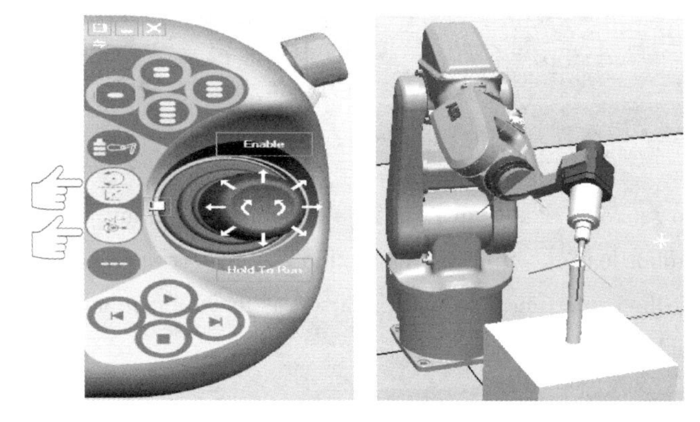


Figure 5-20 Create tool data 'Point 1'

(2) Select 'Point 1', click 'Modify Position' to record the position of 'Point 1', and the tool reference point will lean against the fixed point in this attitude (shown in Figure 5-21).

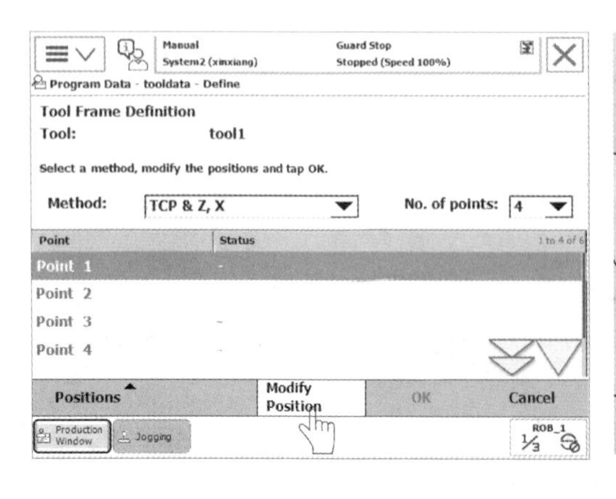


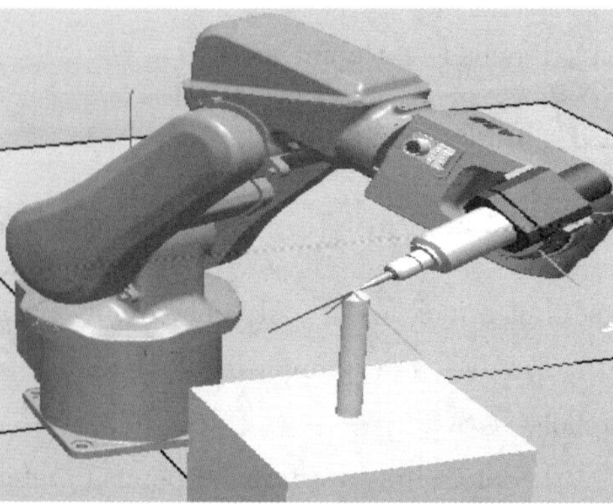

Figure 5-21 Create tool data 'Point 2'

(3) Select ‘Point 2’, click ‘Modify Position’ to record the position of ‘Point 2’, and the tool reference point will lean against the fixed point in this attitude. Select ‘point 3’, click ‘Modify Position’ to record the position of ‘point 3’, and the tool reference point is next to the fixed point in this attitude (shown in Figure 5-22).

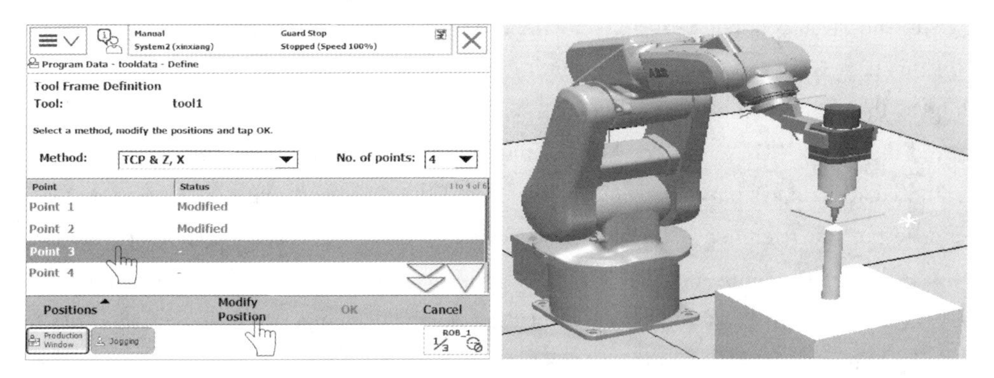


Figure 5-22 Create tool data ‘Point 3’

(4) Select ‘Point 4’, click ‘Modify Position’ to record the position of ‘Point 4’, and the tool reference point will lean against the fixed point in this attitude (shwon in Figure 5-23).

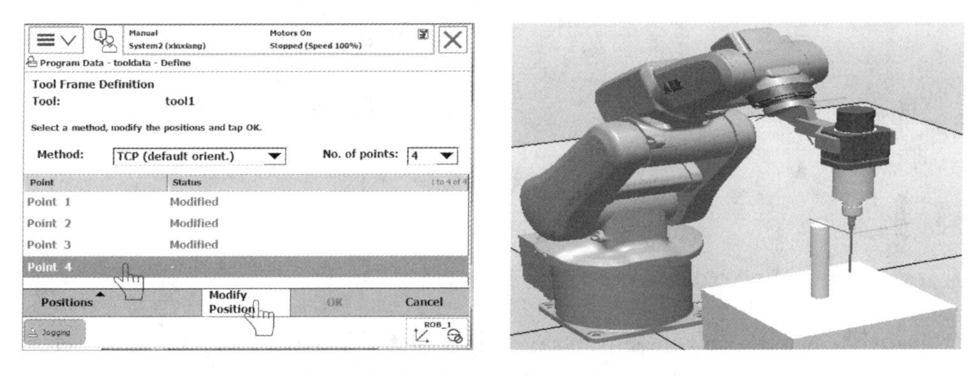


Figure 5-23 Create tool data ‘Point 4’

(5) Select ‘Elongator Point X’, click ‘Modify Position’, record the position of ‘Elongator Point X’, and click ‘Yes’ to complete the setting.

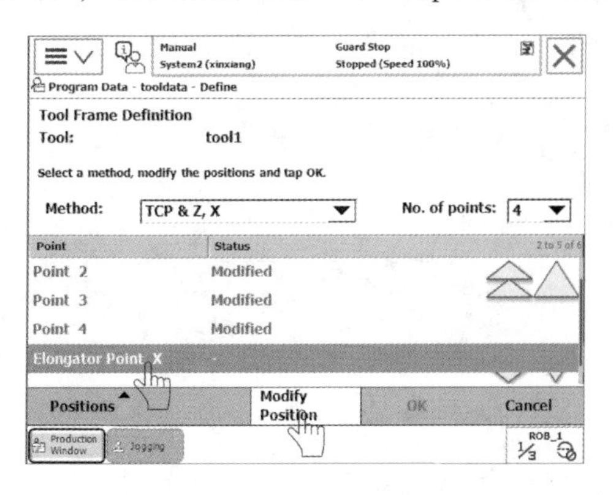

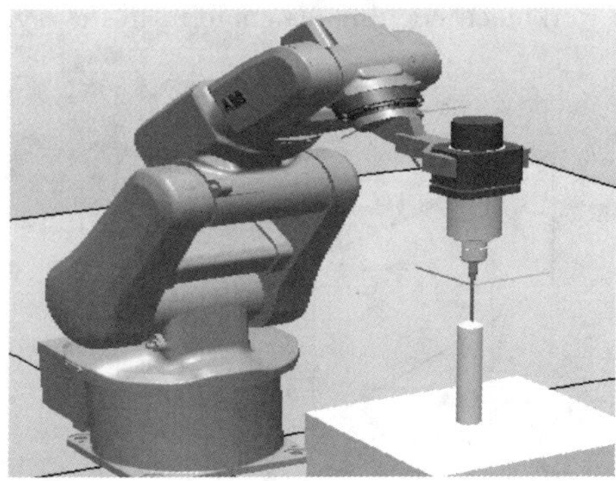

Figure 5-24 Create tool data ‘Elongato point X’

(6) Select ‘Elongator Point Z’ and click ‘Modify Position’ to record the position of ‘Elongator point Z’, and then click ‘OK’ to complete the setting and confirm the error. However, the actual verification effect shall prevail. Next, set the weight and center of gravity of tool1, select ‘tool1’, and then open ‘Edit Menu’ and choose ‘Change Value’. The content displayed on this page is the data generated during the definition of TCP. Click the arrow to turn down the page. On this page, set the weight mass (unit: kg) and center of gravity position data of the tool according to the actual situation (the center of gravity is based on the offset value of tool0 in mm), and then click ‘OK’. Select ‘tool1’ and click ‘OK’. The action mode is selected as relocate, the coordinate system is selected as tool, and the tool coordinate is selected as tool1. Use the rocker to put the reference point of the tool against the fixed point, and then operate the robot manually in the repositioning mode. If the TCP setting is accurate, the reference point of the tool is always in contact with the fixed point, and the robot will change its attitude according to your repositioning operation (shown in Figure 5-25).

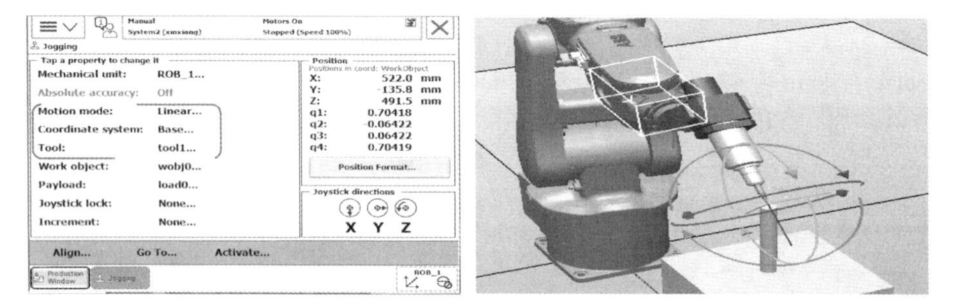

Figure 5-25 Verify new tool data

### 5. 2. 3. 2 Etting of Workpiece Coordinate Wobjdata

In the plane of the object, only three points need to be defined to establish a workpiece coordinate (shown in Figure 5-26), as follows:

(1) X1 and X2 determine the positive direction of workpiece coordinate X.

(2) Y1 determine the positive direction of workpiece coordinate Y.

(3) The origin of workpiece coordinate system is the projection of Y1 on workpiece coordinate X.

The workpiece coordinates conform to the right-hand rule, as shown in Figure 5-27.

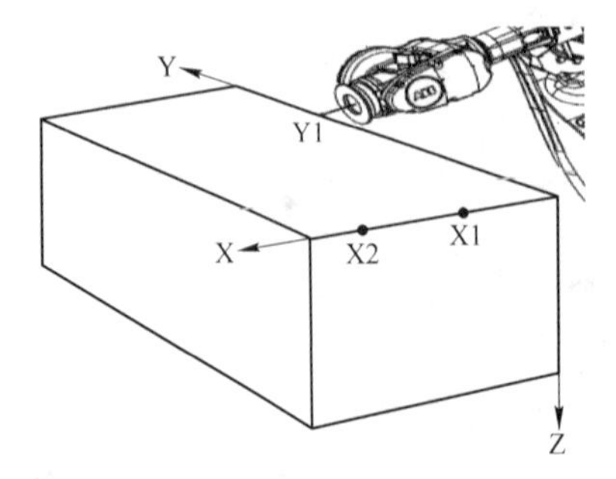

Figure 5-26 Workpiece coordinate system

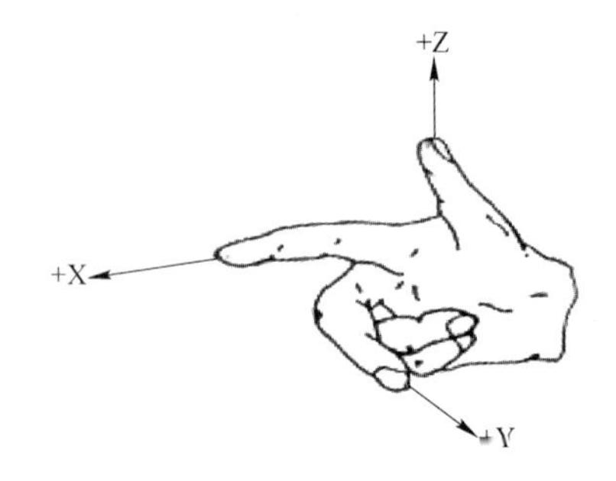

Figure 5-27 Right-hand

The following are the operation steps to establish workpiece coordinates:

(1) Click the main menu button in the upper left corner, select 'Manual Operation' and 'Workpiece Coordinate'. Click 'New' to set the workpiece data attribute, and click 'OK'. Select 'wobj1', click 'Definition' in 'Edit' menu, and the user method selects '3 points'. The tool reference point of the manually operated robot is close to the X1 point defining the workpiece coordinate system. Select 'User Point X1', click 'Modify Position', and then record the position of point X1 (shown in Figure 5-28).

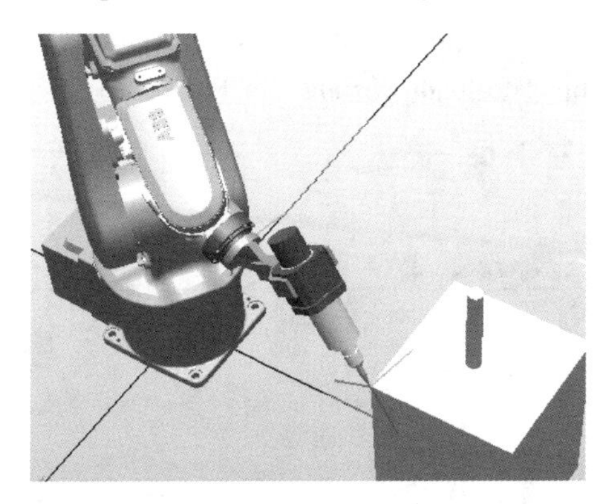

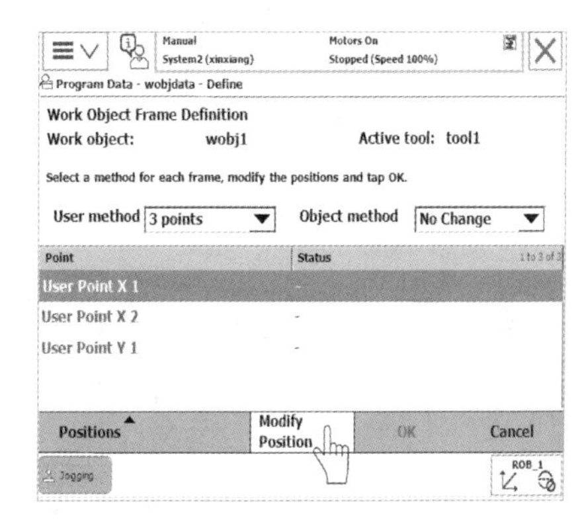


Figure 5-28 Set 'User Point X1'

(2) The tool reference point of manually operated robot is close to point X2 of defined workpiece coordinate system. Select 'User Point X2', click 'Modify Position', and record the position of point X2. The tool reference point of manually operated robot is close to Y1 point of defined workpiece coordinate system. Select 'User Point Y1', click 'Modify Position', and record the position of point Y1 (shown in Figure 5-29).

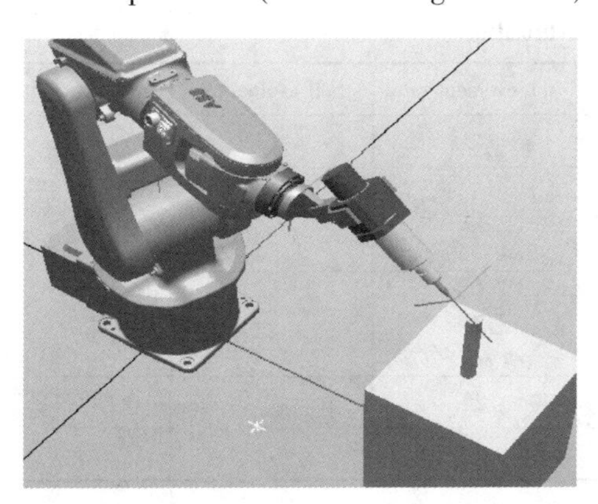

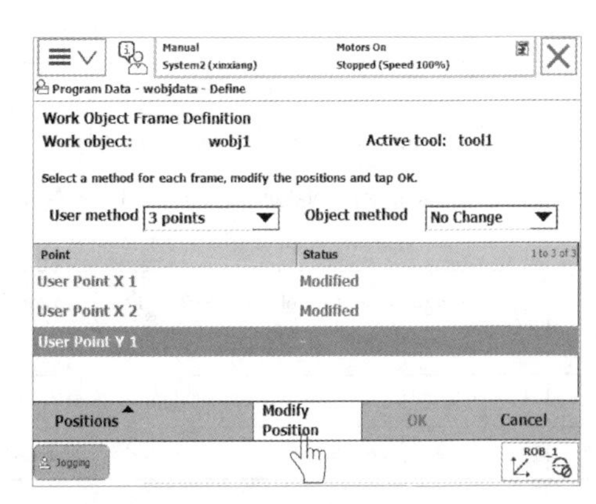


Figure 5-29 Set 'User Point Y1'

(3) Click 'OK' to complete the setting. After confirming the automatically generated workpiece coordinate data, click 'OK', select 'wobj1' and then click 'OK'. The action mode is selected as 'linear', the coordinate system is selected as 'Workpiece Coordinate', and the workpiece coordinate is selected as 'wobj1'. Set the item of manual operation screen, and use

linear action mode to experience the newly established workpiece coordinates, as shown in figure.

#### 5.2.3.3 Setting of payload loaddata data

The creations of loaddata are as follows:

Click the main menu button in the upper-left corner, select 'Manual Control', and select 'Payload'. Click 'New' to set the attributes of the data as required. Generally, it does not need to be modified. Click 'Initial Value' to set the payload data according to the actual situation, and click 'OK'. Clamp and specify the weight and center of gravity of the current carrying object load1, loosen the clamp, and clear the carrying object to load0 (shown in Figure 5-30).

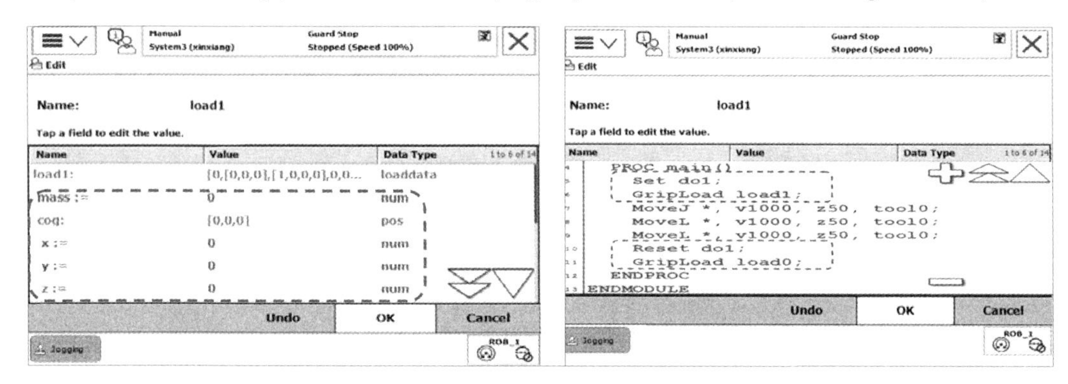


Figure 5-30 Set payload data

### 5.2.4 Evaluation

The table of the task evaluation is shown in Table 5-4.

**Table 5-4 Task evaluation**

| No. | Evaluation content | Achievement ratio | Self evaluation | Teacher evaluation |
|---|---|---|---|---|
| 1 | The meaning of three key programs of industrial robot | 15 | | |
| 2 | The setting method of the first two key programs of industrial robot | 15 | | |
| 3 | The implementation of the first two key program data of industrial robot | 25 | | |
| 4 | Setting and operation of payload data of industrial robot | 25 | | |
| 5 | Safety precautions for operating industrial robots | 20 | | |

(1) The tool center point (TCP) of a tool action is called the tool center point, which is the origin of the tool coordinate system.

(2) The tool center point can be set on or outside the tool.

(3) The default tool center point of the industrial robot body is located in the center of the outer surface of the tool mounting flange plate.

(4) The workpiece coordinate system corresponds to the workpiece, which defines the position

of the workpiece relative to the geodetic coordinate system (or other coordinate system).

(5) Programming industrial robot is to create target and path in workpiece coordinate system.

(6) Before the industrial robot carries out the task, it must correctly set the quality of the fixture, the center of gravity tooldata, the quality of the task object and the center of gravity data loaddata.

(7) The load data of industrial robot often defines the payload of industrial robot or the load of grabbing object (set by instruction gripload or mechunitload), that is, the load held by the human fixture of industrial robot.

### 5.2.5 Development

(1) Please find the original tool coordinates and workpiece coordinates of the robot in the teaching device.

(2) Create a tool data named toolrobot in the teachpad.

(3) Create an artifact data named 'Wobjrobot' in the teachpad.

# Project 6 Programming Practice of ABB Industrial Robots

In this project, the learning goal is to understand the ABB industrial robot programming language, to understand the relationship among tasks, to routines in programming, and to master the usage of common rapid instructions to create a complete rapid program.

## Task 6.1 Buildup of Program Architecture of RAPID

### 6.1.1 Introduction

Rapid language is a high-level programming language based on computer system, which is easy to learn and use, and has strong flexibility. It also has high-level functions such as interrupt, error processing, multi-task processing, etc.

In a rapid program, all instructions are written with a specific vocabulary and syntax, which are called rapid programming language. The instructions can be used to move industrial robots, set output signals, and read input signals. It also can realize decision-making, repeat other instructions, construct programs, communicate with system operators and so on.

The operation task of a robot in the industrial scene usually needs dozens or even tens of thousands of instructions. In order to facilitate the management and reading of programs, instructions are usually packaged into routines, which are modularized and stored in modules.

### 6.1.2 Related Knowledge

The basic architecture of rapid program is shown in Table 6-1.

**Table 6-1 RAPID program architecture**

| RAPID program (Task) | | | |
|---|---|---|---|
| Program module 1 | Program module 2 | Program module 3 | System module |
| Program data | Program data | … | Program data |
| Main routine | Routine | … | Routine |
| Routine | Trap | … | Trap |
| Trap | Function | … | Function |
| Function | — | … | — |

The descriptions of rapid program architecture are as follows:

(1) Rapid program (Tasks): the rapid program is also called a task. This task is composed of

a series of modules, including program modules and system modules. In general, the robot programs is built by creating new program modules, meanwhile system modules are mostly used for system control.

(2) Program module: Multiple program modules can be created according to different purposes, such as the program module for main control, the program module for location calculation, and the program module for storing data, so as to facilitate the classification and management of routine programs and data for different purposes.

Each program module contains four objects: program data, routine procedure, trap procedure and function, but not necessarily in one module. The data, routine, trap procedure and function among the program modules can be called by other.

(3) Main program: In the rapid program, there is only one main program, which can exist in any program module. Like other high-level programming languages, it is the starting point for the execution of the entire rapid program.

### 6.1.3 Implementation

Next, let's take the actual rapid program example to understand the structure level of rapid. The steps of specific operation are as follows:

(1) Click the main menu button in the upper left corner and choose 'Program Editor' (shown in Figure 6-1), and click 'Tasks and Programs' (shown in Figure 6-2).

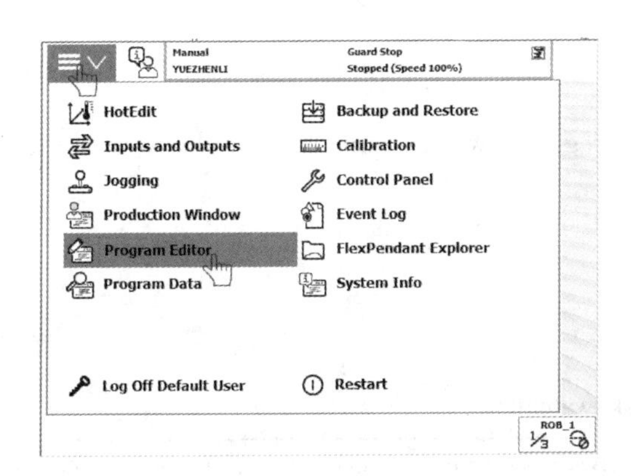


Figure 6-1 Choose 'Program Editor'

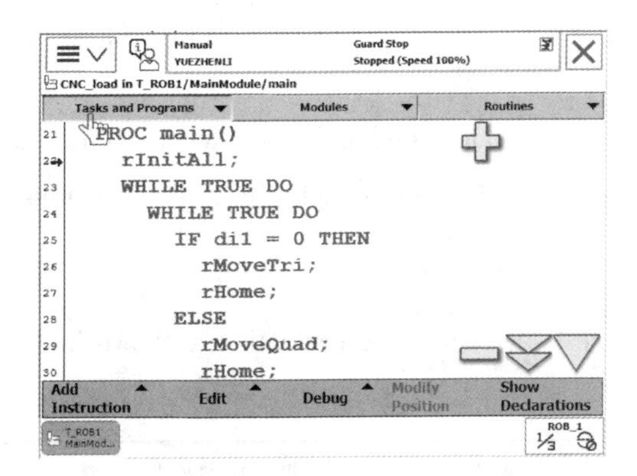


Figure 6-2 Click 'Tasks and Programs'

(2) Then a task can be seen named 'T_ROB1', and click 'Show Modules' (shown in Figure 6-3). There are two system modules named 'BASE' and 'User', and a program module named 'MainModule' in the program. Select 'MainModule' and click 'Show module' to view all routines in the module (shown in Figure 6-4).

(3) As shown in Figure 6-5, 'main ( ) ' is the main program, 'rpick1 ( ) ' is the routine, and 'tpallet1' is the interrupt program. Select a routine and click 'Show Routine' to view the instructions.

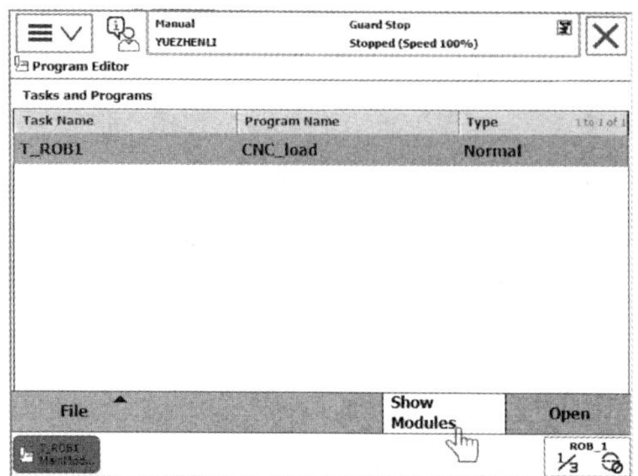


Figure 6-3　Interface of 'Tasks and Progarms'

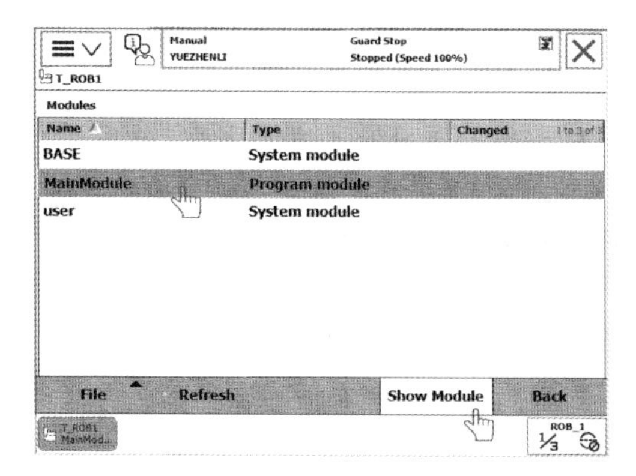


Figure 6-4　Modules

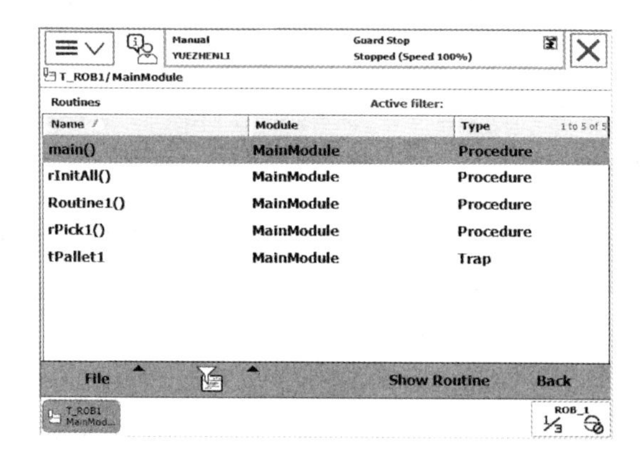


Figure 6-5　Routines

After understanding the architecture of the rapid program, we can quickly grasp the work of the whole program and prepare for program writing.

### 6.1.4　Evaluation

The table of the task evaluation is shown in Table 6-2.

**Table 6-2　Task evaluation**

| No. | Evaluation content | Achievement ratio | Self evaluation | Teacher evaluation |
|---|---|---|---|---|
| 1 | Understand the program composition of rapid | 30 | | |
| 2 | Understand the definition of tasks | 20 | | |
| 3 | Understand the definition of program module | 20 | | |
| 4 | Understand the definition of routine | 30 | | |

(1) A rapid program usually is a task, a task can be composed of several modules, and a module is composed of a number of routine procedures. Interrupt procedures are usually used to store robot instructions.

(2) There is only one main program in rapid program, which is the starting point of the whole

program execution.

Through the study of this task, the basic concepts of ABB industrial robot programming language, the relationship between tasks, modules and routines can be understood.

### 6.1.5 Development

Design a program architecture using ABB robot to load and unload materials for lathe and CNC, with the task name of 'load_unload', and add three program modules, with the names of 'MainModule', 'Lathe' and 'CNC'. 'MainModule' module contains the main routine, 'Lathe' module contains the two routines of 'Lathe_load' and 'Lathe unload', and 'CNC' module contains the two routines of 'CNC_load' and 'CNC_unload'.

## Task 6.2 Learn Common Rapid Programming Instructions

### 6.2.1 Introduction

After building the rapid architecture, the practical things need to be filled in, that is to add specific instructions, so that the rapid program is complete finally. This task is to learn the common rapid instructions in practice.

The rapid program of ABB industrial robot provides abundant instructions for various simple and complex applications. It is the reasonable application of these instructions that can realize the control and operation of industrial robot.

### 6.2.2 Related Knowledge

If ABB industrial robot wants to complete a simple or complex task, it is likely to use a variety of instructions. Common rapid instructions include assignment instruction, robot motion instruction, I/O control instruction, conditional logic judgment instruction, waiting instruction and other common instructions.

### 6.2.3 Implementation

Next, the rapid instructions are learnt from the most commonly used instructions, and the programming convenience is appreciated provided by the rich instruction set of rapid.

#### 6.2.3.1 Assignment Instruction

Assignment instruction (:=) is used to assign value to program data. Assignment can be a constant or a mathematical expression. The use of this instruction is explained by adding a constant assignment: reg1: =5; and a mathematical expression assignment: reg2: =reg1+4; .

The specific operations of reg1: =5; are as follows:

(1) Select ': =' in the common instruction list (shown in Figure 6-6). Select '<VAR>' to highlight it, and click 'Change Data Type...' to select 'Num' numeric data. In the numeric

data ‘Num’, select ‘reg1’. If there is no reg1 in num data, you can choose to click ‘New’ to create reg1.

(2) Select ‘< EXP >’ to highlight it, and open the edit menu and choose ‘Only Selected’. Enter the number ‘5’ through the soft keyboard, and then click ‘OK’. It will add ‘reg1: =5’ instruction in the program (shown in Figure 6-7).

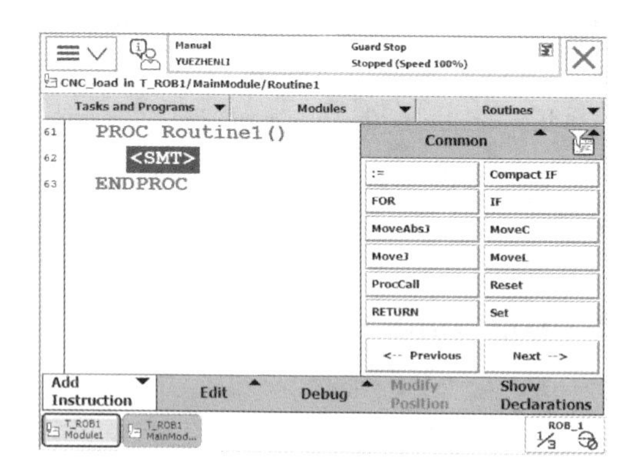


Figure 6-6 The instruction list

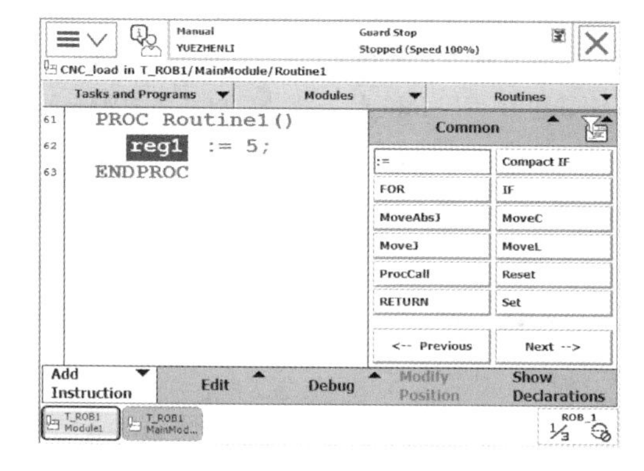


Figure 6-7 Added instruction

Next, add ‘reg2: =reg1+4;’. The specific operations are as follows:

(1) Select ‘: =’ in the instruction list again, click ‘< VAR>’, and select ‘reg2’ in ‘Num’ data. If there is no reg2 in num data, you can click ‘New’ to create reg2.

(2) Then select ‘<EXP>’ on the right, select ‘reg1’, and click ‘reg1’ just added to make it in the selected state. Then click ‘+’ sign in the right of the figure. It will add ‘<EXP>’ on the sight of ‘+’ (shown in Figure 6-8).

(3) Select ‘<EXP>’ and it will behighlighted in blue, then open the ‘Edit’ menu and select ‘Only Selected’. Enter the number ‘4’ through the soft keyboard, and then click ‘OK’.

(4) The added command will be prompted whether it is above or below the cursor. Click ‘Below’.

Figure 6-9 shows the added interface. After the instruction is added successfully, click ‘Add Instruction’ to collect the instruction list.

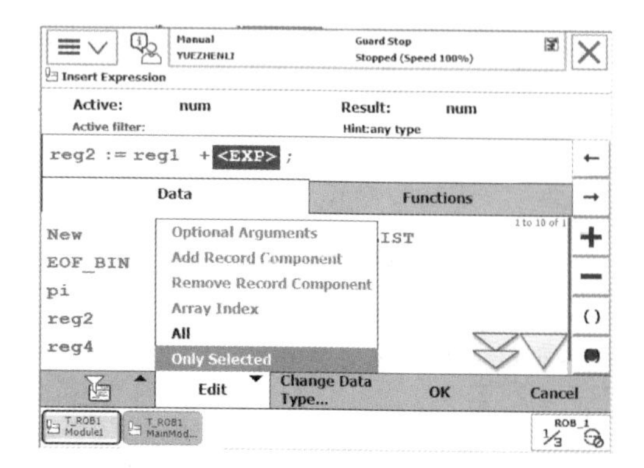


Figure 6-8 Add ‘<EXP>’

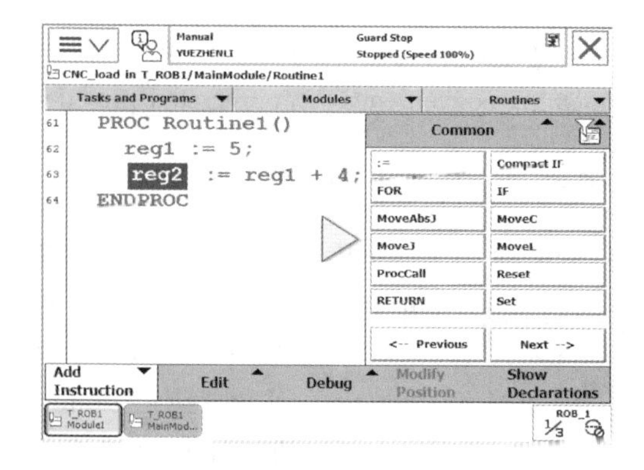


Figure 6-9 Added instruction

### 6. 2. 3. 2 Motion Command of Industrial Robot

There are three main modes of robot motion in space: Joint motion (MoveJ), Linear motion (MoveL), and Circular motion (MoveC).

Note: Before adding or modifying the robot's motion command, it is necessary to confirm the used tool coordinates and work-object coordinates.

**Linear motion command (MoveL)**

First of all, MoveL is learnt. The linear motion is that the TCP path of the robot is always a straight line from the beginning to the end. Generally, welding, gluing and other applications use this command when the path requirements are high. The schematic diagram of linear motion is shown in Figure 6-10.

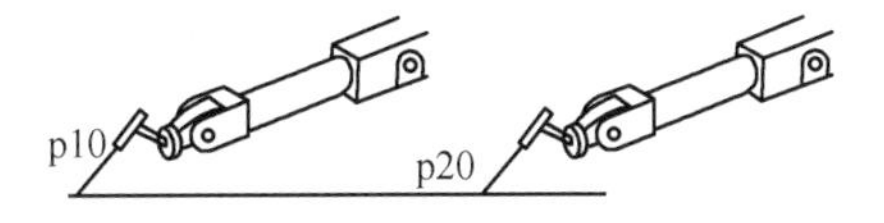


Figure 6-10 The linear motion diagram

The operations of adding MoveL are as follows:

(1) Select '<SMT>' to add the instruction, and select 'MoveL' in the common instruction list. Click ' * ' twice to replace it with robtarget data (shown in Figure 6-11).

(2) Click 'New' to add a Robtarget data. After setting the data attribute of the target point, click 'OK'. At this time, ' * ' has been replaced by the target point named 'p10' and the instruction is added.

(3) Click 'Add Instruction' to collect the instruction list. Clicking '-' to narrow the editing window, the whole motion command can be seen (shown in Figure 6-12). Select 'p10', then click 'Modify Position', and the position information of 'p10' will be stored base on 'tool 1' in and 'wobj1'.

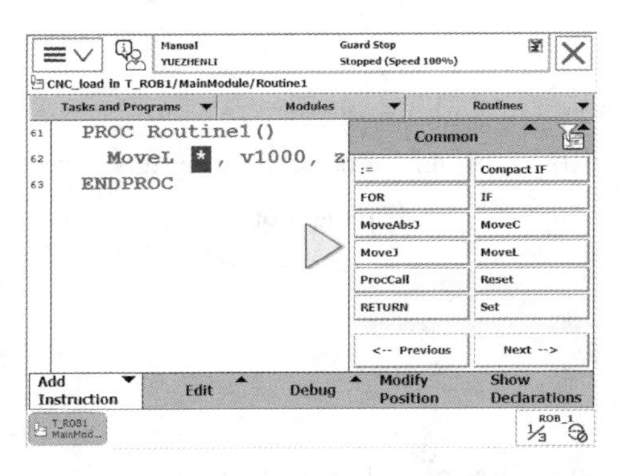


Figure 6-11 Click ' * ' twice

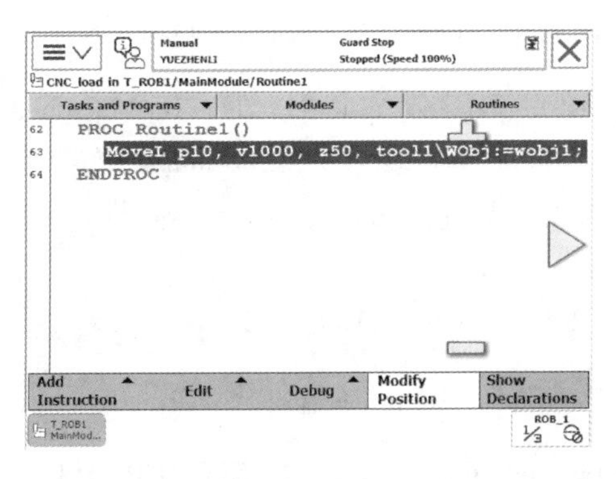


Figure 6-12 Show the whole motion command

Instruction example:

MoveL p10, v1000, Z50, tool1 \ wobj: =wobj1;

Instruction description: The movement mode of the robot is linear movement, the target point is p10, the movement speed data is 1000mm/s, the turning area data is 50mm, and the robot uses tool coordinate data 'tool1' and work-object coordinate data 'wobj1'.

**Joint Motion Command (MoveJ)**

The joint motion command used for low path accuracy requirement. The TCP of the robot moves from a position to another, and the path between the two positions is not necessarily a straight line. The joint motion diagram is shown in Figure 6-13.

The joint motion command is suitable for large-scale motion of robot, and it is not easy for the joint axis to enter the mechanical dead point in the process of motion.

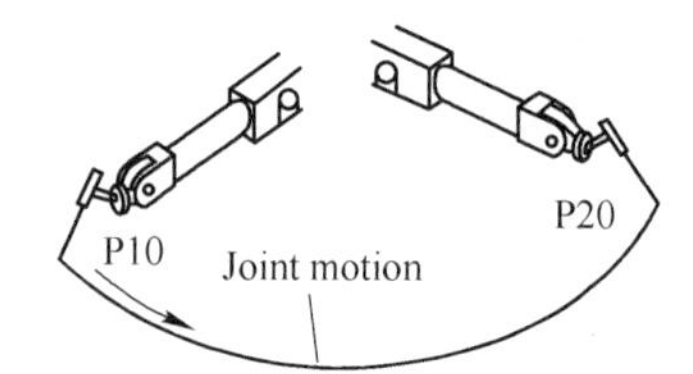


Figure 6-13 Joint motion diagram

A typical example of linear motion and joint motion is shown in Figure 6-14.

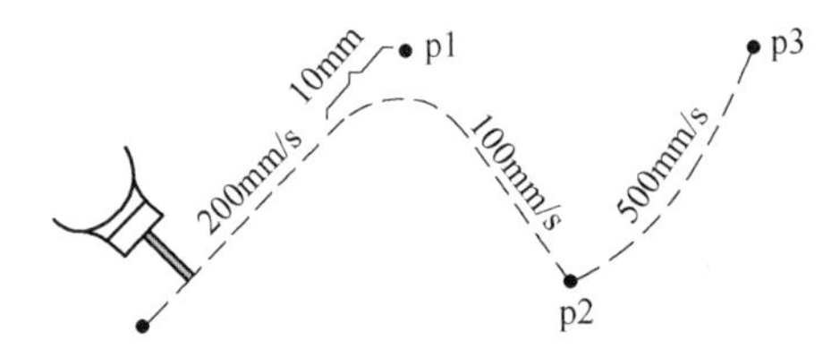


Figure 6-14 Typical example

The motion track can be constituted by three motion instructions as follows:

(1) Instruction example 1:

```
MoveL p1, v200, z10, tool1 \ wobj: =wobj1;
```

Instruction description 1: The TCP of robot moves forward in linear motion from current position to p1 point, with speed of 200 mm/s, turning area data of 10 mm, and starts turning when there is 10 mm from p1 point, using tool coordinate data 'tool1', and using work-object coordinate data 'wobj1'.

(2) Instruction example 2:

```
MoveL p2, v100, fine, tool1 \ wobj: =wobj1;
```

Instruction description 2: The TCP of the robot moves forward in linear motion from p1 to p2, with the speed of 100mm/s. The data of turning area is fine, the robot pauses at p2.

(3) Instruction example 3:

```
MoveJ p3, v500, fine, tool1 \ wobj: =wobj1,
```

Instruction description 3: The steps of adding instruction are the same as MoveL. The TCP of the robot moves forward from p2 to p3 in the way of joint motion, with the speed of 100mm/s, the data of turning area is fine, and the robot stops at p3.

In the turning area, 'fine' refers to the point where the TCP reaches the target point, and the speed of the target point drops to zero. The robot stops and then moves downward. Generally, the last point of the path must be 'fine'. The larger the turning area, the smoother and smoother the robot motion path.

**Arc Motion Command (MoveC)**

The arc path defines three position points that the robot can reach in the space. The first point is the starting point of the arc, the second point is used for the curvature of the arc, and the third point is the end point of the arc. The schematic diagram of the arc motion is shown in Figure 6-15.

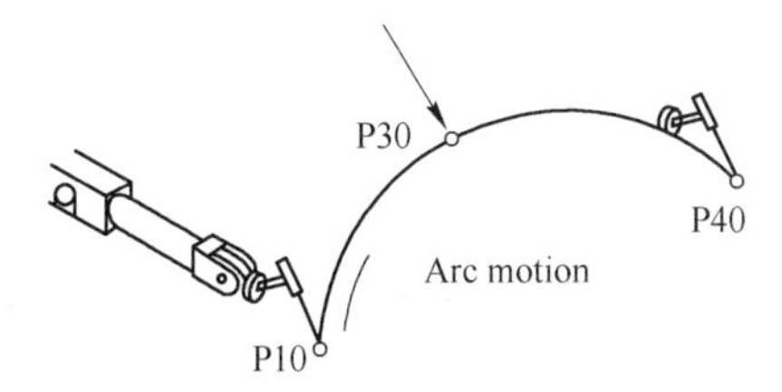


Figure 6-15 Arc motion diagram

Instruction example:

```
MoveL p1, v200, z10, tool1 \ Wobj: =wobj1;
MoveC p30, p40, v1000, z1, tool1 \wobj: =wobj1;
```

Instruction description: p10 is the starting point of the arc; p30 is the second point of the arc, according to which the path radius is determined, and p40 is the end point of the arc.

Note: In subsequent debugging, when debugging a MoveC instruction, you cannot start debugging with this instruction directly because the MoveC instruction requires the first point. The MoveC instruction must be preceded by the MoveL or MoveJ instruction to determine the first point, otherwise the system will take the position of the robot when it starts debugging as the first.

### 6.2.3.3 I/O Control Command

I/O control command is used to control I/O signals to communicate with peripheral equipment. The basic I/O control instructions are described below.

**Digital Signal Set Instruction of Set**

The digital signal set instruction is used to set the digital output to 1 or high level.

Instruction example:

```
Set do1;
```

Instruction description: Set the output signal do1 to 1.

**Digital signal reset command of Reset**

The reset digital signal reset instruction is used to set the digital output to 0 or low level.

Instruction example:

```
Reset do1;
```

Instruction description: Reset the output signal do1 to 0.

Note: If there are motion instructions like 'MoveJ', 'MoveL' and 'MoveC' in front of set or reset instructions, you must use fine in the motion instructions to accurately reach the target point and then change the IO signal state.

**Digital input Signal Judgment Instruction of WaitDI**

Digital input signal judgment instruction is used to determine whether the value of the digital input signal is consistent with the target.

Example instructions:

```
WaitDI di1, 1;
MoveL p1, v200, z10, tool1\wobj:=wobj1;
```

Instruction description: When the program executes, the value of di1 is waitted for to be 1. If di1 is 1, the program will continue to run down and move to p1 in a straight line, otherwise it will wait forever. But if the maximum waiting time reaches 300s (this time can be set according to the actual situation), and the value of di1 is not 1, the robot will alarm or enter the error processing program.

### 6.2.3.4 Logic Judgment Instruction

Conditional logic judgment instruction is used to perform corresponding operations after judging several conditions, which is an important component of rapid.

**Compact Condition Judgment Instruction (Compact IF)**

The compact condition judgment instruction is used to execute an instruction when a condition is met.

Instruction example: if flag1=true Set do1;

Instruction description: If the status of flag1 is true, do1 is set to 1.

**Condition Judgment Instruction of IF**

The condition judgment instruction is to execute different instructions according to different conditions.

Note: The number of conditions determined can be increased or decreased according to the actual situation.

Example instructions:

```
IF num1=1 THEN
        flag1:=TRUE;
ELSE
        flag1:=FALSE;
        Set do1;
```

```
ENDIF
```

Instruction description: If num1 is 1, flag1 is assigned to true. Otherwise, flag1 will be assigned false, and the digital output do1 is set to 1.

**Repeatedly Executing Judgment Instruction of FOR**

Repeatedly execution judgment instruction is used when one or more instructions need to be executed repeatedly for several times. This instruction starts with 'FOR' and ends with 'ENDFOR'.

The usage of the instruction is:

```
FOR i FROM n TO m DO
        <instructions or routines >
ENDFOR
```

The number of cycle is (m−n)+1.

Example instructions:

```
FOR i FROM 1 TO 10 DO
        Routine1;
ENDFOR
```

Instruction description: The routine "Routine1" repeated 10 times.

**Condition Judgment Loop Instruction of WHILE**

Condition judgment loop instruction is used to repeatedly execute the corresponding instruction when the given condition is met. This instruction starts with 'WHILE' and ends with 'ENDWHILE'.

The usage of the instruction is:

```
WHILE <conditions> DO
        <instructions or routines>
ENDWHILE
```

Example instructions:

```
WHILE num1 > num2 DO
        num1: =num1-1;
ENDWHILE
```

Instruction description: When the condition of 'num1>num2' is satisfied, the operation of 'num1: =num1−1' is executed repeatedly until the condition is not satisfied.

**Waiting Instruction 'WaitTime'**

Waiting instruction is used to wait for a specified time before the program continues to execute downward. The usage of this instruction is WaitTime in s.

Example instructions:

```
WaitTime 4;
Reset do1;
```

Instruction description: After 4s, the program executes 'Reset do1' instruction.

### 6.2.3.5 Other Common Instructions

**Call Routine Instruction of ProcCall**

Other routines at the specified location are called by using this instruction. The specific steps are as follows:

(1) Select '<SMT>' as a confirmed location to call other routine, and select the 'ProcCall' instruction in the instruction list (shown in Figure 6-16). Select the 'Routine1' routine we want to call, and then click 'OK' (shown in Figure 6-17).

The result ofthe call routine instruction is shown in Figure 6-18.

Instruction description: If the input signal 'di1' is 1, Routine1 will be executed, otherwise, the program skips this statement and executes others in sequence.

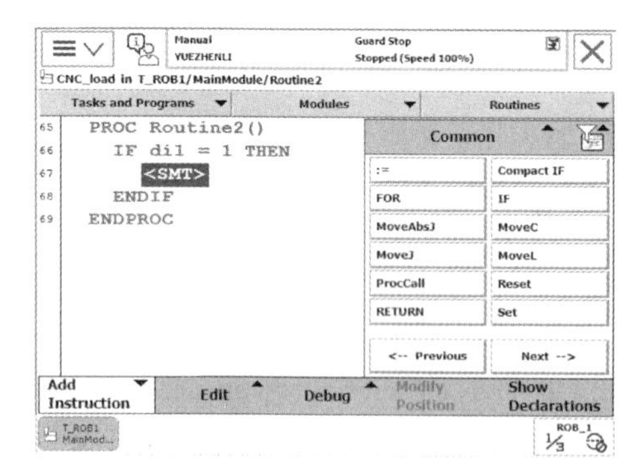


Figure 6-16 Select 'ProcCall'

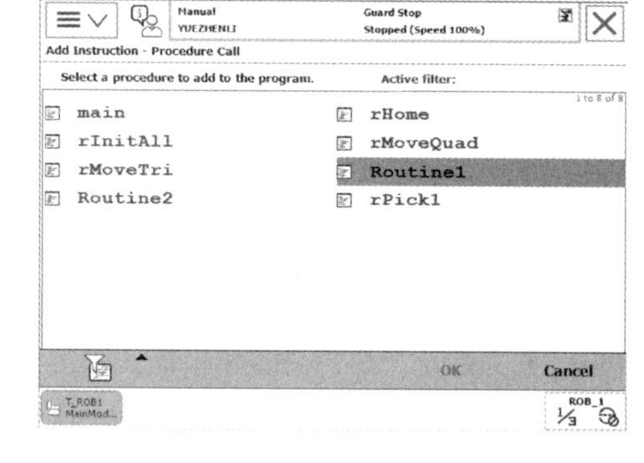


Figure 6-17 Routines

**Routine Exit Instruction of RETURN**

When the instruction is executed, the routine will be terminated immediately, and the program pointer will be returned to the place where the routine is called.

An example instruction is shown in Figure 6-19.

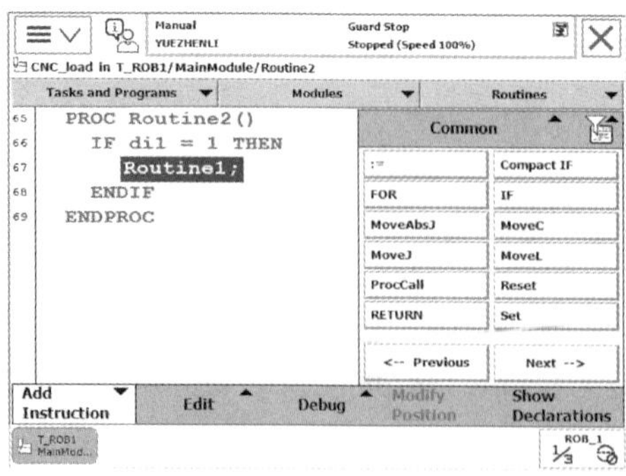


Figure 6-18 Added instruction

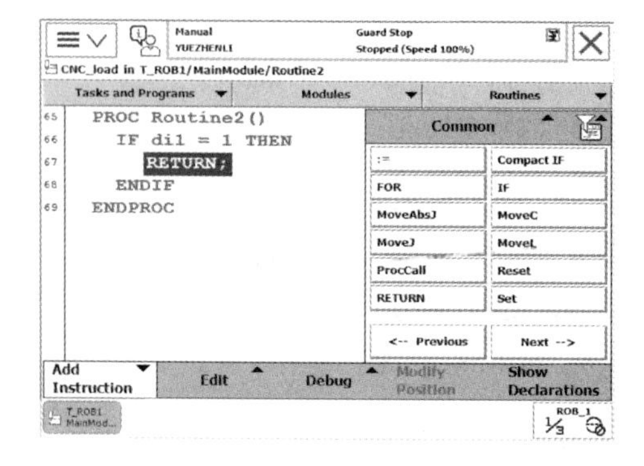


Figure 6-19 Example instruction

Instruction description: When di1 = 1, the program executes the return instruction, and the program pointer returns to the position where routine2 routine is called.

### 6. 2. 4 Evaluation

The table of the task evaluation is shown in Figure 6-3.

**Table 6-3 Task evaluation**

| No. | Evaluation content | Achievement ratio | Self evaluation | Teacher evaluation |
|---|---|---|---|---|
| 1 | Assignment instruction | 5 | | |
| 2 | Motion command | 35 | | |
| 3 | I/O control instruction | 20 | | |
| 4 | Conditional logic judgment instruction | 20 | | |
| 5 | Wait instruction | 5 | | |
| 6 | Other common instructions | 15 | | |

(1) The common motion instructions in RAPID programming system are MoveJ, MoveL and MoveC.

(2) The common I/O control instructions in rapid programming system include digital signal setting instruction of Set, digital signal reset instruction of Reset, and digital input signal judgment instruction of WaitDI.

(3) Conditional logic judgment is the core of thinking and an important part of the program. Common signals include Compact IF, IF, FOR and WHILE.

(4) ProcCall and RETURN are often used in complex programs.

### 6. 2. 5 Development

(1) Add a routine named 'rTrain' to the module named 'MainModule'.

(2) In the 'rTrain' routine, create a new num type data 'count', and edit of count: =count +1;

(3) In the 'rTrain' routine, the MoveL and MoveJ instructions are used to walk a square track. The four points are p1, p2, p3 and p4.

(4) In the 'rTrain' routine, add 'Main' program, call the 'rTrain' routine to it, and debug the whole program.

# 项目1　工程实践创新技术的认知

创新是一个民族进步的灵魂，也是一个国家兴旺发达的不竭动力。创新涵盖众多领域，包括政治、军事、经济、社会、文化、科技等领域。本项目介绍了1000多个组件，并按照结构件、连接件、传动件和电气组件四大类分别进行说明，以及对常见传动方式的搭建方法和搭建技巧进行介绍。同时能认识VJC图形化交互开发系统并使用VJC图形化交互开发系统设计出程序，实现对复杂机电系统的逻辑控制。

## 任务1.1　认识工程实践创新与能力源套件

### 1.1.1　任务引入

一百多年前，有位发明家外出旅行，看见一位老太太携带的袋子袋口被人挤坏了，东西撒了一地。回到家中，他联想起了老太太那天的遭遇，最终发明了人类历史上第一根拉链，解决了口袋拉链问题。大多数人只是认为这是一件生活小事，不值得大惊小怪，但发明家却有比较强烈的好奇心和探索精神，现在，几乎所有人都用到他的发明成果。这个故事告诉我们要留意生活中许多不起眼的小事，勤于思考、勇于创新，就会有所收获。

在开启能力源创新实践之旅之前，需要了解下创新所用的工具——能力源创新课程套件。能力源创新课程套件的组件及特点可以通过一些典型案例感受出能力源创新课程套件的强大功能，将套件中的组件以结构件、连接件、传动件和电气组件四大类进行呈现，对套件的基本组成有一个清晰的认识。同时，本项目还介绍了一些常见的搭建方法与技巧，为下一阶段的项目实践打下基础。

### 1.1.2　相关知识

#### 1.1.2.1　认识工程实践创新

工程是指将自然科学的理论应用到具体工农业生产部门中形成的各学科的总称，如水利工程、化学工程和土木建筑工程。工程还可以用较多的人力、物力来进行较大而复杂的工作，需要一个较长时间周期内来完成，如城市改建工程和高速铁路工程。关于工程的研究称为“工程学”，关于工程的立项称为“工程项目”，一个全面的、大型的、复杂的包含各子项目的工程称为“系统工程”。

**工程**

工程是指将自然科学原理应用到工农业生产部门中去而形成各学科的总称。狭义工程概念为：以某组设想的目标为依据，应用有关的科学知识和技术手段，通过一群人的有组织活动将某个（或某些）现有实体（自然的或人造的）转化为具有预期使用价值的人造产品过程；广义工程概念为：由一群人为达到某种目的，在一个较长时间周期内进行协作活动的过程。

**实践**

实践是人类改造客观世界的一切物质性活动，具有客观物质性和主观能动性和社会历史性；实践是人类所特有的活动，与动物消极适应自然的本能活动不同，是客观的物质性的活动；实践是改变客观事物的活动，不是纯主观的思维活动，必然会引起客观对象的变化；实践具有客观性和物质性，有目的、意识的能动性，以及社会性和历史性的活动。

**创新**

创新是指以现有的思维模式提出有别于常规或常人思路的见解为导向，利用现有的知识和物质，在特定的环境中，本着理想化需要或为满足社会需求，从而改进或创造新的事物、方法、元素、路径和环境，并获得一定有益效果的行为。

创新的类型包括：

（1）突破性创新。其特征是打破陈规，改变传统和大步跃进。

（2）渐进式创新。其特征是采取下一逻辑步骤，让事物越来越美好。

（3）再运用式创新。其特征是采用横向思维，以全新的方式应用原有事物。

#### 1.1.2.2 认识能力源套件

**能力源创新课程套件**

能力源创新课程套件涉及机械、电子、传感器、计算机软硬件、控制等各方面的专业知识和技术技能。采用项目式教学，以小组的形式，通过实施精心设计的由浅入深的工程实践创新项目，让学生了解实际工程，并在了解、学习真实工程项目的基础上，提炼出真实工程项目中机电的核心技术。以能力源创新课程套件为载体，激发学生的创新意识，强化学生的团队协作能力，利用学生所学的各种专业知识去解决实际工程问题，通过动手协作，把所学的知识点再现，帮助学生建构一个广域的工科知识体系能力源套件如图 1-1 所示。

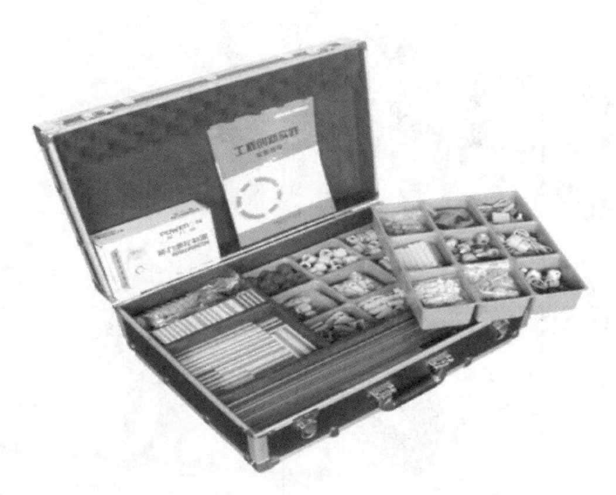

图 1-1　能力源套件

能力源创新课程套件可以仿真出工程项目，不仅可以通过书本去学习和了解这些工程项目，而且可以在动手实践的过程中进行学习和体验，深刻理解工程组成结构和工作原理，还可以按照自己的想法对各种工程项目做无限的创新和拓展。

**能力源创新课程套件典型组件说明**

根据不同的功能，能力源创新课程套件的组件分为结构件、连接件、传动件和电气组件四大类，即：

（1）结构件——三维结构的基础组件。结构件就如盖房子用的砖头，是构建工程项目最基础的组件。能力源创新课程套件的结构件分为点、线、面三种类型（见图1-2），彼此可以直接连接或者借助连接件连接、三维扩展，构建三维空间里的工程模型，同时可以根据项目的需求灵活使用。

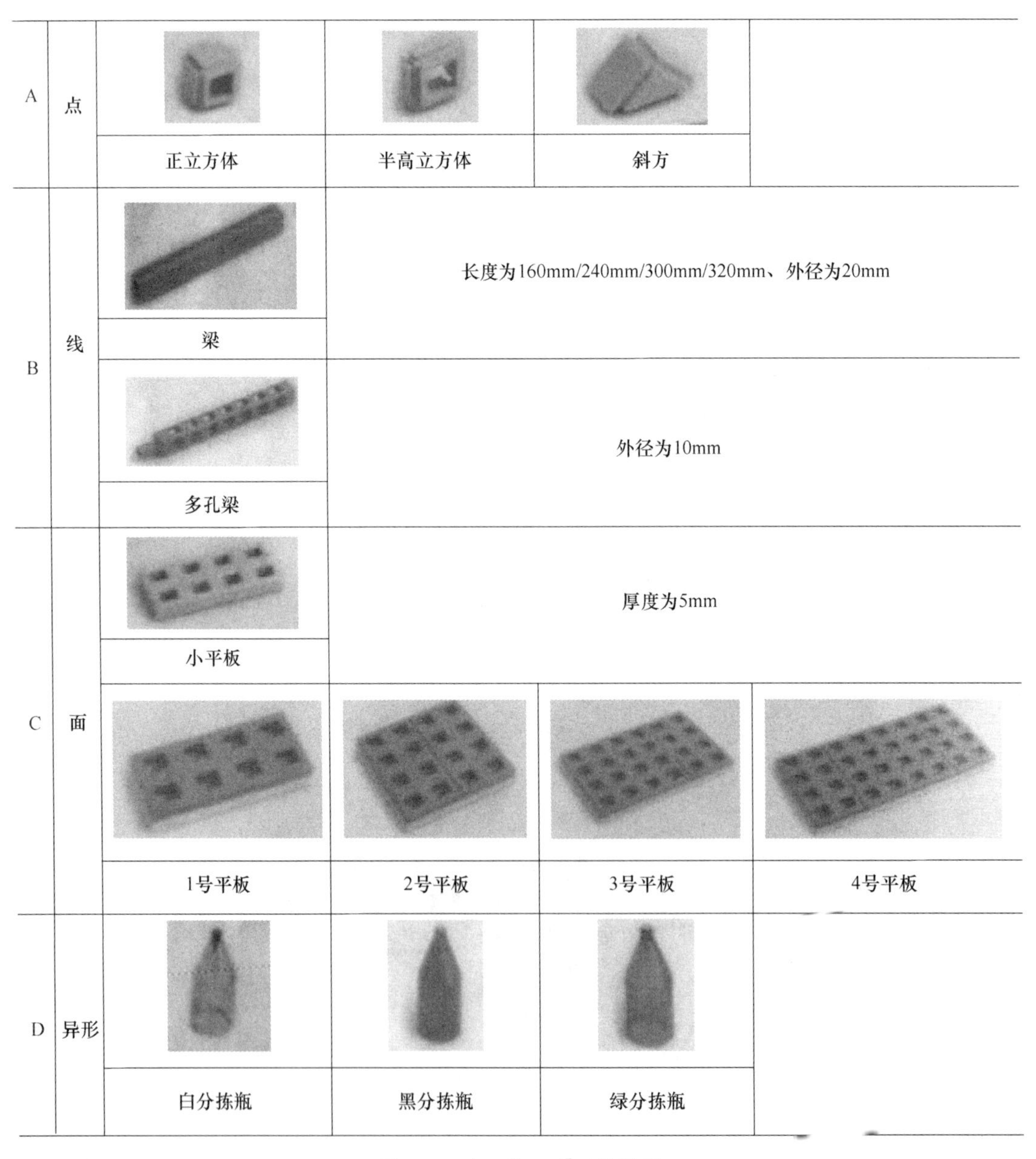


图1-2　点、线、面三种类型

（2）连接件——三维扩展的协助者。连接件类似于盖房子用的水泥或者灰土，可以提供一种合适的方式将结构件彼此连接。

一些结构件不需要额外连接组件就能够相互连接，如立方体和梁。大部分结构件要借助连接件才能连接。结构件有点、线、面三种类型，相对应的连接方式分为点与点、点与线、点与面、线与线、线与面和面与面之间的连接，如图 1-3 所示。

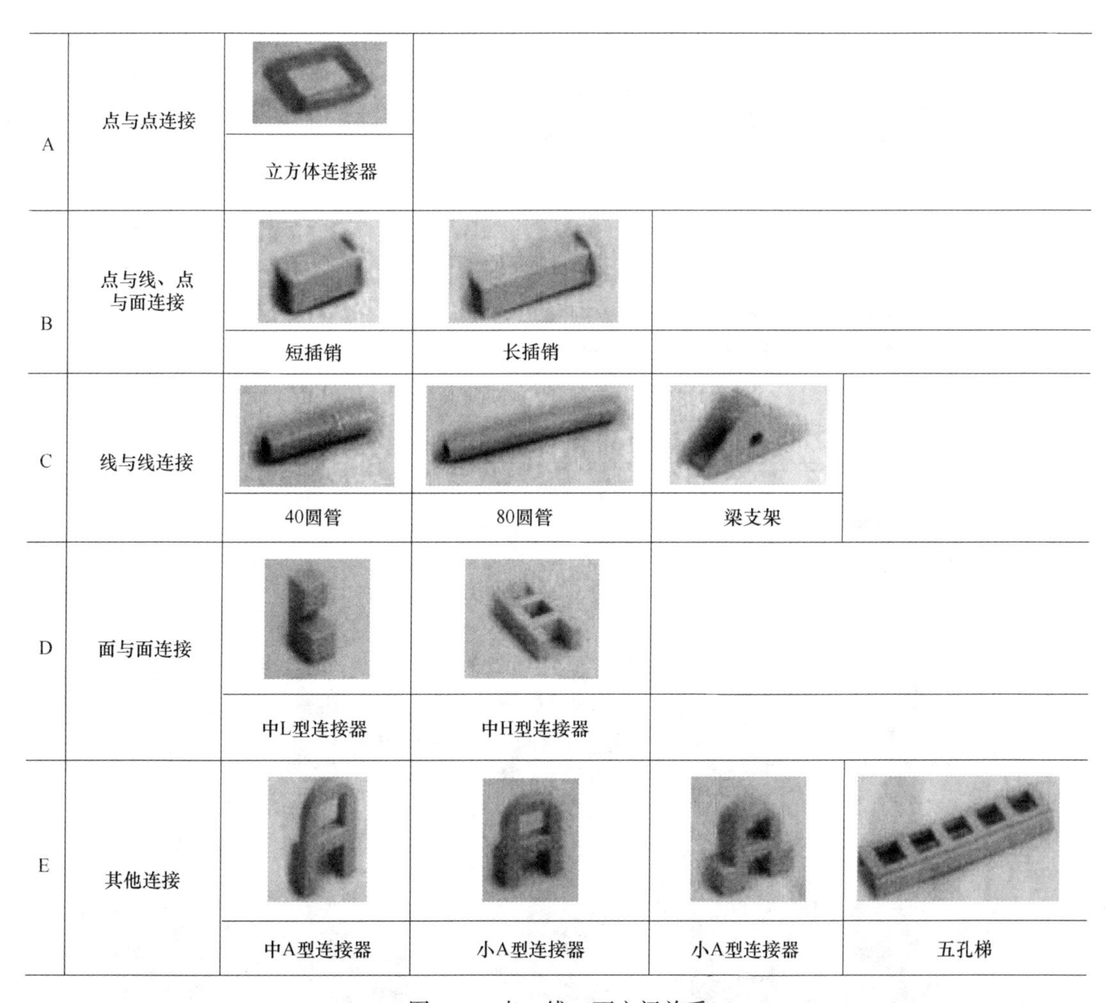

| | | | | | |
|---|---|---|---|---|---|
| A | 点与点连接 | 立方体连接器 | | | |
| B | 点与线、点与面连接 | 短插销 | 长插销 | | |
| C | 线与线连接 | 40圆管 | 80圆管 | 梁支架 | |
| D | 面与面连接 | 中L型连接器 | 中H型连接器 | | |
| E | 其他连接 | 中A型连接器 | 小A型连接器 | 小A型连接器 | 五孔梯 |

图 1-3　点、线、面之间关系

（3）传动件——动力的传递手。传动件主要是指传递动力或者改变运动方向和形式的组件，这些组件在设计时同时考虑了灵活性与易用性，组合效率高。常见的传动方式有齿轮传动、齿轮齿条传动、蜗轮蜗杆传动、皮带传动和螺旋传动等，如图 1-4 所示。

（4）电气组件——自动化的基础。能力源创新课程套件中的电气组件是指各种传感器、执行器和电机线，套件中所有的传感器和执行器集成在立方体中，便于实现三维搭建和扩展，无须使用任何固件就可以完成与其他组件的连接。

传感器作为采集信息的主要组件，将输入信息传递给控制器。例如，AGV 小车利用灰度传感器能够检测环境光的功能来实现巡线，自动门通过触碰开关操作门的开启与关

| | | | | | | |
|---|---|---|---|---|---|---|
| A | 模块化减速器 | 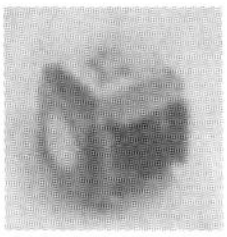 |  | 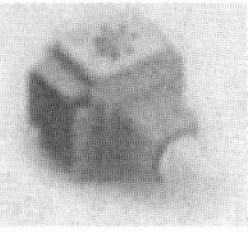 | 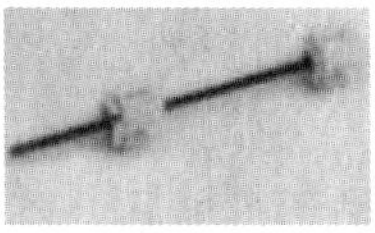 | |
| | | 5:1减速 | 1:1转向 | 1:1带轴 | 丝杠组件 | |
| B | 齿轮 | 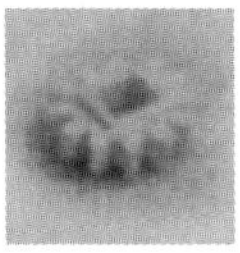 | 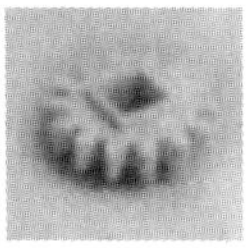 | 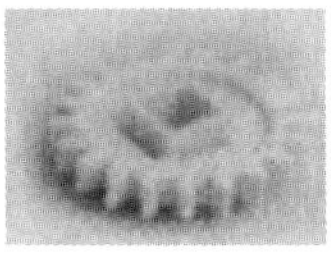 | 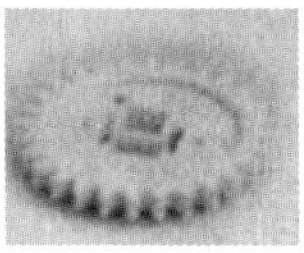 | 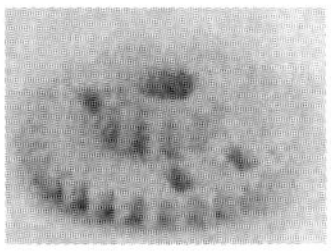 |
| | | 12齿 | 14齿 | 20齿 | 28齿 | 12/28齿 |
| C | 蜗轮蜗杆 | 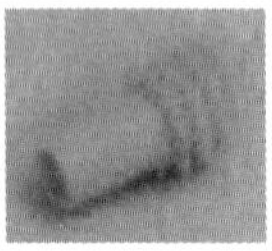 | 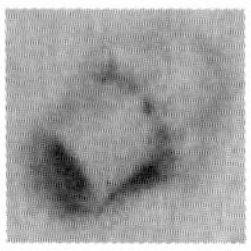 | | | |
| | | 蜗杆 | 12齿蜗轮 | | | |
| D | 齿条 | 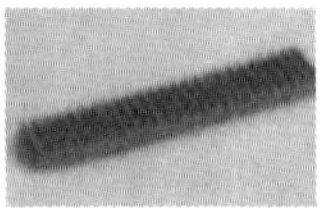 | | | | |
| | | 齿条 | | | | |
| E | 轴承及轴 | 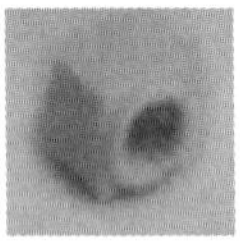 | 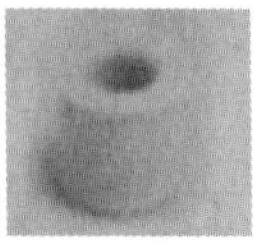 | 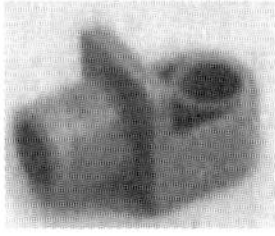 | 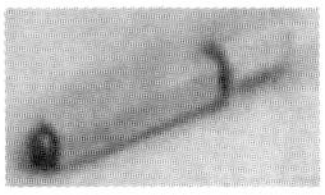 | 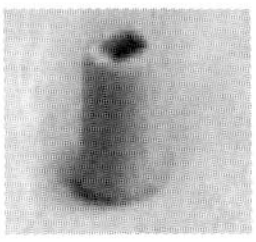 |
| | | 轴承 | 滑动轴承 | 关节件 | 带台阶轴 | 外圆内方管 |
| F | 轮毂 | 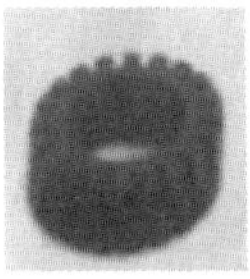 | 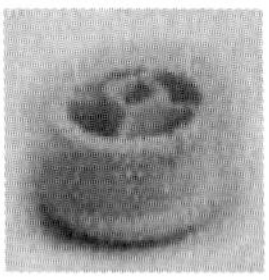 | 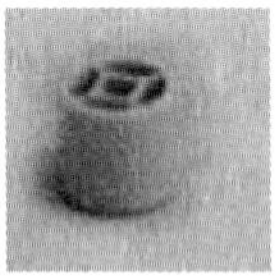 | 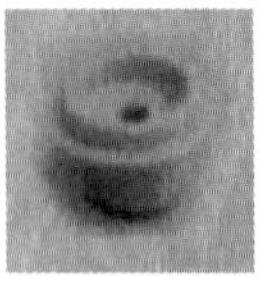 | 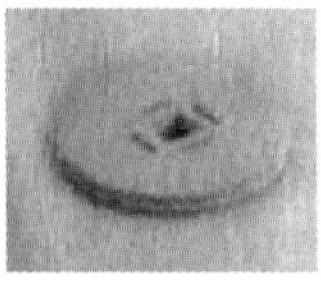 |
| | | 驱动轮胎 | 驱动轮毂 | 皮带轮 | 导向轮 | 滑轮 |

图 1-4　传动件

闭，同时通过磁敏开关和磁铁配合来控制门扇的极限位置，数控机床通过旋转计数器检测丝杠的精确转数，从而控制每个自由度的运动范围。

常用的执行器包括不同颜色的 LED 灯（数字输出型执行器）、电磁铁以及电机（可调输出型执行器）。执行器在工程项目中也有广泛的使用，例如利用工业机械手通电与否来实现“抓手”取、放分拣瓶，用不同颜色的 LED 灯反馈不同的工作状态等。电机导线是连接电机与控制器的专用导线。

总之，电气组件（见图 1-5）是实现工程项目自动化控制的主要组件。

| | | | | | |
|---|---|---|---|---|---|
| A | 模拟量传感器 |  | 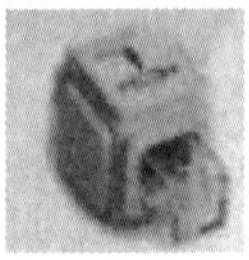 | | |
| | | 光敏传感器 | 温度传感器 | | |
| B | 数字量传感器 | 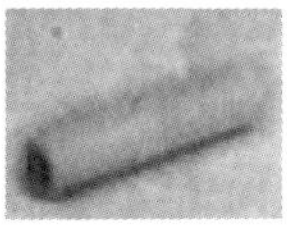 | 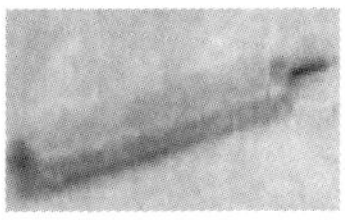 |  |  |
| | | 磁铁 | 磁敏开关 | 旋转计数器 | 触碰开关 |
| C | 可调输出型执行器 | 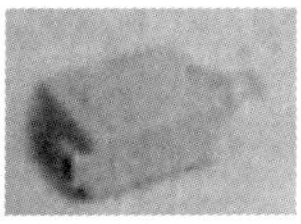 |  | | |
| | | 电机 | 电磁铁 | | |
| D | 数字输出型执行器 | 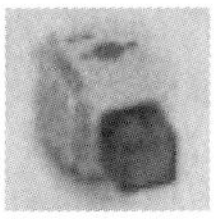 | 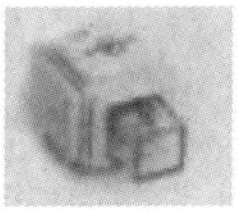 | 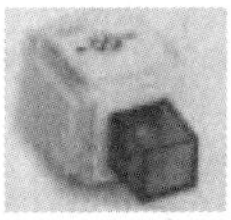 |  |
| | | 红灯 | 黄灯 | 绿灯 | 蓝灯 |
| E | 连接导线 | 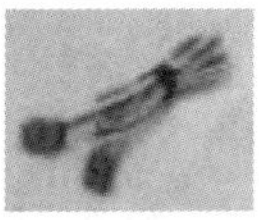 | | | |
| | | 电机线 | | | |

图 1-5　电气组件

**控制器**

控制器（见图 1-6）是能力源创新课程套件的控制部分，让 1000 多个能力源的各类组件彼此配合成一个项目整体（相当于人类的“大脑”）。内部处理器是一个 32 位 ARM 处理器，主频为：72MHz，64K SRAM，用户存储器 3.96Mb。能力源创新课程套件的控制器包含 12 路 I/O 接口和 4 路直流电机接口。12 路 I/O 接口支持模拟输入、数字输入、数字输出、计数器、485 通信等功能。

控制器的详细参数如下：

（1）USB 下载口。控制器与上位机之间通过 USB 下载口通信，在上位机 VJC 系统中写好的程序借助下载线通过 USB 下载口下载到控制器中，下载程序的过程中需要打开控制器电源。

（2）运行键（〈Enter〉）。用于程序的执行，程序只有下载到能力源控制器中，并按下

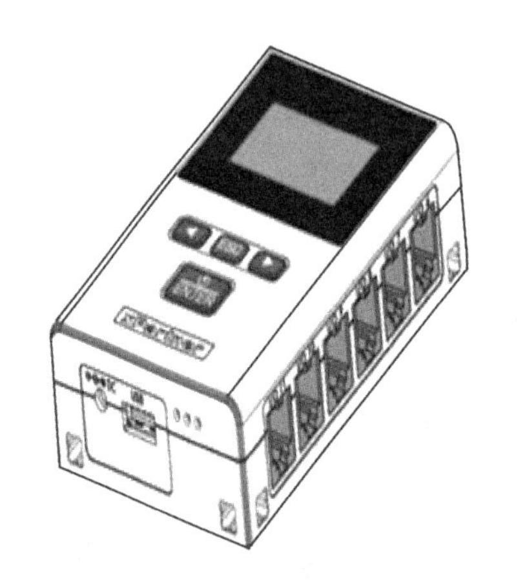

| 序号 | 定义 | 序号 | 定义 |
|---|---|---|---|
| A | 电源+,<br>电池电压 | D | AI/DI |
| B | +5V<br>电源 | E | RS485 D- |
| C | 电源-<br>/地/DO<br>输出 | F | RS485 D- |



图 1-6　控制器

<Enter>键后才能运行。

（3）电机口。共 4 路，可以控制电机的速度和转向，支持普通直流电机和闭环电机，输出电压为电池电压，单路支持最大电流为 1.5A。

（4）电源。自带 8.4V 1A 适配器，接在电源口上可以直接给控制器供电。另外也可以使用专用锂电池供电，专用锂电池为 8.4V 1500MAH，最大放电电流为 6.5A，自带保护电路。

（5）液晶屏。128×64 点阵液晶显示屏，带背光，可以显示图形和字符；另外通过设置界面可以读取和调整 EEPROM 的数值、设置控制器屏幕背光板和蜂鸣器的开关状态。

（6）退出键（〈Esc〉）。用于能力源控制器的复位。

（7）其他。Ccon102 控制器采用 U 盘方式下载，提供高达 3.96M 的用户程序存储空间，可通过左右按键在程序界面选择不同程序来并运行。

控制器包括：

（1）主界面。控制器开机后进入主界面，界面内容包含电量显示和主菜单。电量显示图标可以粗略显示当前电池电量，主菜单包含 4 个项目，可以通过左右按键在各个项目之间切换，黑色方框框起的图标是当前选择的项目，按<Enter>键进入该项目的下一级界面，如图 1-7 所示。

（2）程序界面。控制器采用 U 盘方式下载，提供高达 3.96M 的用户程序存储空间。当控制器中没有程序时，无法进入程序界面。如图 1-8 所示的控制器中有 2 个程序，分别为 PROGRAMA 和 PROGRAMB。前面的序号 1 和 2 是程序的编号，这些程序是按照下载时间进行排序编号，从而方便用户查找。

用户在该界面上可以通过左右键选择不同的用户程序，被选中的用户程序会反色显示，按<Enter>键后被选中的用户程序将会被读入内存并开始运行。用户在程序运行过程中，按<Esc>键用户程序终止运行，界面返回主界面。

（3）端口界面。进入端口界面后，左右键可以依次在 AI、DO、DI、舵机和计数 5 个界面之间切换，AI、DO、DI、舵机和计数5个标识会在屏幕左下角显示。在AI模拟输入端口的检测界面（见图1-9）中，0~11代表I/O口号，后面的数字“0”表示对应端口实时采集的AI输入值，图片显示的0表示没有传感器接入。控制器AI的返回值范围是0~4095。端口的AI功能是传感器输入测试功能，只需读取数值，没有下一步操作，如图1-9所示。

在DO数字输出端口测试界面（见图1-10）中，0~11代表I/O口号，后面的数字表示DO的当前开关状态，0表示断开，1表示接通。通过<Enter>键和左右键可以实现对单个端口的DO状态进行单独控制。为保证项目模型的安全性，当退出DO控制时，DO口会恢复到断开状态。

图1-7　主界面

图1-8　程序界面

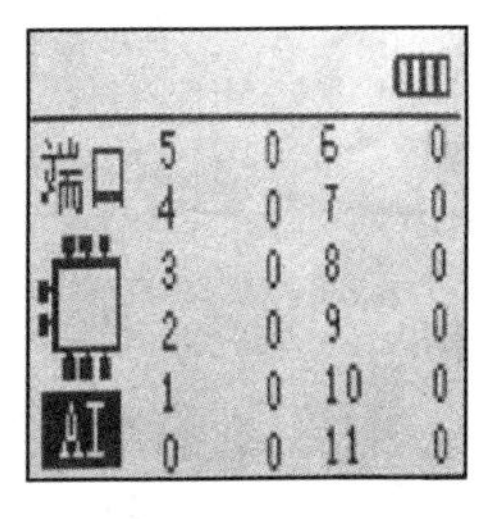

图1-9　AI界面

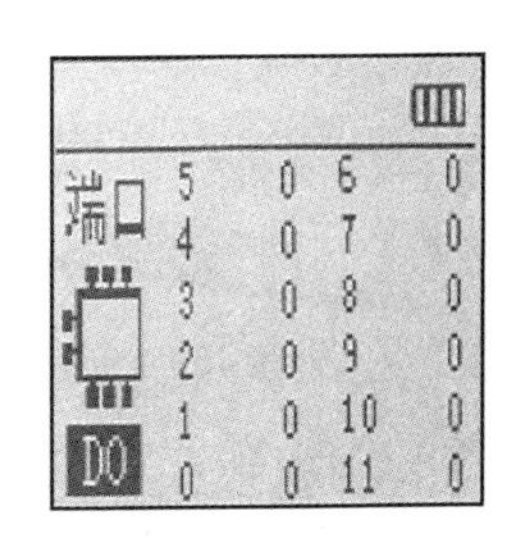

图1-10　DO界面

在DI数字输入端口检测界面（见图1-11）中，0~11代表I/O口号，后面的数字“0”表示对应端口实时采集的DI输入值，数字量传感器的返回值只有“接通”和“断开”两种状态，所以对应的返回值只有0和1。显示为0时会有两种情况，没有接传感器或者所接的数字量传感器是断开的。端口的DI功能是数字量传感器输入测试功能，只读取数值，没有下一步操作。

在数字舵机检测界面（见图1-12）中，该功能只能控制1个舵机，右侧下方是数字舵机的控制参数。IO表示数字舵机所接的I/O口，目的是打开该端口的485电源；ID表示所要控制的数字舵机的ID号；A表示读取和设置数字舵机角度。

在旋转计数器检测界面（见图1-13）中，该功能只能读取1个旋转计数器的返回值，因为旋转计数器需要电机带动，所以在该界面下还需要控制电机的运动，右侧是旋转计数器的控制参数。其中，IO表示旋转计数器所接的I/O端口号；DC表示带动该旋转计数器的电机所接的DC端口号；S表示控制器电机转向，在S反显时，按<Enter>键进入电机控制功能，默认是在中间的停止符上，可以通过左右键在正转符（左箭头）、停止符（方块）、反转符（右箭头）之间切换，同时电机的当前速度会在S后面显示出来；CNT表示显示计数器的返回值，通过按键操作使该值反显时，按<Enter>键可以进行清零操作。

（4）电机界面。在DC测试界面（见图1-14）中，DC一列中M0~M3对应电机口DC0~DC3上的电机；Speed一列中的数值代表对应DC口的电机速度（取值为-100~100），Encode一列中的数值代表闭环电机编码器返回的数值（仅在接通闭环电机时可用，按照电机正反转增减，数值变化范围为0~65535）。

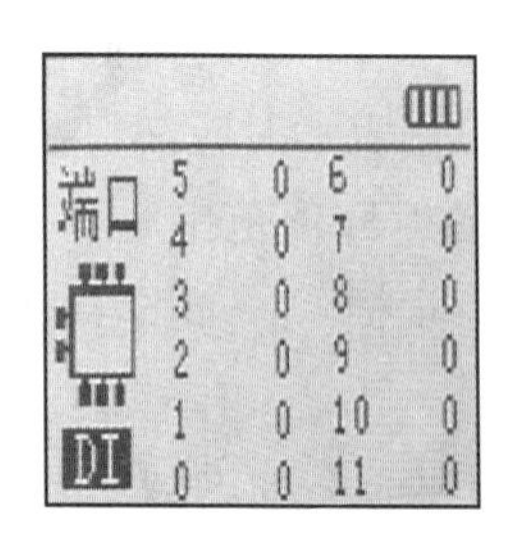


图 1-11　DI 界面

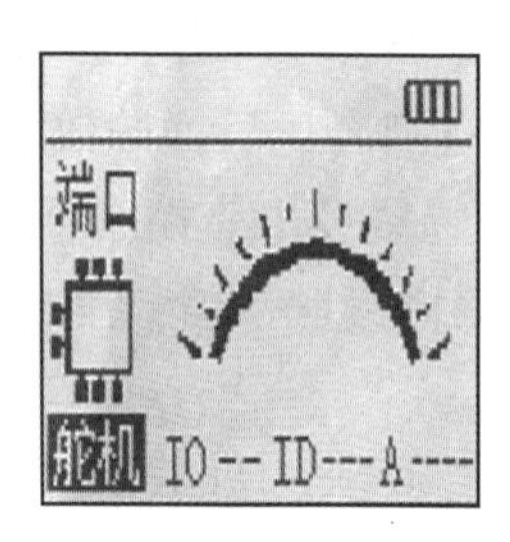


图 1-12　舵机界面

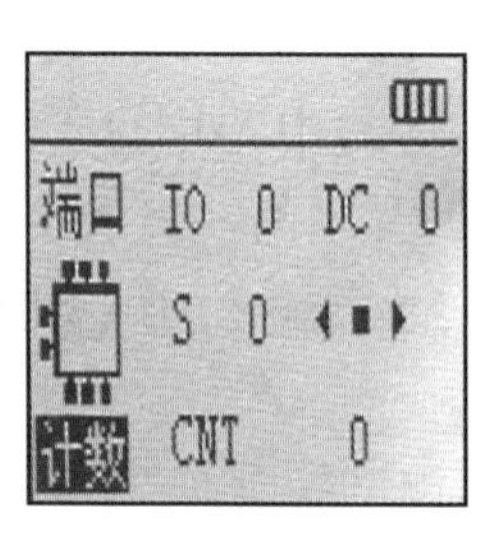


图 1-13　计数界面

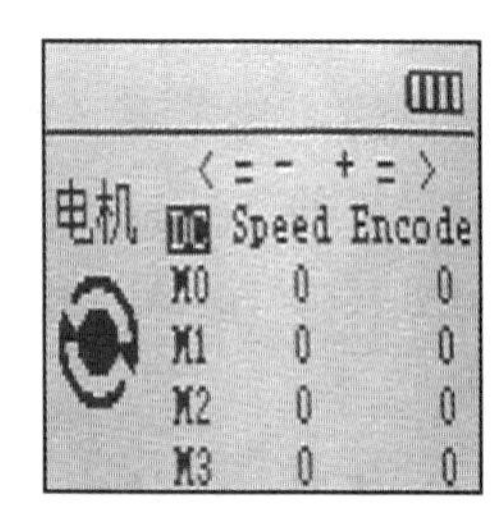


图 1-14　电机界面

### 1.1.3　任务实施

熟知能力源创新套件后要能够使用组件搭建简单的结构掌握套件常用的搭建方法和技巧。

#### 1.1.3.1　结构件与连接件搭建方法

接下来可以利用套件进行一些常见套件的连接。首先举例说明点、线、面三者之间如何直接连接、如何借助连接件连接。例如：立方体（点）与立方体（点）之间直接相互连接（见图 1-15），立方体（点）借助立方体连接器连接（见图 1-16），借助中 L 型连接器连接立方体（或半高立方体）（点）与平板（面）（见图 1-17），借助矩插销连接立方体（点）和平板（面）（见图 1-18），借助长插销连接立方体（点）与梁（线）（见图 1-19），借助梁支撑架连接两根梁（线）（见图 1-20），借助长插销连接梁（线）与平板（面）（见图 1-21），以及借助中 L 型连接器直连接两个平板（面）（见图 1-22）。

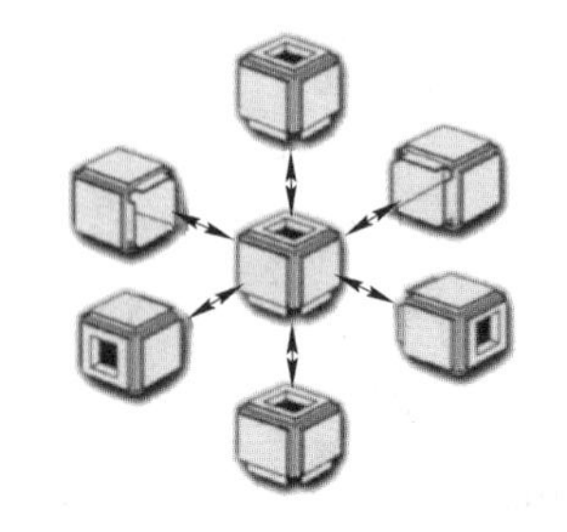

图 1-15　点与点（直接连接）

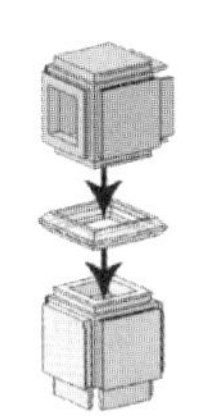

图 1-16　点与点（借助立方体连接器）

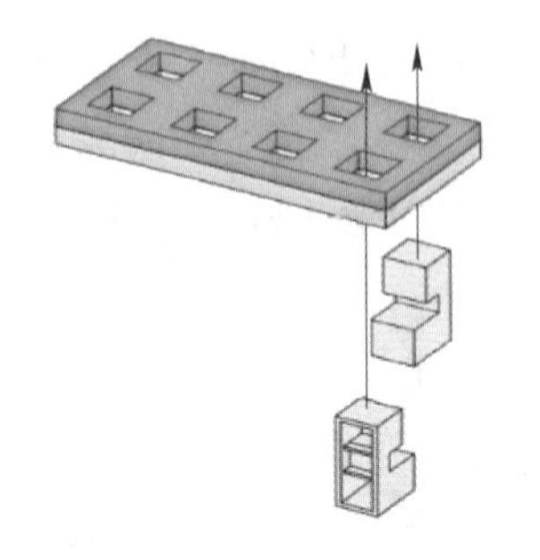

图 1-17　点与面（借助 L 连接器）

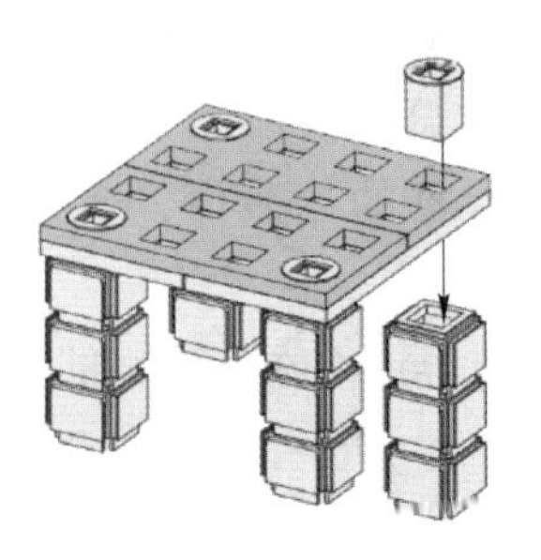

图 1-18　点与面（借助短插销）

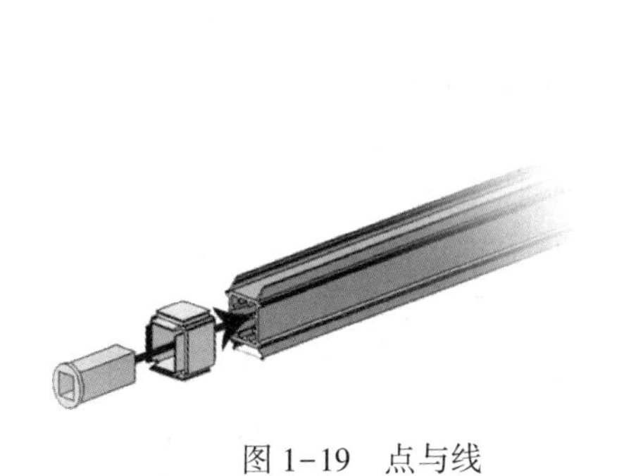

图 1-19　点与线

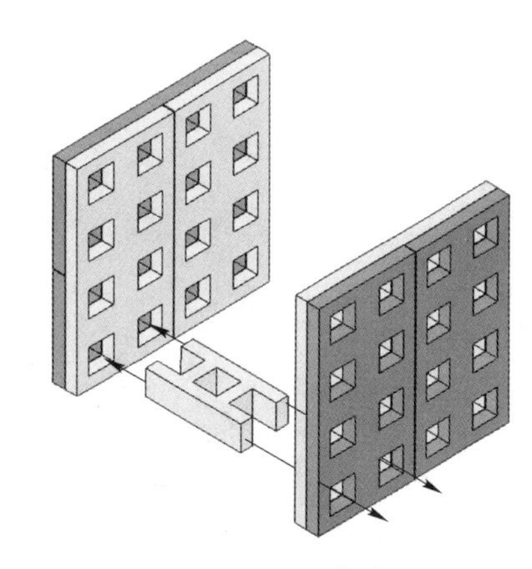

图 1-20　线与线

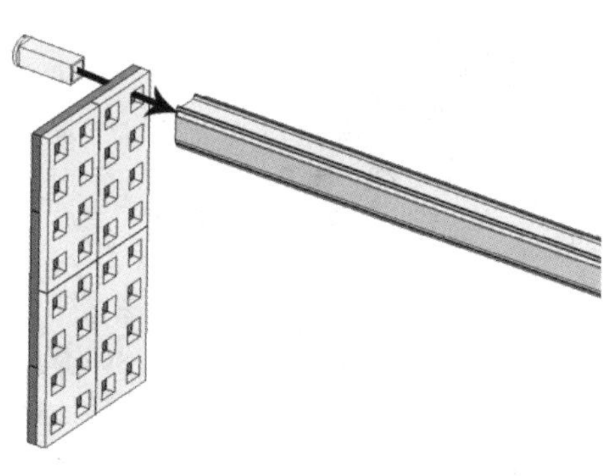

图 1-21　线与面

图 1-22　面与面

这里仅列举了部分常见的搭建案例，机械结构的搭建是很巧妙的。各部分连接件的使用非常灵活，同一种功能的模型其连接方式以及结构件的使用是多种多样的。

1.1.3.2　电气组件的搭建方法

电气组件包括电机、传感器和传动件，其各组件的搭建方法分别为：

(1) 电机的连接方法。电机是指能力源创新组件中的动力之源，所有的运动都要借助于电机。电机在与其他传动件连接时通常要借助立方体或者平板。电机最常用的连接方式如图 1-23 所示。

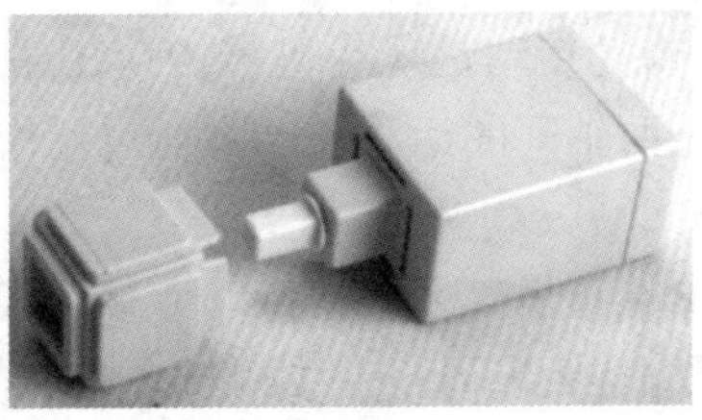

图 1-23　电机

电机的相距是指发动机从曲轴端输出的力矩。在功率固定的条件下它与发动机转速成反比，转速越快扭矩越小，反之越大。为了增加电机的扭矩，常常需要给电机增加减速机构，以提高它的扭矩。

在能力源创新课程套件中常常采用减速箱和齿轮组两种方法来达到减速箱的效果，如图 1-24 和图 1-25 所示。

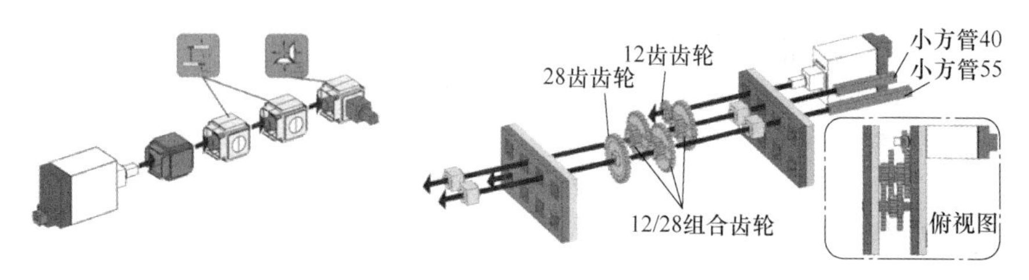


图 1-24　借助 5∶1 减速箱

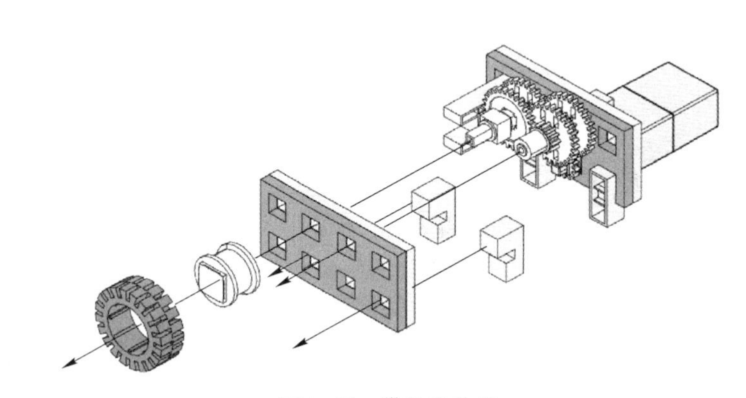

图 1-25　借助齿轮组

(2) 传感器的连接方法。能力源创新课程套件的传感器（除磁敏开关之外）全部集成在立方体当中，用法与立方体完全一样；磁敏开关集成在小方管内，用法同小方管一样。例如：磁敏开关可以通过短插销与立方体进行连接（见图 1-26），或者通过两个小 A 连接器与五孔梯进行连接（见图 1-27）；触碰开关与 LED 灯或者温度传感器、光敏传感器等可以直接相互连接，或者借助立方体连接器连接（见图 1-28）；磁铁通过短插销与立方体进行连接（见图 1-29）。

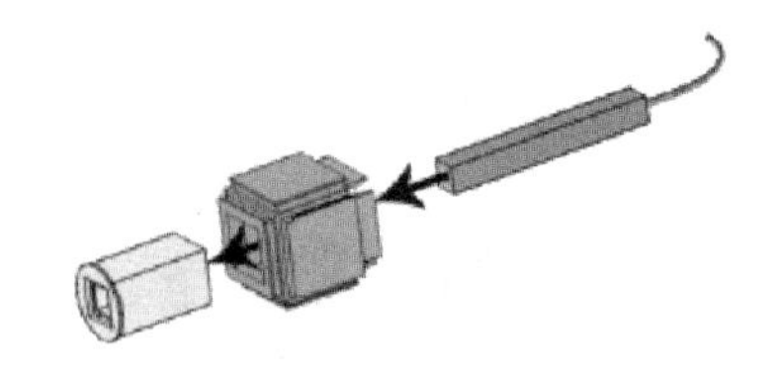

图 1-26　磁敏传感器的连接（1）

(3) 传动件的搭建方法。传动件的传动方式包括：

1) 齿轮传动。齿轮传动是现代机械中应用最广泛的一种机械传动，它通过齿的相互啮合来传递空间任意两轴间的运动和动力，也可用来改变运动和速度。传动组中的齿轮有主齿轮和从齿轮之分。

齿轮传动结构中，直齿轮之间的传动有减速和增速两种形式，减速形式主齿轮的齿数多于从齿轮，增速形式从齿轮的齿数多干主齿轮，两种形式的传动组输出的扭矩是不同的。

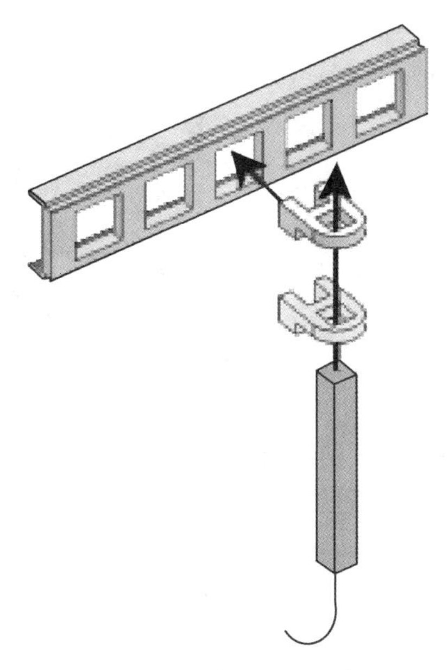

图 1-27　磁敏传感器的连接（2）

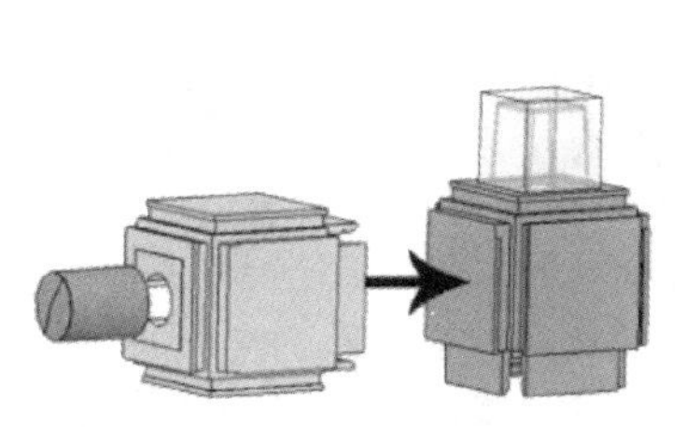

图 1-28　灯与触碰开关的连接

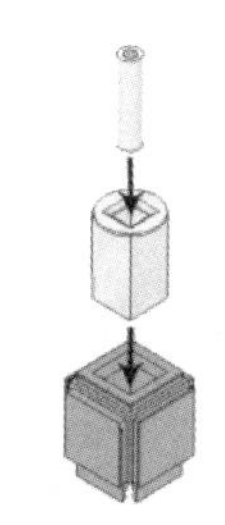

图 1-29　磁铁的连接

2）齿轮与齿条传动。齿条在机械上属于半径无限大的齿轮，齿轮齿条传动严格意义上属于齿轮传动的一种特殊形式，是直线运动与圆周运动两种运动形式的相互转换。典型的应用是自动门项目中走轮系统的结构（见图 1-30）。

图 1-30　自动门的走轮系统

3）蜗杆传动。蜗杆传动由蜗轮和蜗杆组成，用于传递空间交错轴间的运动和动力，两轴间的交错角度为 90″，其中蜗杆主动，蜗轮从动。蜗轮蜗杆传动装置如图 1-31 所示。

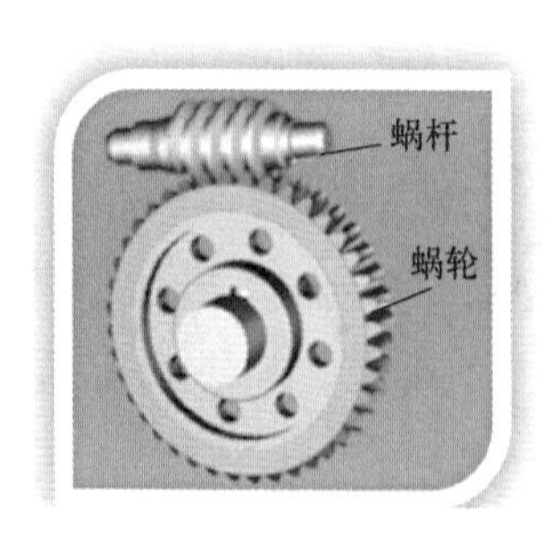

图 1-31　蜗轮蜗杆传动装置

4）带传动。带传动是一种常用的、成本较低的动力传动装置（见图 1-32），具有运动平稳、清洁（无须润滑）、噪声低的特点，同时具有缓冲、减振、过载保护的作用，维修方便。

带传动是由主动带轮，从动带轮和传动带组成，是利用环状的传动带紧箍两个带轮。在动带与带轮之间产生摩擦力，将主动带轮的运动和动力传递给从动轮。

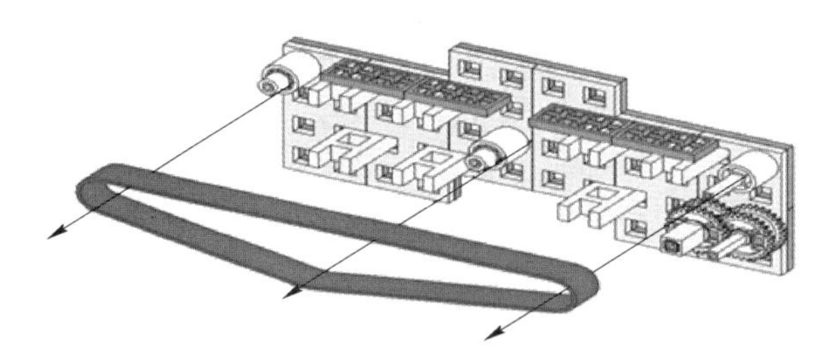

图 1-32　带传动装置

5）螺旋传动。螺旋传动是由螺杆和旋合螺母组成的机械传动，主要用于将旋转运动转换成直线运，将转矩转换成推力。在能力源创新课程套件中，丝杠是螺旋传动的典型代表（见图1-33），螺旋传动在工业机械手项目中有应用。

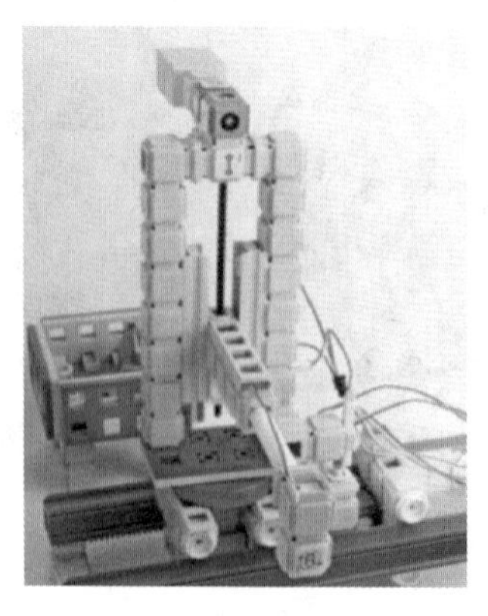

图 1-33　采用丝杠实现的螺旋传动装置

6）链传动。链传动适用于不宜使用带传动，且要求两轴平行、距离较远、功率较大、比较准确的场合。

链传动由主动链轮、从动链轮和绕在链轮上并与链轮啮合的链条组成，是通过链条将具有特殊齿形的主动链轮的运动和动力传递到具有特殊齿形的从动链轮的一种传动方式。在能力源创新课程套件中，智能电梯项目中有链传动，采用滑轮和棉线模拟。

### 1.1.4 任务评价

任务评价表见表 1-1。

**表 1-1 任务评价表**

| 序号 | 评价内容 | 成绩占比 | 自评 | 师评 |
|---|---|---|---|---|
| 1 | 工程实践创新的认知 | 10 分 | | |
| 2 | 结构件、连接件、传动件和电气组件的功能特点 | 30 分 | | |
| 3 | 常见的传动方式 | 25 分 | | |
| 4 | 结构件与连接件、电气组件及传动件的搭建方法 | 35 分 | | |

通过以上内容的学习，可以让学生对这些套件的使用和作用有个初步的认知。在传动方式上有齿条传动、蜗轮蜗杆传动、带传动、螺旋传动和链传动等，同时也介绍了电气组件和常见传感器的搭建方法。通过学习，对能力源创新课程套件有了一个初步的整体认识，并掌握了常用组件的基本搭建方法和搭建技巧，为下阶段的项目实践打下基础。

### 1.1.5 任务拓展

（1）工程实践创新概念是什么？
（2）能力源套件的分类有哪些？
（3）能力源套件常见组件的基本搭建方法有哪些？

## 任务 1.2 认识 VJC 图形化交互开发系统

### 1.2.1 任务引入

在能力创新课程套件中各功能部件好像是人的身体，控制器好像是指挥人体的大脑，而程序就好像是大脑中的思想。通过使用 VJC 图形化交互开发系统可以很方便编制出程序，实现对复杂机电系统的逻辑控制。可以说 VJC 图形化交互开发系统为我们搭建的套件系统赋予了灵魂，使其能独立自主地完成既定工作。

### 1.2.2 相关知识

VJC 图形化交互式开发系统（以下简称 VJC）是能力源创新课程套件的专用软件，同时支持流程图编程和交互性 JC 语言编程，编写好的程序下载到控制器中可以直接运行。

流程图采用模块化编程的形式，接近人类自然语言（见图 1-34），流程图程序的形式与标准流程图完全一致，不会造成学生后期学习的错误理解。

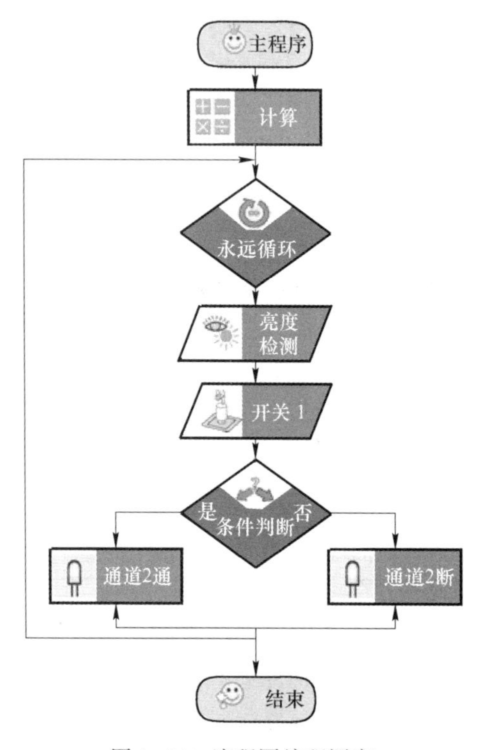


图 1-34　流程图编程语言

1.2.2.1　VJC4.3 安装过程

可以通过网站（http://www.abilix.com）获得 VJC4.3 的安装程序，安装程序名称为“VJC _4.3._CH_SetUp_bi * * * * * * （* * * * * * 为年月日）.exe”。双击该安装文件，根据提示选择“确定”或者“下一步”即可顺利完成安装，安装过程用户可以选择安装目录，具体安装内容如下：

（1）安装程序会首先验证用户的电脑上是否已经安装“.NET2.0”（或以上版本），如果没有安装，则进入“.NET 2.0”的安装；如果已经安装则直接进入 VJC 主程序的安装。

（2）VJC 的默认安装目录为 C 盘的根目录，用户可以在安装过程中按照提示更改 VJC 的安装目录，但建议目录路径中不要包含中文。

（3）安装完成后会在桌面和快速启动栏各生成 1 个名为“VJC_4.3_CH”的快捷方式。双击该快捷方式即可以打开 VJC 4.3。

注意：VJC4.3 的推荐运行环境是 Win 7。

1.2.2.2　首次使用

安装成功后，第一次进入 VJC 时会显示新建界面（见图 1-35），根据要新建的程序格式选择对应选项。如果需要打开一个已有的程序，可以选择打开菜单进行操作。

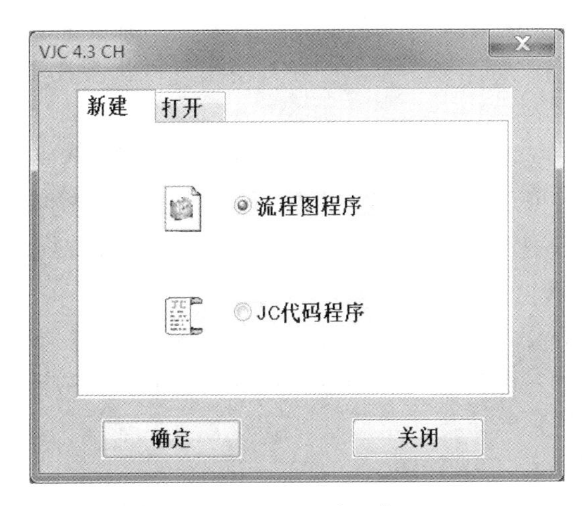

图 1-35　新建程序

1.2.2.3　菜单栏与工具栏

VJC 的菜单栏和工具栏非常符合用户习惯，和常用的 Office 软件相似，本手册将对一些不常见和常用的功能做详细介绍。

**流程图界**

流程图界面下的部分菜单包括：

(1) “文件（F）→输出 JC 程序（J）...”：将当前流程图程序保存为 JC 程序。

(2) “编辑（E）→主程序（M）”：从子程序编辑界面返回到主程序界面。该功能仅在编辑子程序时有效。

(3) “编辑（E）→新建子程序（N）...”：新建一个子程序，会弹出新建子程序对话框（见图 1-36）。点击子程序设置下的“其他程序...”可以调用其他程序中的子程序。在“子程序名称”后面注明该子程序名称，它会显示流程图的该子程序模块上，所以建议根据该子程序的实际功能命名。可以写上用户的名字，以方便其他人阅读遇到问题时咨询。

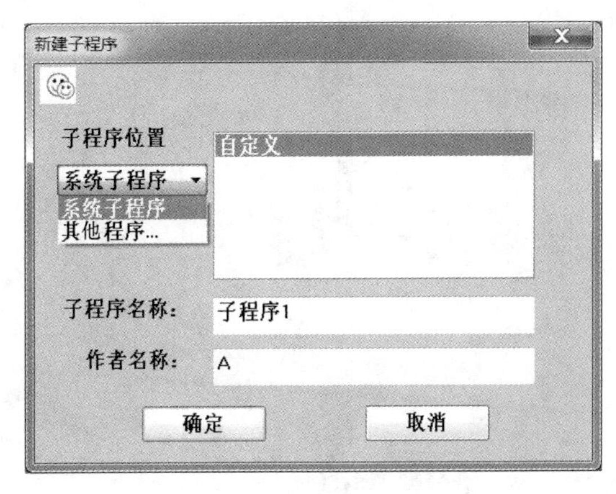

图 1-36　新建子程序

(4)“编辑（E）→删除当前子程序（D）”：删除当前正在编辑的子程序。该功能仅在子程序编辑界面下有效，如果该子程序已经被使用（即出现在流程图编辑窗口中），则无法删除。

(5)“视图（V）→显示 JC 代码（C）”：流程图右侧 JC 代码视窗的显示开关。

(6)“视图（V）→模块注释”：流程图界面下模块内参数提示窗口的显示开关。

(7)“视图（V）→巡线模块库”：勾选后左侧会显示巡线模块库，这是针对特定装配方式的 5 灰度或者 7 灰度巡线小车而设计的模块库。

(8)“视图（V）→高级模块库”：勾选后左侧会显示高级模块库。

(9)“工具（T）→下载当前程序（D）”：将当前流程图程序下载到控制器，可以使用快捷键<F6>（下载当前 JC 程序的快捷键是<F5>）。

注意：下载前请连接好 USB 线缆。

(10)“工具（T）→窗口截图”：将当前显示的流程图窗口截图，会跳出对话框提示用户设置保存目录。如果想要白底的图片，可以在视图菜单中取消流程图背景。

(11)“工具（T）→流程图实时保存”：勾选该选项后，可以自动保存用户编写的流程图程序。首先用户要手动保存一次，自动保存功能能够提高用户编写程序时的安全性，不会因为断电或遗忘而前功尽弃。

(12)“帮助（H）→帮助主题（H）”：调出帮助文档，VJC 的帮助文档即本文档，但可能会包含更新的内容。更新内容一般会放在文档的末尾。

(13)“帮助（H）→检查更新（U）”：从 VJC4.0 版本起，就开始提供网络升级功能，该菜单即是帮助用户检查是否有可用更新（并完成更新）。这个功能需要用户的计算机连接 Internet。

**JC 界面**

JC 界面如图 1-37 所示。其中，JC 界面下的部分菜单包括：

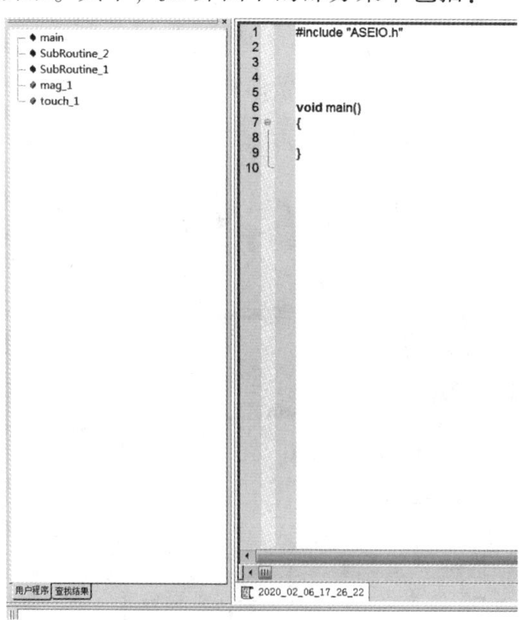


图 1-37　JC 界面

（1）“编辑（E）→转到行（G）”：弹出转到行对话框。在 JC 界面下，每行 JC 代码都有对应行号。

（2）“视图（V）→交互信息窗口（I）”：交互信息窗口的显示开关。交互信息窗口是指 JC 界面最下面的窗口，一般会显示编译错误、警告、下载过程等信息。支持窗口位置重置。

（3）“视图（V）→多功能提示窗口”：多功能提示窗口的显示开关。多功能提示窗口是最左侧的窗口，包含用户程序和查找结果两个窗口。用户程序窗口会显示当前程序中的主程序、子程序和变量，双击可直接定位。在用户程序中选择一个对象（如变量），右键点击“全部查找”按钮，结果会显示在“查找结果”窗口，双击“可定位”。支持窗口位置重置。

（4）“工具（T）→实时编译”：实时编译的功能开关。打开实时编译功能后，VJC 会在“交互信息窗口”中显示编译结果，能够第一时间找到语法错误、警告。

（5）“窗口（W）”：JC 界面下可以同时打开多个窗口，可以将窗口任意排序。

**工具栏**

JC 编程界面下的工具栏如图 1-38 所示。

图 1-38　JC 编程界面下的工具栏

流程图编程界面下的工具栏如图 1-39 所示。

图 1-39　流程图编程界面下的工具栏

当用户将鼠标移动到工具栏上的图标时，会显示该图标的功能。

Ccon102 控制器能够保存多个用户程序。为了方便区分，允许用户自定义程序名称。用户填写在“主程序名字”后面的内容将会作为程序名称显示在 Ccon102 控制器的屏幕上，需要注意的是仅能显示英文。

**快捷按键**

程图编程界面下的快捷键见表 1-2。

**表 1-2　程图编程界面下的快捷键**

| <F2> | 全屏显示 |
| --- | --- |
| <Esc> | 全屏状态下退出全屏，大部分对话框打开时该键相当于取消键 |
| <F6> | 下载流程图程序 |
| <F1> | 帮助，调出帮助文档 |
| <F9> | 在流程图上做反色标记 |
| <F12> | 流程图与代码编辑器切换，注意不是将流程图转换成 C 代码的功能 |

码编辑器中的部分快捷按键见表 1-3。

表 1-3 码编辑器部分快捷键

| <F2> | 找标记 |
| --- | --- |
| <Ctrl>+<+> | 代码放大 |
| <Ctrl>+<-> | 代码缩小 |
| <Ctrl>+<W> | 查找全部内容 |
| <Ctrl>+<F3> | 向下查找 |
| <Shift>+<F3> | 向上查找 |
| <Ctrl>+<F> | 查找对话框 |
| <Ctrl>+<A> | 全选 |
| <Ctrl>+<H> | 替换 |
| <Ctrl>+<Z> | 撤销 |
| <Ctrl>+<C> | 复制 |
| <Ctrl>+<V> | 粘贴 |
| <Ctrl>+<X> | 剪切 |
| <F5> | 代码编辑器程序下载 |

### 1.2.2.4 程序下载

下载当前正在编辑的流程图程序（或 JC 程序）有 3 种方法：

（1）点击工具栏的下载按钮。

（2）使用菜单栏“工具（T）”菜单下的“下载当前程序（D）”。

（3）使用快捷方式，按<F5>键下载当前 JC 程序，按<F6>键下载当前流程图程序。

程序下载前需要先连接 USB 下载线缆。Ccon102 控制器连接好 USB 线缆后控制器的显示屏上会进入下载界面。如果没有连接好 USB 线缆，而直接下载，VJC 的下载界面也会有提示。

程序中如果有语法错误，会无法通过编译，也会导致程序无法下载。流程图程序的语法错误一般出现在自定义模块中。JC 代码的语法错误会显示在“交互信息窗口”中，双击错误信息，会在 JC 代码窗口进行自动定位，方便用户的修改。请尽量不要在下载过程中拔掉 USB 线缆，可能会造成死机。下载完成后拔掉 USB 线缆，即可运行程序。

### 1.2.2.5 软件在线升级

VJC 提供了在线升级功能，用户将使用中发现的一些问题的改善方法作为补丁共享给大家，通过在线升级可以消除这些问题；另一方面，该功能还会把一些 VJC 新功能、例程等资料共享给用户。在流程图界面的帮助菜单中，点击“检查更新（U）”会弹出如图 1-40所示的对话框。确定保存已编好的程序，以免造成不必要的损失。

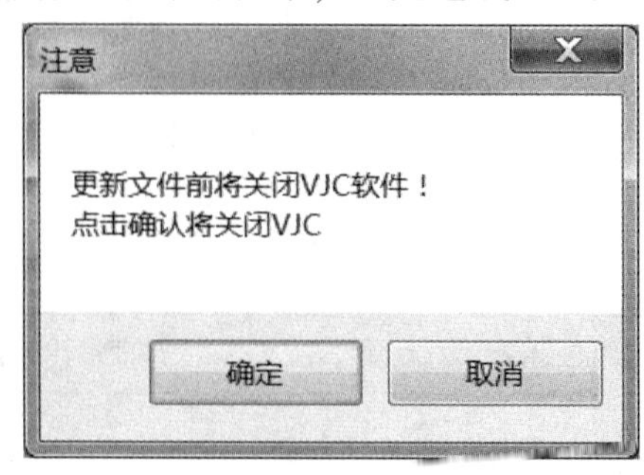


图 1-40 更新文件提示

点击“确定”按钮后，会弹出升级管理对话框，如图 1-41 所示。

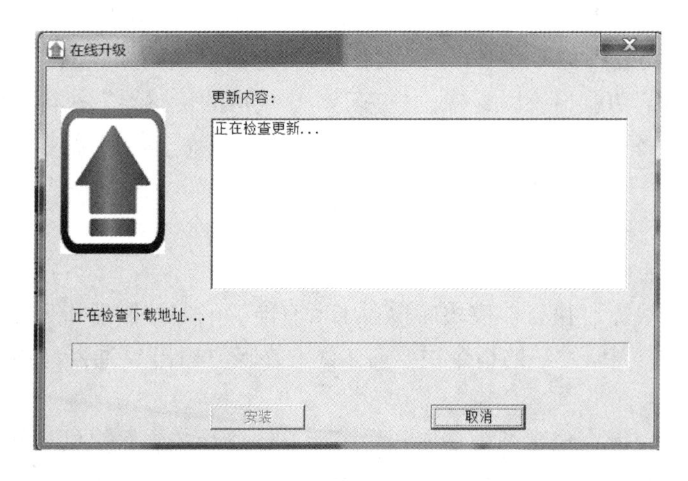


图 1-41　检查更新

确认计算机已经连接到网络上后，点击“检查更新”按钮，即开始下载服务器上没有下载的文件。如果有文件需要更新，会有提示，此时点击“有可用更新，点击开始”按钮，然后点击“开始安装”就可以安装新的软件。安装完成后退出即可。

1.2.2.6　流程图界面

**流程图界面介绍**

如图 1-42 所示为流程图程序编辑窗口。左侧为模块库，中间是程序编辑窗口，右侧为 JC 代码栏（可以控制是否显示），该栏显示内容为流程图程序自动生成的，不可更改，能方便用户学习 C 语言结构和读取各模块参数。

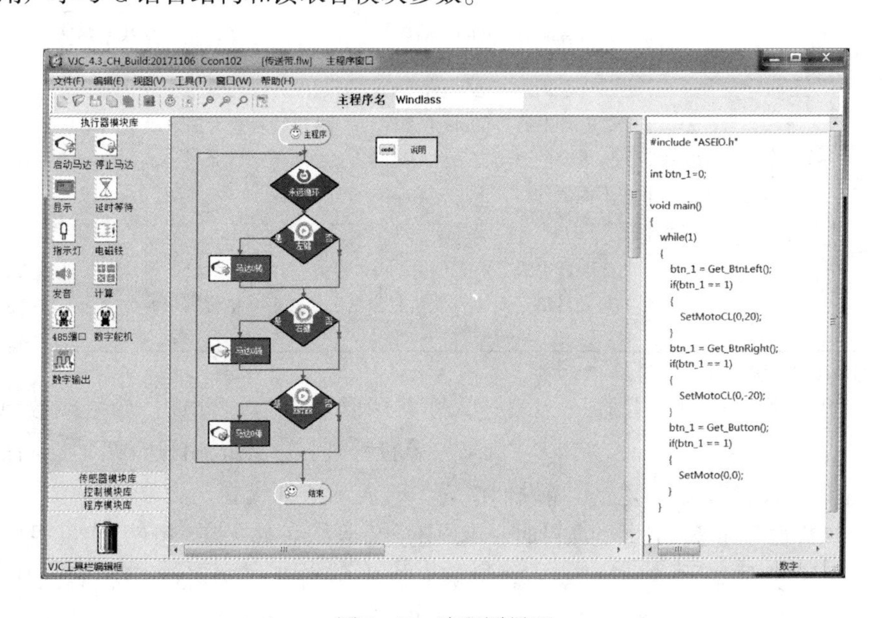


图 1-42　流程图界面

一台机器人主要包含控制器、传感器、执行器和用户程序，传感器和执行器都是接在控制器的各端口上的，用户程序在控制器中运行。可以理解为用户程序通过控制器的端口采集传感器的值，经过各种计算后再通过端口给执行器发送指令让其动作。Ccon102 控制器还具有 RS485 通信功能，所以支持一些基于 485 接口的传感器、执行器和通讯。

VJC4.3 的模块库进行了重新分类，分为执行器模块库、传感器模块库、控制器模块库和程序模块库，另外还可以在工具菜单中调出其他一些非常用的模块库。

**流程图各模块名称和功能**

流程图模块包括：

（1）执行器模块库。执行器模块库包含了所有能够响应控制器指令而有所动作的电子器件的控制模块。控制器发送的指令主要有 DO（数字值输出）通断、DC（电机）运动、声音、显示等。

（2）传感器模块库。传感器模块库包含了所有能够为控制器提供外部环境数据的采集模块。控制器通过端口的 AI、DO、计数等功能采集环境信息。

（3）控制模块库。在用户程序中，读取各端口传感器的返回值一般有储存和判断这两种用途，其中用于判断的情况居多。在 VJC 中提供了“While”语句、“If…eles…”语句和“For”语句这三种判断方式的流程图模块，它们都在控制模块库中。

如果要做判断必须有被比较的对象和比较参考值。被比较的对象一般是传感器的返回值，或者是更新后的变量值，所以在传感器模块库中所有具备“读取”功能的模块都可以直接装换成条件判断模块。控制模块库中各模块的图标、名称、作用见表 1-4。

**表 1-4　控制模块库**

| 序号 | 图标 | 名称 | 功　能 |
|---|---|---|---|
| 1 | | 多次循环 | C 语言中的“For”语句，参数为循环次数，功能是让循环体内的语句循环运行用户指定的次数 |
| 2 | | 永远循环 | C 语言中的“While（1）”语句，没有参数，功能是让循环体内的语句重复执行下去 |
| 3 | | 条件循环 | C 语言中的“While（条件）”语句，参数为用户设置的条件，功能是让循环体内容的语句在条件满足时重复执行下去，条件不满足时跳出循环，继续执行下面语句 |
| 4 | | 条件判断 | C 语言中的“If（条件）…else…”语句，参数为用户设置的条件，功能是如果条件满足则执行模块左分支语句，否则执行模块右分支语句 |

表 1-4 中提到“条件”是指一个表达式，共两侧可以是运算式、变量或数值，中间使用“==”“！=”“>”“<”“>=”“<=”等符号连接。该语句只有两种返回值 0、1，当返回值为 0 时表示条件不成立，当返回值为 1 时表示条件成立。

在设置条件循环和条件判断的条件时，用户还会看到“条件一”“条件二”（见图 1-43），一般情况下用户只使用“条件一”。点击“条件二”并勾选“有效”后会看到如图 1-43 所示的界面，外观上和“条件一”基本一致，但是多了“有效”和“条件逻辑关系”两项，即：

1）只有勾选了“有效”，“条件二”才生效。

2）条件逻辑关系表示“条件二”和“条件一”的逻辑关系，包含“与”“或”“非”三种运算关系（在C语言中使用“&&”“‖”“&&!”表示）。与或非运算的两侧依旧可以是表达式或数值，计算结果为：

①“条件一”与“条件二”：仅当两个条件都成立时结果为1，其他情况结果为0；

②“条件一”或“条件二”：只要有一个条件成立结果就为1，都不成立时结果为0；

③“条件一”非“条件二”：仅当两个条件返回值不同时结果为1，其他情况结果为0。

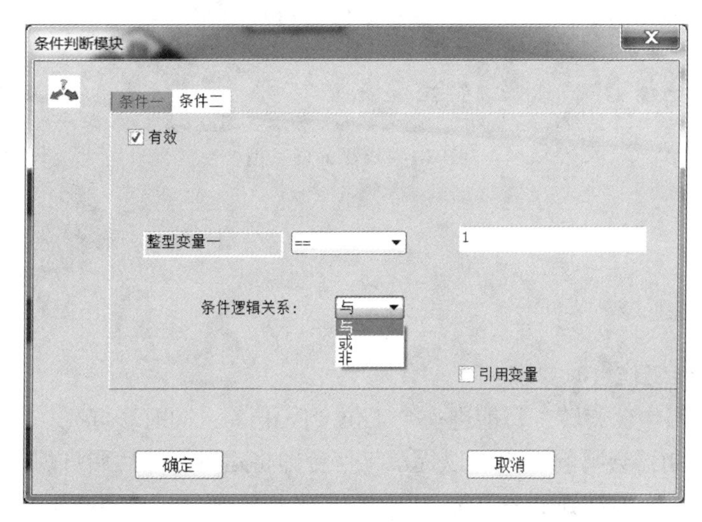


图1-43　条件判断模块

（4）程序模块库。程序模块库中各模板的图标、名称、用途见表1-5。

**表1-5　程序模块库**

| 序号 | 图标 | 名称 | 用　　途 |
|---|---|---|---|
| 1 | | 新建子程序 | 新建一个子程序，还可以通过该模块调用其他程序的子程序，子程序可以理解成把多个模块打成一个包（做成1个模块实现某一特定功能），方便以后直接使用，还可以让主程序不显得太长 |
| 2 | | 子程序返回 | 子程序中的结束模块，使流程图外观完整，无实际意义 |
| 3 | | 结束模块 | 主程序中的结束模块，使流程图外观完整，无实际意义 |

（5）变量百宝箱。在流程图各模块上，用户会看到有“引用变量”或“某某变量”的可选择或可点击的地方，操作后看到的界面就是变量百宝箱（见图1-44）。

变量百宝箱中集成了所有在流程图中的变量，通过选择下半部分的各个图标可以切换变量类型，通过选上面的变量菜单，可以切换变量编号。

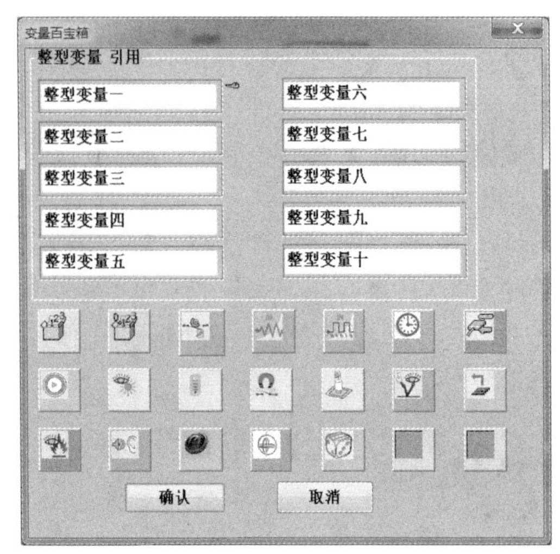


图 1-44　变量百宝箱

## 1.2.3　任务实施

### 1.2.3.1　任务描述

搭建一个旋转工作台时要求：

（1）运行后工作台旋转，每前进一个工位会停止 3s，同时彩灯亮。

（2）3s 后控制器会鸣响 0.5s，以提示工作台即将运行，然后彩灯熄灭。

（3）工作台继续前进到下一工位。

### 1.2.3.2　结构搭建

旋转工作台模型如图 1-45 所示。其中，旋转工作台搭建的四个步骤模型分别如图 1-46~图 1-49 所示。

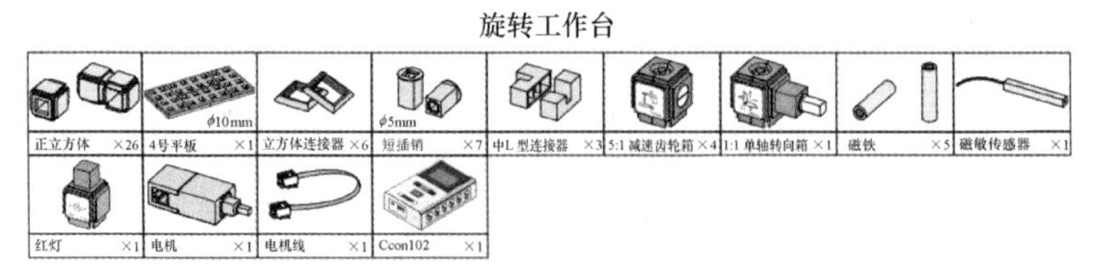


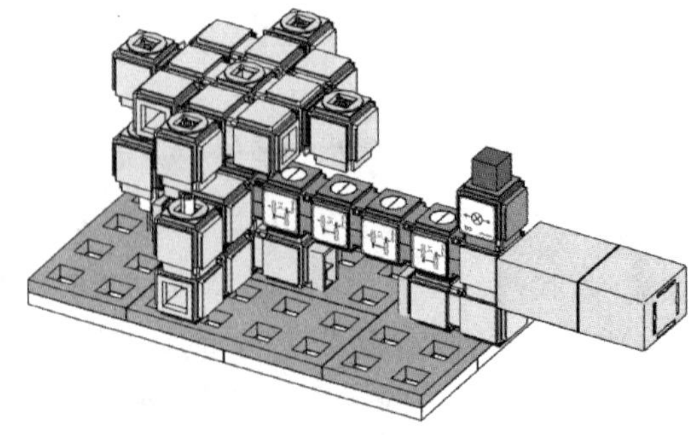

图 1-45　旋转工作台

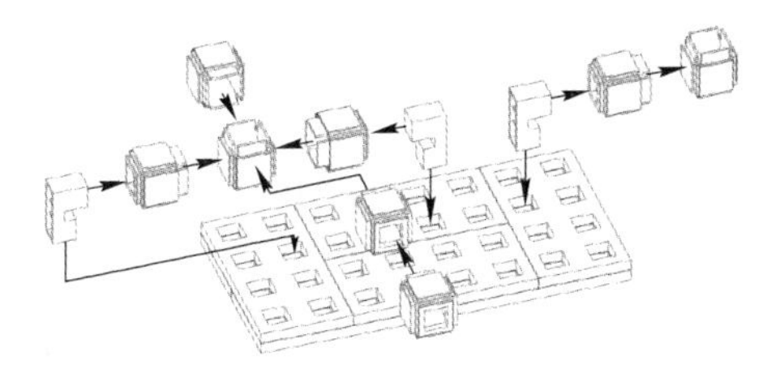

图 1-46　旋转工作台搭建步骤（1）

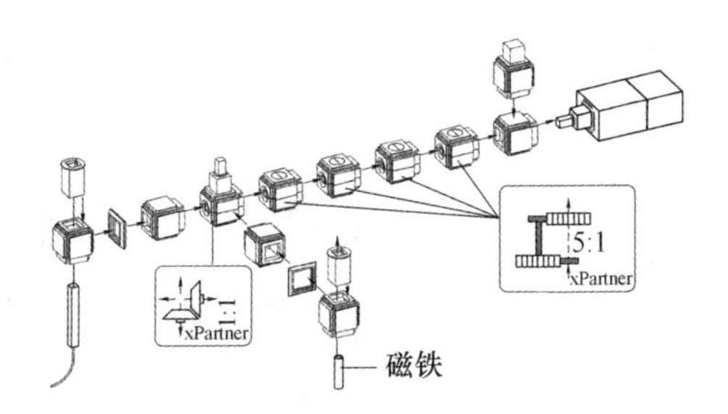


图 1-47　旋转工作台搭建步骤（2）

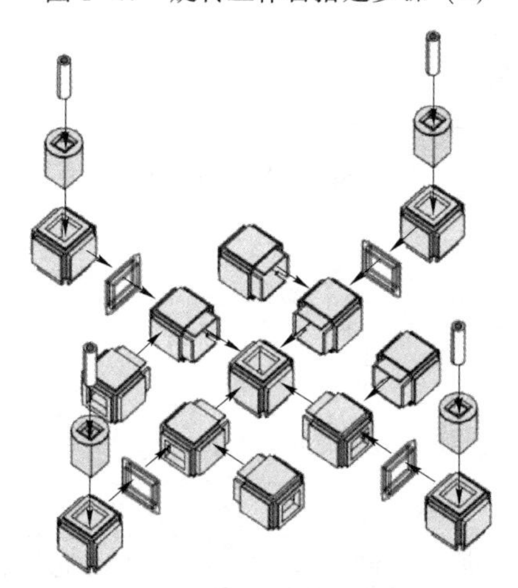

图 1-48　旋转工作台搭建步骤（3）

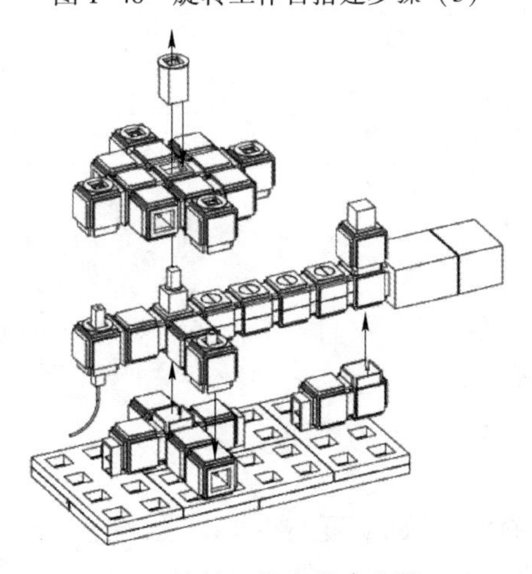

图 1-49　旋转工作台搭建步骤（4）

#### 1.2.3.3 程序设计

工位到达是磁敏传感器返回值从 0 到 1 的过程，工位离开是磁敏传感器返回值从 1 到 0 的过程。旋转工作台程序如图 1-50 所示。

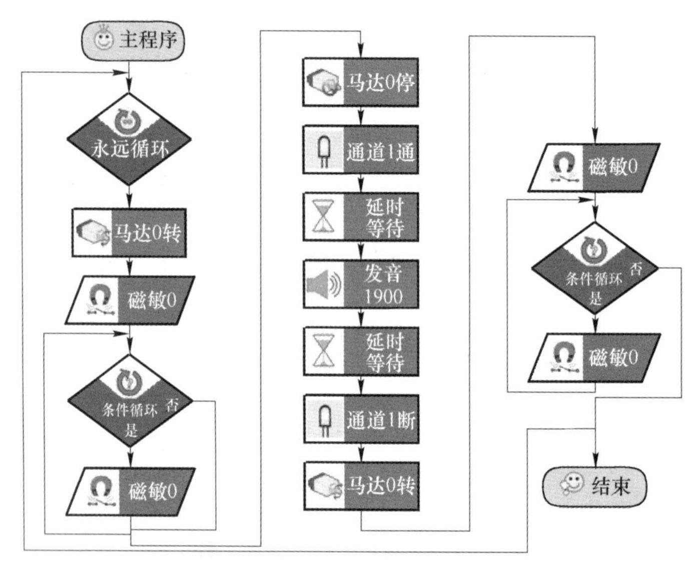


图 1-50 旋转工作台程序

### 1.2.4 任务评价

任务评价表见表 1-6。

**表 1-6 任务评价表**

| 序号 | 评价内容 | 成绩占比 | 自评 | 师评 |
| --- | --- | --- | --- | --- |
| 1 | 磁敏传感器工作原理 | 10 分 | | |
| 2 | 条件循环模块的原理 | 10 分 | | |
| 3 | 旋转工作台的搭建 | 30 分 | | |
| 4 | 旋转工作台的编程 | 30 分 | | |
| 5 | 旋转工作台的调试 | 20 分 | | |

### 1.2.5 任务拓展

针对一些特殊功能和高级用户，我们做了一些针对性较强的模块库，这些模块库在视图菜单下，选择后才会出现。

#### 1.2.5.1 巡线模块库

巡线模块库是专门针对使用地面灰度巡线小车设计的用于巡线的模块库。使用该模块时要求地面灰度成排安装在小车的前方，灰度数量为 5 个。巡线模块库如图 1-51 所示。

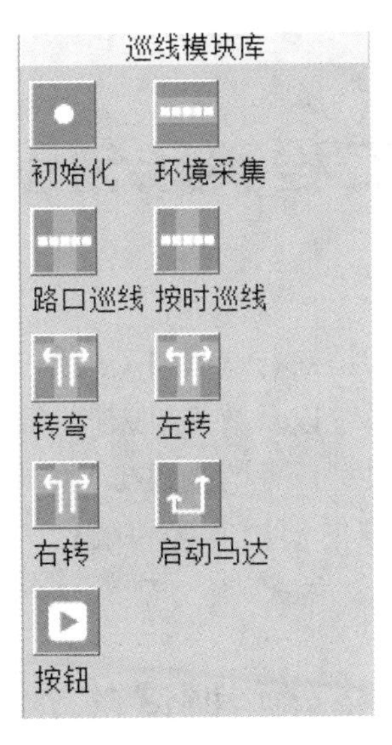


图 1-51　巡线模块库

#### 1.2.5.2　高级模块库

Ccon102 控制器还支持一些不常用的传感器、执行器或其他功能，因此可以把其放在高级模块库中。高级模块库中的模块会根据功能更新和元器件更新不断变化。该模块库包含：

（1）编码电机。针对 RS485 通讯的编码器电机而设计的模块，该电机包含电机 ID 和电机速度两个参数。1 个编码电机模块可以最多同时控制 4 个电机，如果用户需要控制更多的编码电机，可以使用多个编码电机模块。

（2）舵机角度。在执行器模块库中有数字舵机模块，它是让舵机转到对应角度，而舵机角度模块是读取数字舵机的当前角度。

（3）数码管。针对 RS485 通讯的数码管而设计的模块，它可以控制数码管显示的数值。需要注意的是，数码管仅能显示 1 位 0~9 的值，所以在使用“引用变量”功能时，被引用的变量不能超出该范围。

（4）读 EEPROM。读取控制器某一地址的 EEPROM 中的数值，关键参数是 EEPROM 的地址值（0~31），一般将程序中的临界值存在 EEPROM 中。

（5）写 EEOROM。更改控制器某一地址的 EEPROM 中的数值。在 Ccon102 控制器上，该功能还可以通过界面操作完成。

# 项目2 工程实践创新项目应用

本项目通过五个典型任务，即搭建与调试自动门控制系统、智能电梯控制系统、AGV小车控制系统、工业机械手控制系统以及数控机床控制系统，由简单到复杂地详细介绍利用能力源套件进行工程实践创新典型任务的搭建与VJC软件编程调试的过程，通过这些任务可以让学生熟练掌握能力源套件的使用方法和编程调试过程，提高学生的创新与实践能力。

## 任务2.1 搭建与调试自动门控制系统

### 2.1.1 任务引入

自动门从理论上理解应该是门的使用观念的延伸。自动门是指用各种信号控制自动开关的单扇、双扇或多扇的门。它是用电力驱动门开启和关闭的，由控制系统发出指令，并通过电机及减速传动系统带动门体启闭的成套装置。自动门工程是综合性的边缘产业，它综合了建筑技术、电子控制技术、机械设计和制造技术、计算机技术及建筑装饰技术，具有广阔的发展空间。

### 2.1.2 相关知识

#### 2.1.2.1 自动门的原理

自动门是应用先进的感应技术，通过控制系统控制机电执行机构使门体自动开启闭合的一种门系统。当人或其他活动目标进入传感器检测工作范围，门扇会自动开启；当人或其他活动目标离开感应区，门扇会自动关闭。

#### 2.1.2.2 自动门分类

**自动平移门**

自动平移门主要应用于公司、政府机关、饭店、银行、医院、写字楼以及商场等主要出入口通道或工厂的车间通道。平移门的优点在于日常维护费用低廉，配件价格低，不占用纵向空间，外观开阔。缺点是它需要较大的横向空间，安装时要做整体工程配套施工，维护保养较不方便，开门时不能形成绝对的密闭空间。

**自动平开门**

自动平开门主要应用于高级办公室、时尚住宅大楼入口、医疗通道、老年公寓、洗手间、残疾人通道以及防火门或横向空间较小的场所。其优点在于安装、维护方便；可以在原有门体上安装，施工时间短；单开门在造价上相对于平移门要低一些。但是由于双开门的平开门要使用两套设备，在造价上相对平移门要高一些，开门时不能形成绝对的密闭空间。

**其他自动门**

自动弧形门、安防门、自动旋转门、消防门、隔音门、防辐射门、防静音门、银行金库门等。

### 2.1.2.3 自动门结构

自动门结构主要包括：

(1) 主控制器。主控制器是自动门的指挥中心，是通过内部编有指令程序的大规模集成块，发出相应指令，指挥马达或电锁类系统工作；同时人们通过主控器调节门扇开启速度、开启幅度等参数。自动门的专用主控制器如图 2-1 所示。

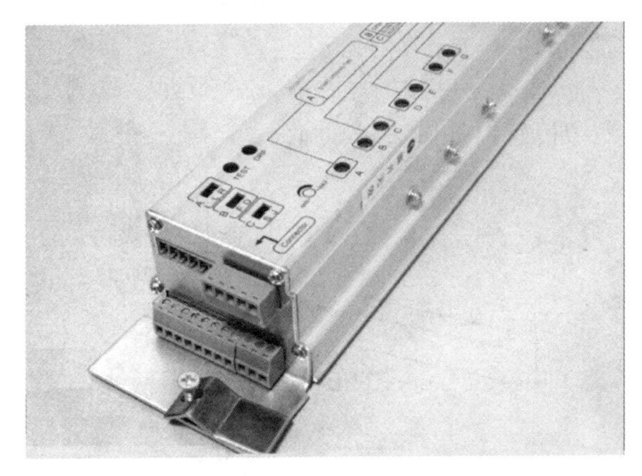

图 2-1 主控制器

(2) 感应探测器。主要负责采集外部信号，如同人们的眼睛，当有移动的物体进入它的工作范围时，它就给主控制器一个脉冲信号。感应探测器外观如图 2-2 所示。

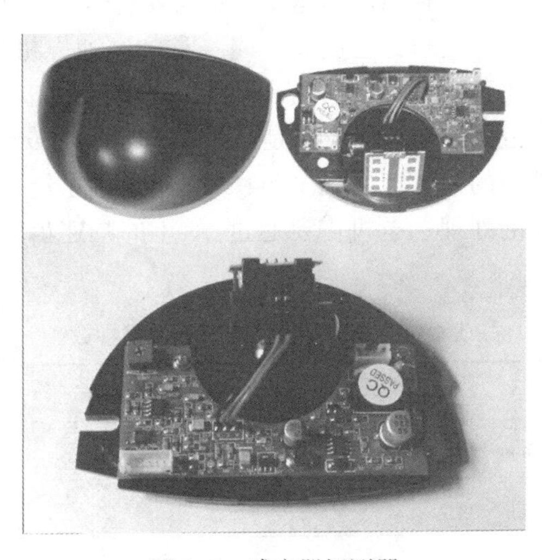

图 2-2 感应器探测器

（3）动力马达。主要提供开门与关门的主动力，控制门扇加速与减速运行。

（4）门扇行进轨道。门扇行进轨道就像火车的铁轨，约束门扇的吊具走轮系统，使其按特定方向行进。

（5）门扇吊具走轮系统。主要用于吊挂活动门扇，同时在动力牵引下带动门扇运行。

（6）同步皮带。主要用于传输马达所产动力，牵引门扇吊具走轮系统。

（7）下部导向系统。下部导向系统是门扇下部的导向与定位装置，用来防止门扇在运行时出现前后门体摆动。

#### 2.1.2.4 自动门核心技术

**自动门的传感技术**

传感技术是指高精度、高效率、高可靠性的采集各种形式信息的技术。自动门常用的传感技术包括行程开关、光电开关、红外传感器和视觉传感器。本任务中使用磁敏传感器代替自动门传感器。

磁敏传感器为数字型传感器，可检测是否有磁铁靠近，监测到磁铁返回值为 1，未检测到磁铁返回值为 0。磁敏传感器如图 2-3 所示。

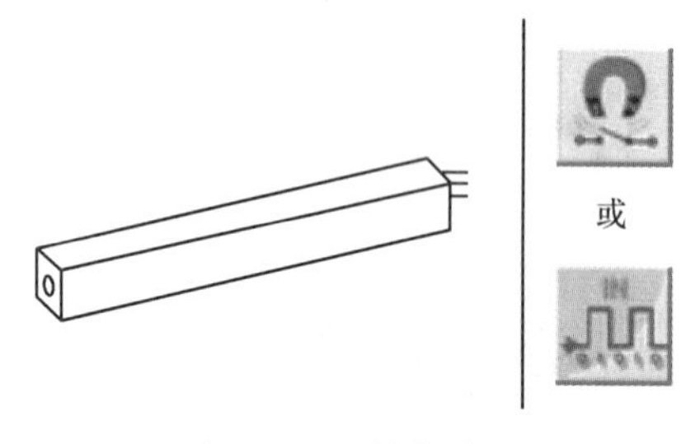


图 2-3　磁敏传感器

**自动门驱动技术**

自动门的驱动装置由电机和减速器或电机和液压系统组成，按照设定的指令程序工作，带动门体开关。从驱动装置到门体的运动，中间要有传动机构，减速器是传动机构的一部分。目前，减速器和电机大部分已连成一体（称为减速电机），减速电机使得设备简化，使设备容易做到标准化、小型化。除减速器外，还要根据具体需要设计传动机构。传动机构可以是齿轮传动，链条传动，也可以是高效率的同步带传动。

能力源创新课程套件用的是直流电机（见图 2-4）。为了能方便地与其他组件进行连接，外形上做了一些特殊的处理。电机需要通过 5 : 1 的减速齿轮箱达到减速的功能，其具体的技术参数见表 2-1。

**表 2-1　能力源直流电机的技术参数**

| 工作电压/V | 6.5~11.2 |
|---|---|
| 空载转速/$r \cdot min^{-1}$ | >5800 |
| 最大效率点力矩/g · cm | 163 |

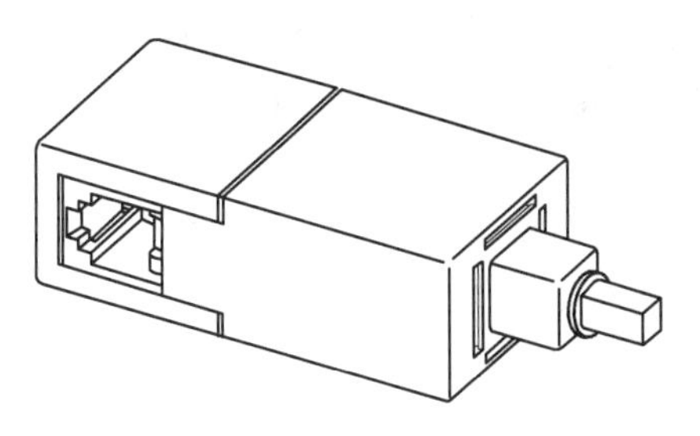

图 2-4　能力源直流电机

**自动门控制技术**

通用控制器。通常采用 PLC（见图 2-5）作为自动门的控制器。PLC 具有抗干扰能力强，可靠性高，控制系统结构简单，通用性强，编程方便，易于实现的优点。常用的 PLC 品牌有三菱、西门子等。

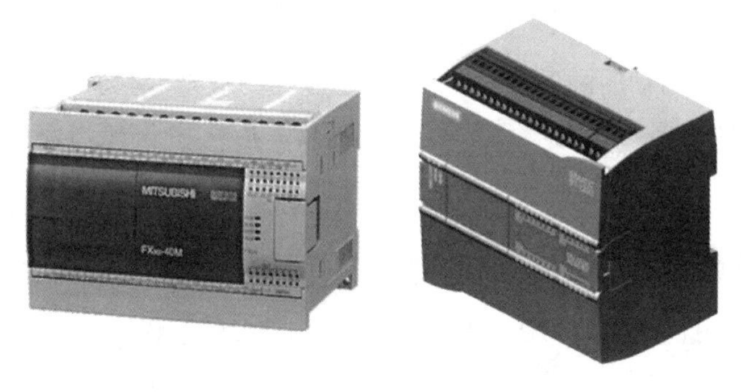


图 2-5　PLC 控制器

专用控制器如图 2-6 所示。它通常采用流行的 32 位数字处理器（DSP）作为控制中枢，外围结合性能优越的通信和数字解码技术，使自动门在性能、可靠性等方面都大大提高。

自动门控制器支持手持 PDA 等智能终端通信的功能，在自动门的调试和维护上更加方便、安全。自动门所有参数可以矢量化显示，运行状态可以通过 PDA 的参数一目了然，控制程序可以升级。手持 PDA 智能终端如图 2-7 所示。

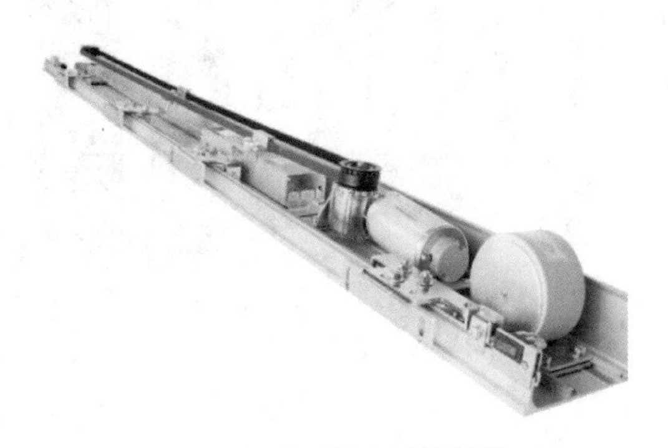

图 2-6　自动门专用控制器

图 2-7　手持 PDA 智能终端

搭建的能力源自动门采用能力源控制器（见图 2-8）作为控制单元。该控制器有 4 路电机接口（接口编号为 DC0-DC3）和 12 路 I/O 接口（接口编号为 I/O0-I/O11），其中 I/O 接口是 A/D 复用的。通过编程，该控制器可以实现自动门的各种基本功能。

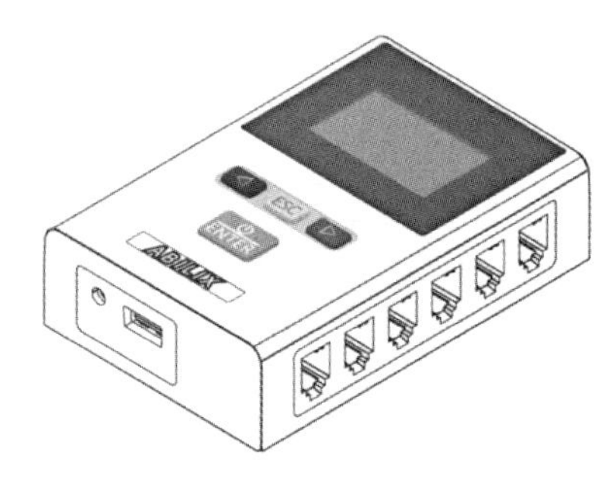

图 2-8　能力源控制器

### 2.1.3　任务实施

选择合适的自动门类型，完成自动门设计，要求自动门制系统检测到开门命令时开门，检测到关门命令时关门，并且用三种颜色的灯，分别示“关闭状态禁止通行”“运动状态注意安全”和“打开可以通行”。

#### 2.1.3.1　方案设计

设计方案包括：

（1）自动门采用移动门的形式；

（2）自动门通过直流电机驱动，通过齿轮减速后再由齿轮齿条驱动自动门开启和闭合。

#### 2.1.3.2　材料准备

参考配套项目用料清单如图 2-9 所示。

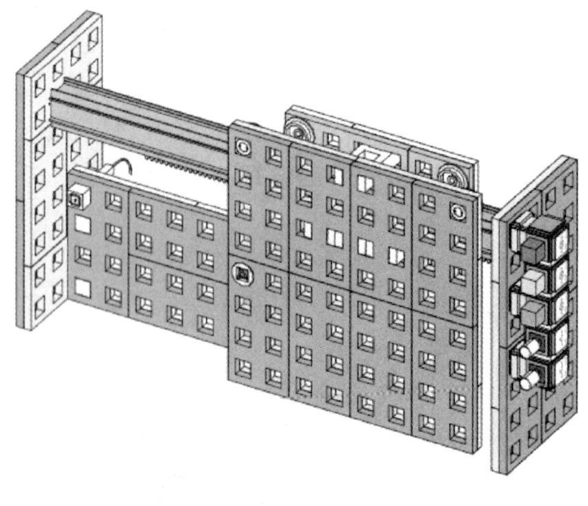

图 2-9　自动门

#### 2.1.3.3 结构搭建

模拟自动门系统的结构搭建模型分别如图 2-10~图 2-16 所示。

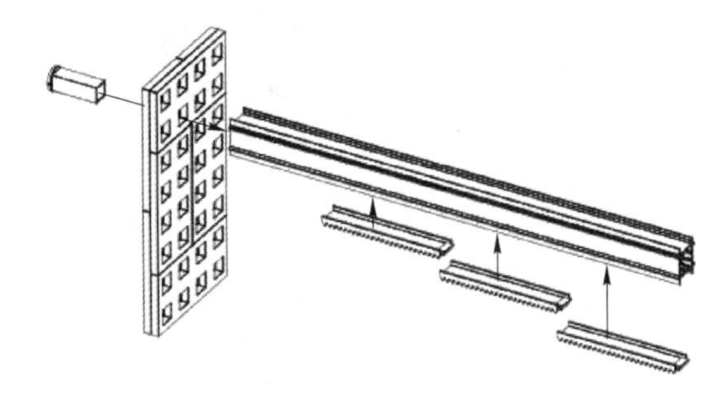

图 2-10 安装门扇行进轨道

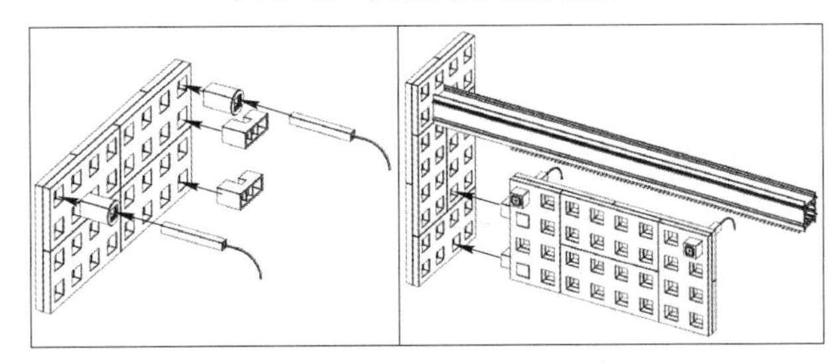

图 2-11 安装传感器和固定支架

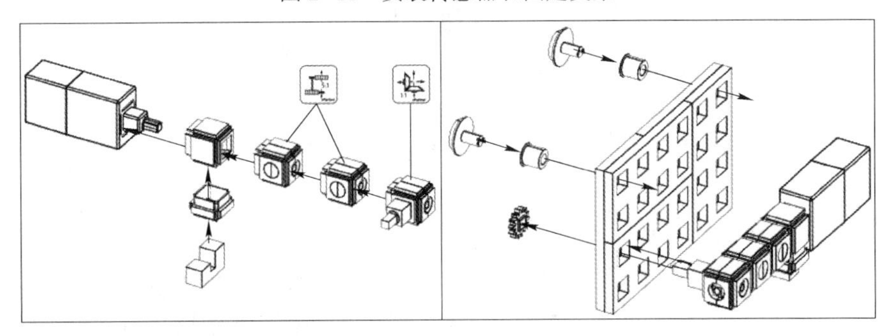

图 2-12 安装传动系统

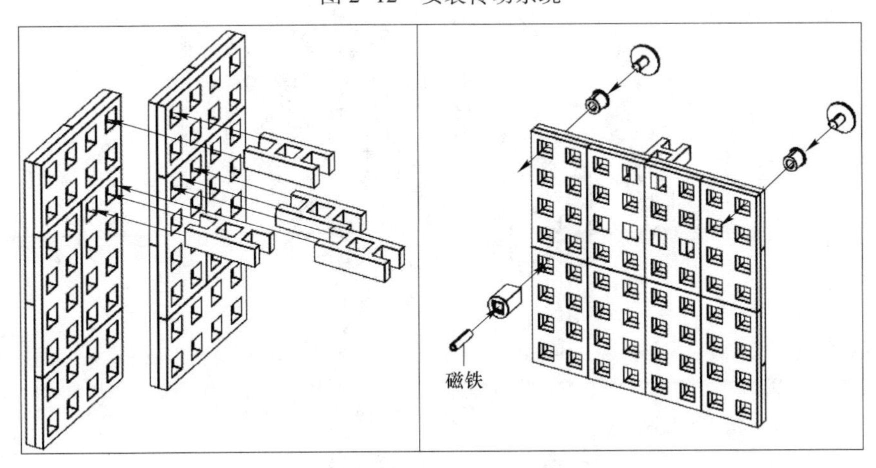


图 2-13 安装吊具走轮

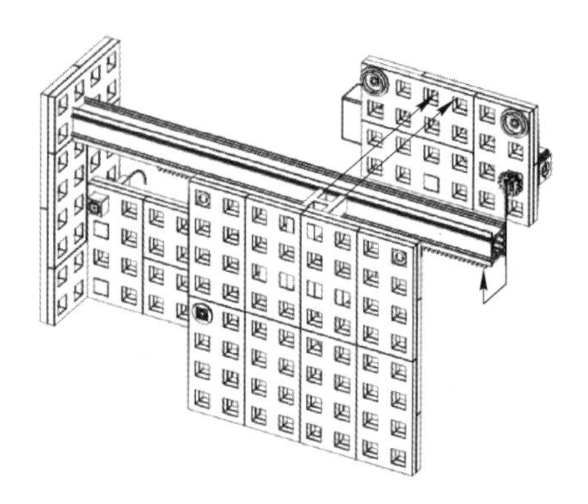

图 2-14　安装门扇

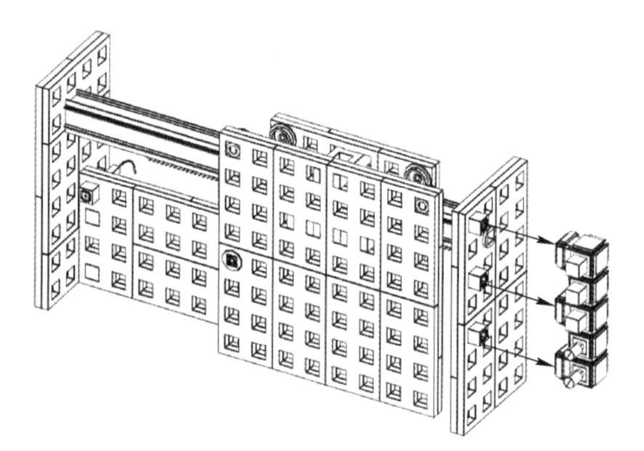

图 2-15　安装按钮及指示灯

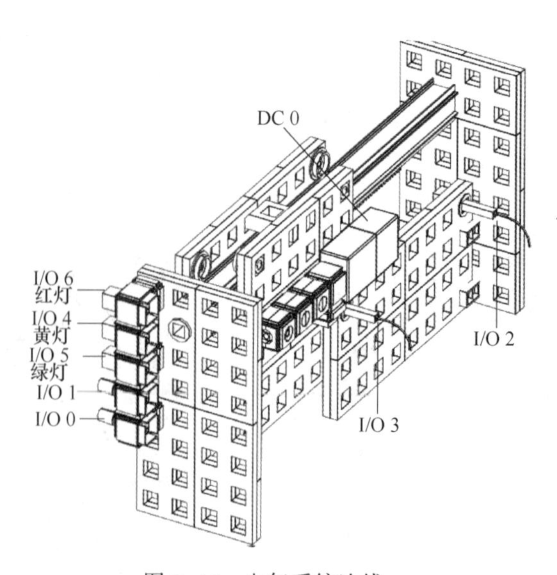


图 2-16　电气系统连线

2.1.3.4　程序设计

自动门程序流程如图 2-17 所示。

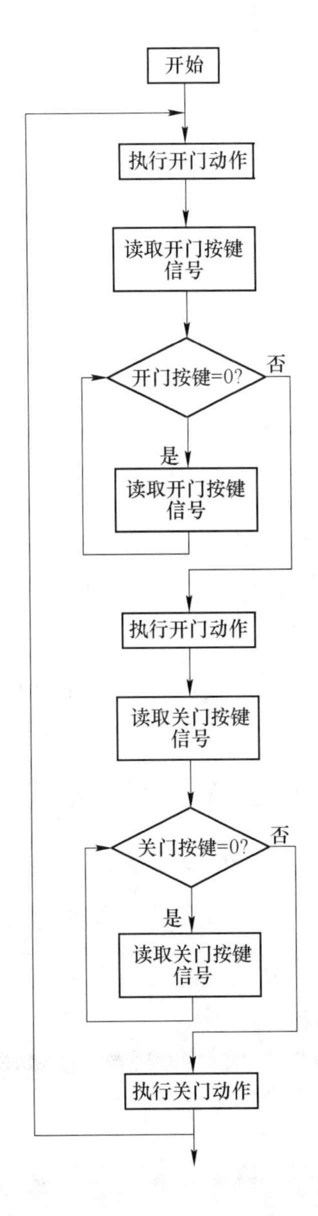


图 2-17　自动门程序流程图

新建流程图程序的对话框如图 2-18 所示；流程图编程界面如图 2-19 所示；新建子程序的对话框如图 2-20 所示；子程序流程图界面如图 2-21 所示；开门子程序中的 VJC 流程图和 JC 语言见表 2-2。

图 2-18　新建流程图程序

图 2-19　流程图编程界面

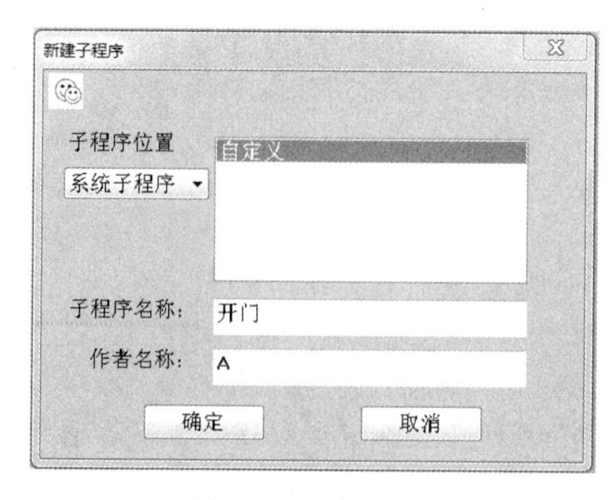


图 2-20　新建子程序

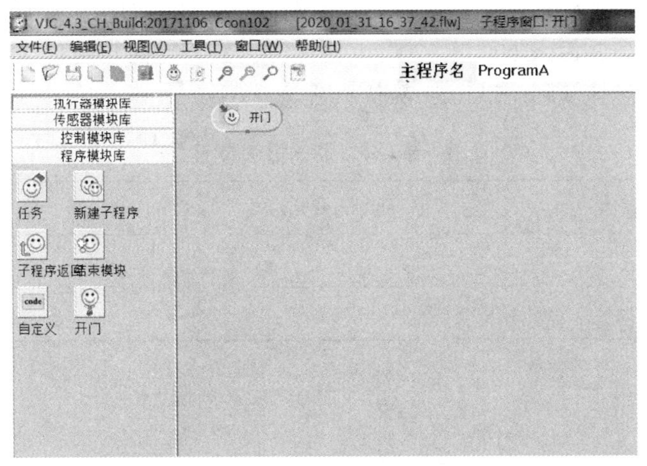


图 2-21 子程序流程图界面

**表 2-2 开门子程序**

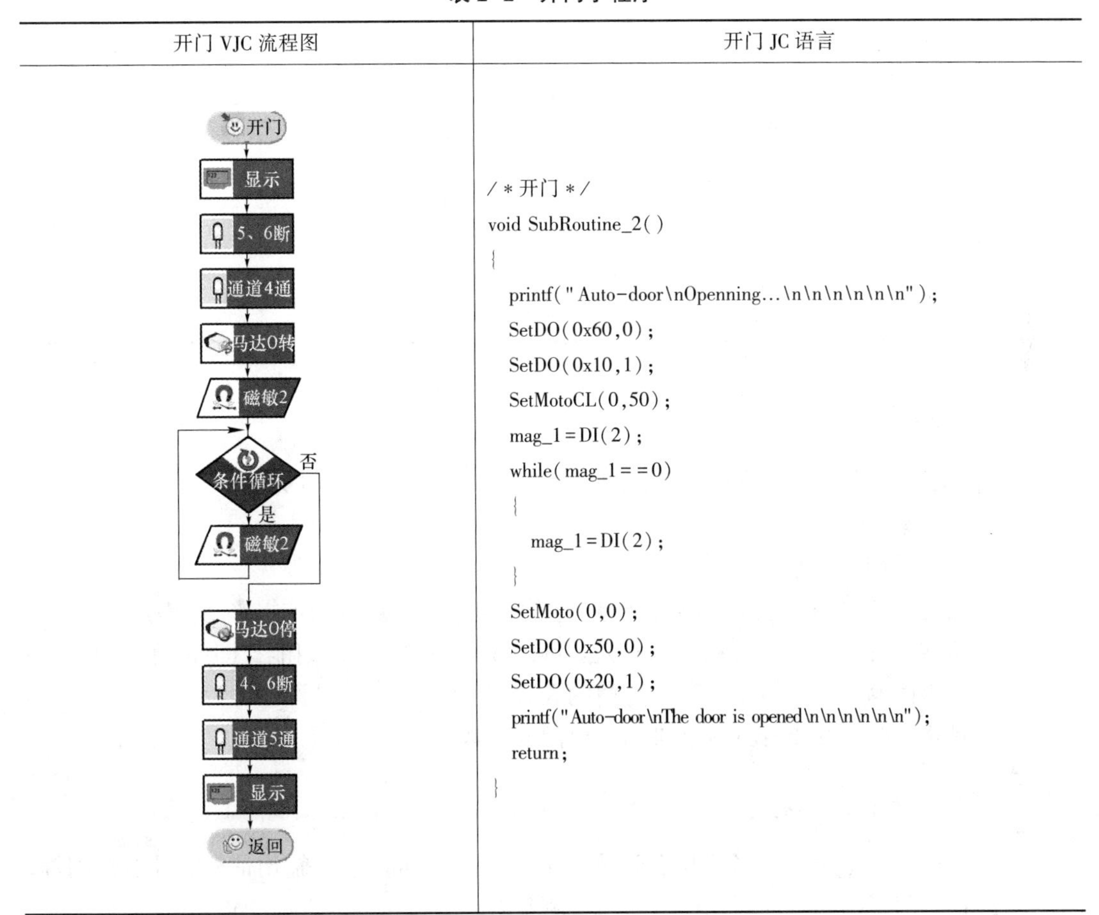


| 开门 VJC 流程图 | 开门 JC 语言 |
| --- | --- |

```
/ * 开门 * /
void SubRoutine_2( )
{
    printf( " Auto-door\nOpenning...\n\n\n\n\n\n" ) ;
    SetDO( 0x60 ,0 ) ;
    SetDO( 0x10 ,1 ) ;
    SetMotoCL( 0 ,50 ) ;
    mag_1 = DI( 2 ) ;
    while( mag_1 = = 0 )
    {
        mag_1 = DI( 2 ) ;
    }
    SetMoto( 0 ,0 ) ;
    SetDO( 0x50 ,0 ) ;
    SetDO( 0x20 ,1 ) ;
    printf( "Auto-door\nThe door is opened\n\n\n\n\n" ) ;
    return ;
}
```

### 2.1.3.5 调试记录

利用 VJC 中的在线检测功能，依次测试各个端口，保证端口对应正确且各元器件工作正常，电机正转时是关门动作。磁敏开关无输入，一般是检测距离过远，可以适当调近磁敏与磁铁之间的距离；齿条下面的 12 齿齿轮打滑，检查导向轮组件装配，是否从梁上滑落。最后把调试信息填写到调试记录表中（见表 2-3）。

**表 2-3　调试记录表**

| 调试项目 | 调试过程记录 | 调试人员 |
|---|---|---|
| | | |
| | | |

### 2.1.4　任务评价

完成任务后对自动门任务进行评价。任务评价表见表 2-4。

**表 2-4　任务评价表**

| 序号 | 评价内容 | 成绩占比 | 自评 | 师评 |
|---|---|---|---|---|
| 1 | 条件模块的原理 | 10 分 | | |
| 2 | 子程序模块的原理 | 10 分 | | |
| 3 | 自动门的搭建 | 30 分 | | |
| 4 | 自动门的编程 | 30 分 | | |
| 5 | 自动门的调试 | 20 分 | | |

### 2.1.5　任务拓展

能不能增加几个组件，进一步丰富自动门的功能？并思考如何实现。

**想一想，练一练**

（1）可以采用两扇平移门增加门的宽厚吗？

（2）当门开启时间超过 1min 后，黄灯闪烁进行报警，可以增加报警功能吗？

（3）门在闭合过程中，若检测到有人要进入或出去时，自动门能立即停止闭合吗？

（4）是否能够运用计数来统计通过门的人数？

（5）是否能够进一步分别统计进门和出门的人数？

**动动手**

（1）自己设计两扇平移门。

（2）采用一个定时器，当门开启后进行计时，超过 1min 后，输出信号，让黄灯闪烁。

（3）在关门子程序中，增加检测，当检测到有人通过时，立即停止输出。

（4）需要增加一个传感器，对进出的人数进行检测。

（5）再增加一个传感器，根据两个传感器动作的先后，是否可以判断人是进还是出？（见图 2-22）

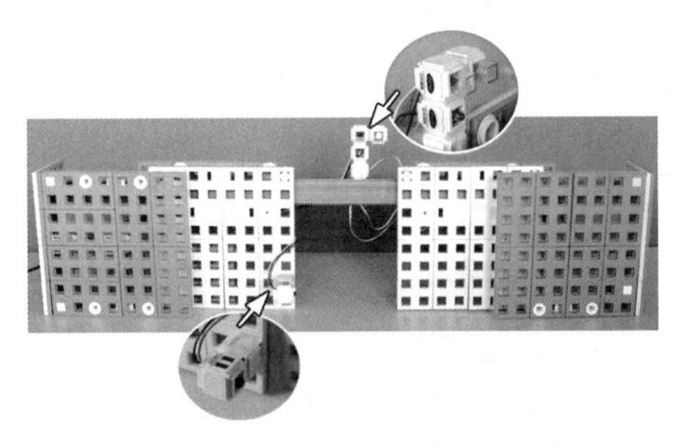

图 2-22　带红外检测的两扇平移门

# 任务 2.2　搭建与调试智能电梯控制系统

## 2.2.1　任务引入

很久以前，人们就开始使用原始的升降工具来运送人和货物，并大量采用人力或者畜力作为驱动力。19 世纪初，随着工业革命进程的发展，蒸汽机成为重要的原动力，人们开始使用蒸汽机作为升降工具的动力，并不断地得到创新和改进。电梯问世已经有 100 多年，经历了无数次改进和提高，电梯已由最早的简陋、不安全、不舒适变成了如今便捷、安全的电梯，其技术发展是永无止境的。21 世纪，随着人口数量与可利用土地面积之间的矛盾进一步激化，将会大力发展多用途、全功能的高层塔式建筑，超高速电梯继续成为研究方向。

## 2.2.2　相关知识

### 2.2.2.1　了解智能电梯

智能电梯设备是由机械和电气两大系统组成，配有独立的标准电气控制系统，机械系统由驱动系统、轿厢及对重装置、导向系统、层门和轿门及开关门系统、机械安全保护系统组成。电气控制系统主要是由拖动控制部分、使用操作部分、井道信息采集部分、安全防护部分等组成。控制系统包含了可编程控制器、变频器、传感器、电机传动和低压电气设备等，同时具有测视频监控、视频显示、消防、电话呼叫等功能，能实现按钮控制、信号控制、集选控制、人机对话等功能，并可进行智能群控、远程监控和故障设置、诊断检测等内容考核，从而体现了现代电梯的主流技术。

### 2.2.2.2　智能电梯的定义与分类

**电梯的定义**

电梯是一种以电动机为动力的垂直升降机构，装有箱状吊舱，用于多层建筑乘人或者载运货物，服务于规定楼层间的固定式升降设备。也有台阶式，其踏步板装在履带上连续运行，俗称自动电梯。

**智能电梯的分类**

智能电梯可以从不同的角度进行分类：

（1）按用途分类。按用途分类，智能电梯可以分为乘客电梯、载货电梯、医用电梯、杂物电梯、观光电梯、车辆电梯、船舶电梯、建筑施工电梯和特殊用途的电梯。

（2）按驱动方式分类。按照驱动方式分类，可以分为交流电梯、直流电梯、液压电梯、齿轮齿条电梯和螺杆式电梯。

（3）按操纵控制方式分类。按照操纵控制方式来分，电梯可分轿内手柄开关控制电梯、按钮控制电梯、信号控制电梯、集选控制电梯、下集合（选）控制电梯、并联控制电梯、梯群控制电梯和梯群智能控制电梯。

### 2.2.2.3 了解几大电梯品牌

常用的电梯品牌包括 OTIS（奥的斯）、Schindler（迅达）、KONE（通力）、GUANGRI（广日）、Mitsubishi（三菱）、FUJIElevator（富士达）、HITACHI（日立）、TOSHIBA（东芝）、ThyssenKrupp（蒂森克虏伯）等。

### 2.2.2.4 智能电梯的结构与组成

电梯是机电一体化产品，其机械部分好比人的躯体，电气部分相当于人的神经，控制部分相当于人的大脑。机械部分和电气部分通过控制部分得以调度，密切协同，使电梯可靠运行。尽管智能电梯种类繁多，但目前使用的智能电梯绝大多数为电力拖动、钢丝拽摇引式结构，其结构如图 2-23 所示。

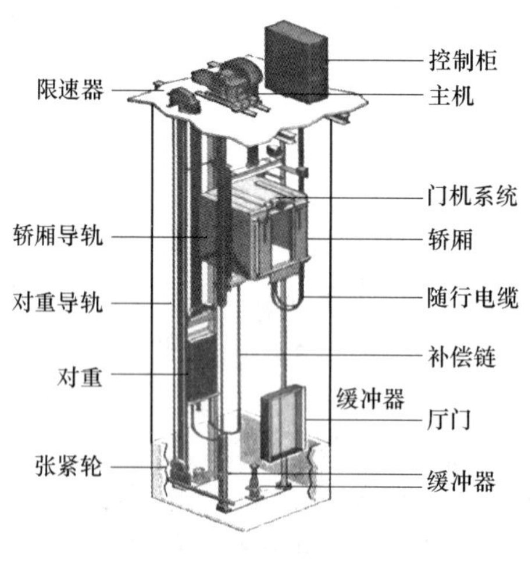


图 2-23　电梯结构图

从智能电梯空间位置看，有以下四个组成部分：依附建筑物的机房和井道，运载乘客或者货物的空间（轿厢），乘客或者货物出入轿厢的地点（层站），即机房、井道、轿厢、层站。

从电梯各构件部分的功能上看，可分为曳引系统、导向系统、轿厢系统、门系统、重量平衡系统、电力拖动系统、电机控制系统和安全保护系统。电梯各部分的功能见表2-5。

**表 2-5　电梯各部分的功能**

| 系统 | 功能 | 主要构件与装置 |
|---|---|---|
| 曳引系统 | 输出与传递动力，驱动电梯运行 | 曳引机、曳引钢丝绳、导向轮、反绳轮等 |
| 导向系统 | 限制轿厢和对重的活动自由度 | 导轨、导轨支架 |
| 轿厢 | 用以运送乘客和货物的组件 | 轿架、轿厢体 |
| 门系统 | 乘客或货物的进出口，运行时层、轿门必须封闭，到站时才能打开 | 轿门、厅门、门机、门锁 |
| 重量平衡系统 | 相对平衡轿厢重量以及补偿高层电梯中曳引绳长度的影响 | 对重、补偿链 |
| 电力拖动系统 | 提供动力，对电梯实行速度控制 | 电动机、供电系统、速度反馈装置调速装置等 |
| 电气控制系统 | 对电梯的运行实时操纵和控制 | 控制柜、平层装置、操纵箱、召唤盒、操纵装置 |
| 安全保护系统 | 保证电梯安全使用，防止一切危及人身安全的事故发生 | 限速器、安全钳、缓冲器、端站保护装置、超速保护装置、断相错相保护装置、上下极限保护装置、门锁联锁装置 |

### 2.2.3　任务实施

模拟一个三层楼房的电梯系统，每层都可以实现“呼唤”服务，且电梯到达楼层是要有亮灯信号。

#### 2.2.3.1　方案设计

方案设计中智能电梯各部分的功能及仿真项目见表 2-6。

**表 2-6　智能电梯各部分的功能及仿真项目**

| 系统 | 功能 | 仿真项目 |
|---|---|---|
| 曳引系统 | 输出与传递动力驱动电梯运行 | 用棉线仿真 |
| 导向系统 | 限制轿厢和对重的自由度 | 导向轮仿真 |
| 轿厢 | 运送乘客和货物的组件 | 平板，横梁搭建 |
| 电力拖动系统 | 提供动力、对电梯实行速度控制 | 电机，齿轮箱仿真 |
| 电气控制系统 | 对电梯的运行实施操纵和控制 | 开关，磁敏传感器等 |

#### 2.2.3.2　材料准备

设计智能电梯时所需材料如图 2-24 所示。

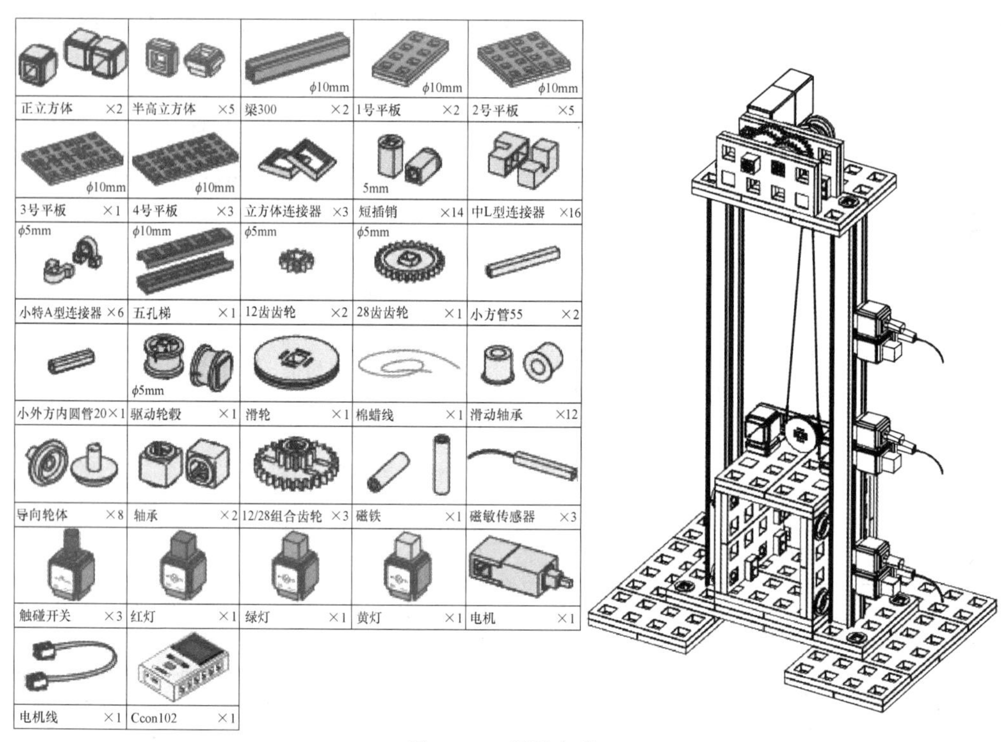


图 2-24　智能电梯

## 2.2.3.3　结构搭建

模拟智能电梯系统的结构搭建模型分别如图 2-25~图 2-30 所示。

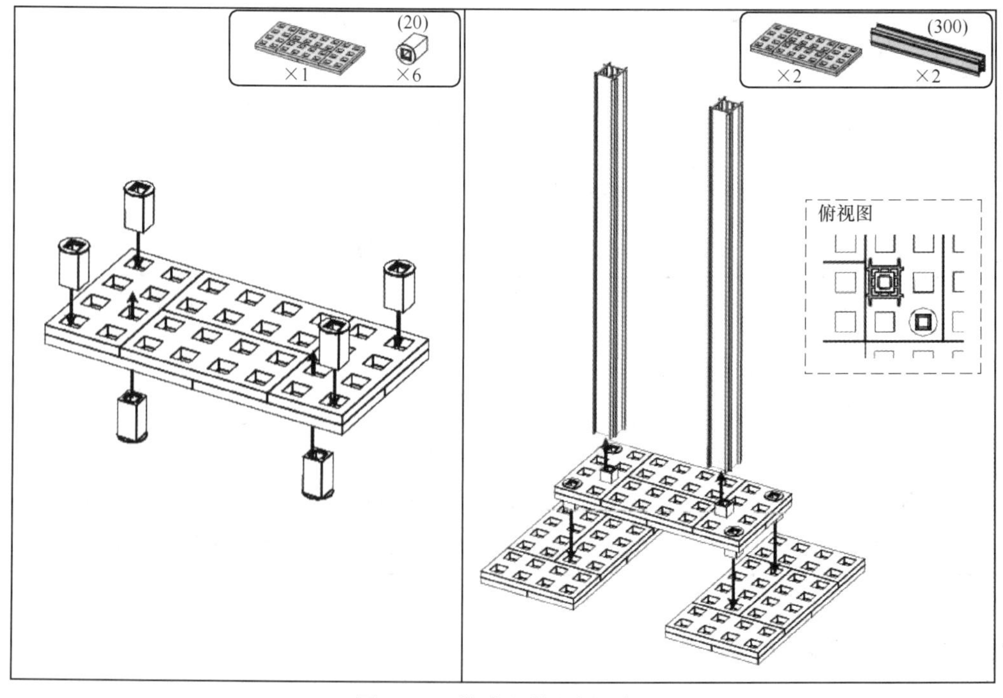


图 2-25　搭建电梯的轿厢架

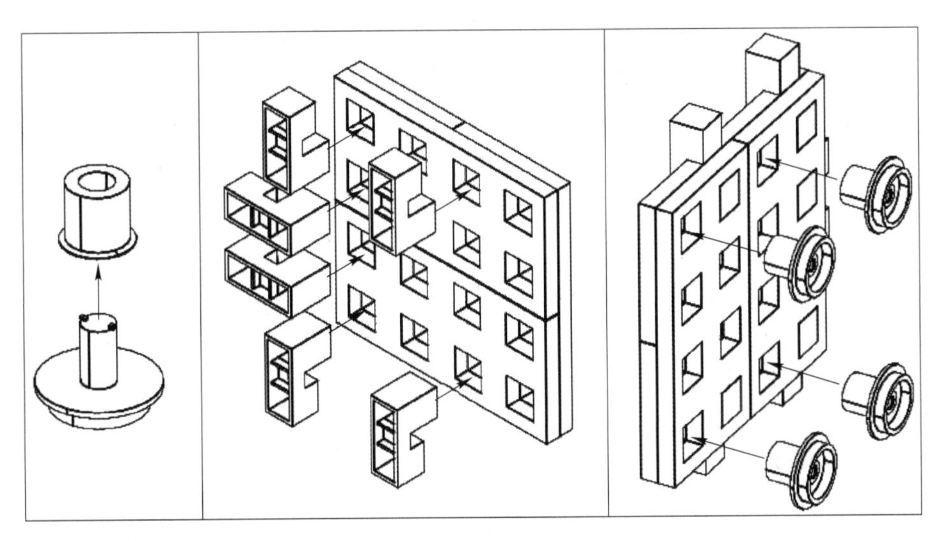

图 2-26　搭建智能电梯轿厢的侧板

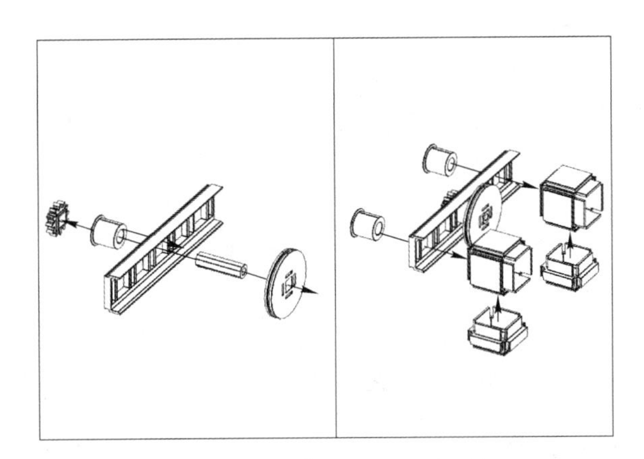

图 2-27　搭建电梯的曳引机构

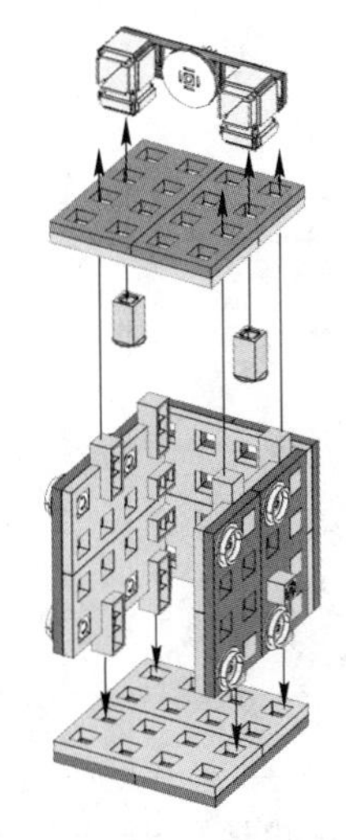

图 2-28　搭建电梯轿厢整体

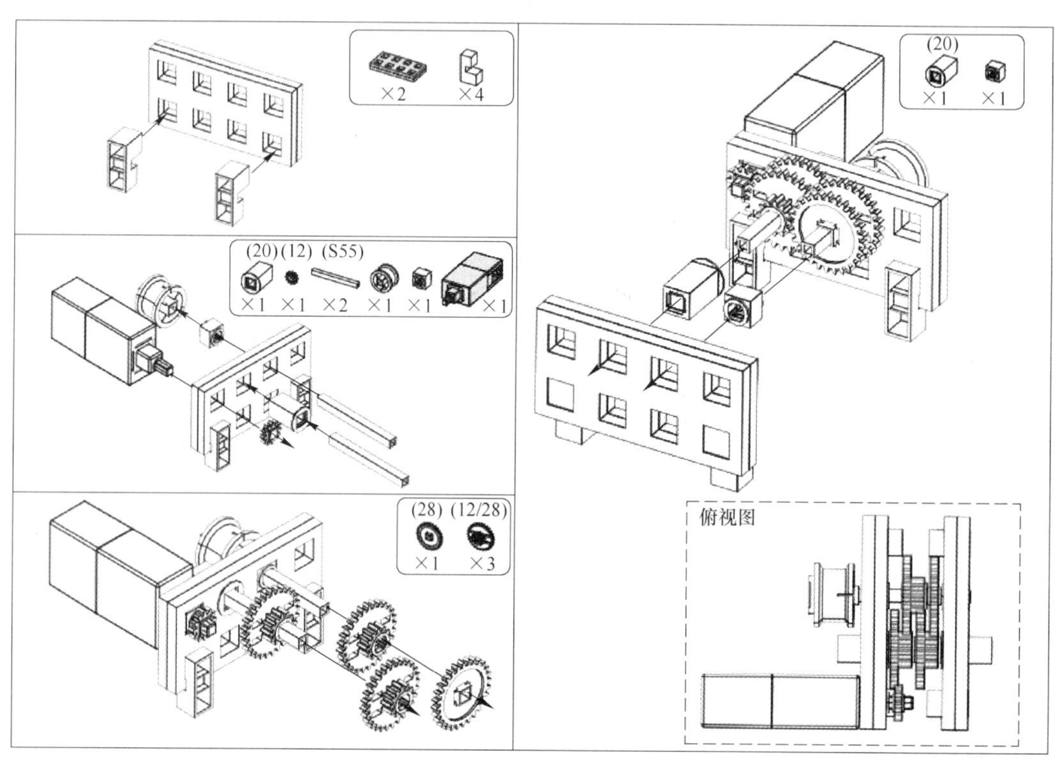


图 2-29　搭建电梯的驱动系统

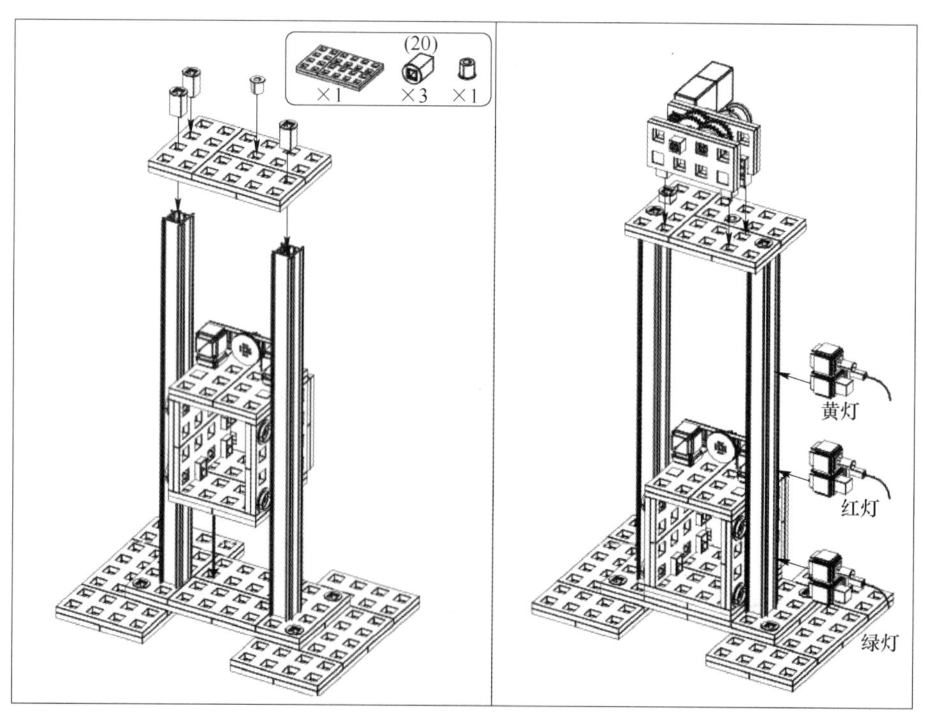


图 2-30　将电梯的各部分组合为整体

#### 2.2.3.4 程序设计

智能电梯程序流程如图 2-31 所示。

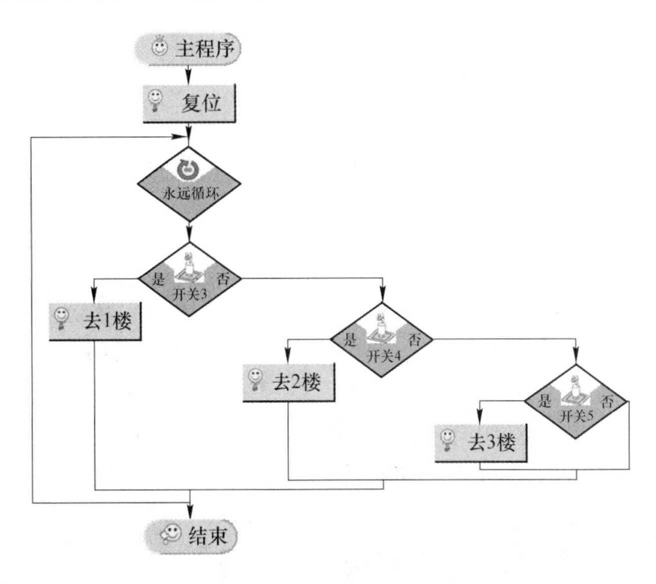


图 2-31 智能电梯程序流程图

#### 2.2.3.5 调试记录

电梯调试记录表见表 2-7。

**表 2-7 调试记录表**

| 调试项目 | 调试过程记录 | 调试人员 |
| --- | --- | --- |
| | | |
| | | |

### 2.2.4 任务评价

任务评价表见表 2-8。

**表 2-8 任务评价表**

| 序号 | 评价内容 | 成绩占比 | 自评 | 师评 |
| --- | --- | --- | --- | --- |
| 1 | 磁敏传感器的应用 | 10 分 | | |
| 2 | 子程序模块的应用 | 10 分 | | |
| 3 | 电梯的搭建 | 30 分 | | |
| 4 | 电梯的编程 | 30 分 | | |
| 5 | 电梯的调试 | 20 分 | | |

### 2.2.5 任务拓展

(1) 如何实现四层电梯结构?

(2) 能否在轿厢中实现选择要到达的楼层?

(3) 并行的多个电梯如何实现调度控制?

# 任务2.3 搭建与调试AGV小车控制系统

### 2.3.1 任务引入

20世纪50年代初，世界上第一台自动导引小车（Automated Guided Vehicle，AGV）开发成功。它是牵引式小车系统，可以十分方便地与其他物流系统自动连接，从而提高劳动生产率，极大地提高装卸、搬运的自动化程度。1954年研制出电磁感应导向的自动导引小车，由于它的显著特点，迅速得到了应用和推广，小车实物图片如图2-32所示。本次任务的目的是完成AGV小车设计和程序设计，重在培养自主创新能力。

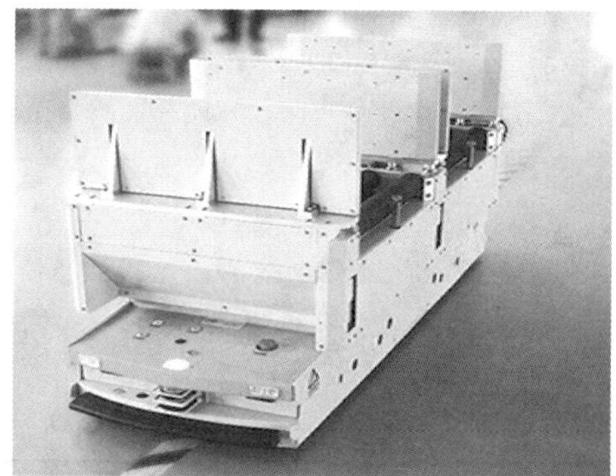

图2-32 AGV实物图片

### 2.3.2 相关知识

#### 2.3.2.1 AGV小车的定义

AGV小车（自动导引小车）是采用自动或者人工的方式装载货物，按设定的路线运动行驶或牵引着载货车至指定地点，再用自动或人工的方式装载货物的工业车辆。AGV是以电池为动力源的一种自动操纵的工业车辆。

#### 2.3.2.2 AGV小车的分类

按照导向原理的不同，自动导引小车可分为：

(1) 外导式。在车辆的运行路线上设置导向信息媒体（如导线、磁带、色带等），由车上的导向传感器接收线路媒体的导向信息（如频率、磁场强度、光强度等），信息经实时处理后控制车辆沿正确路线行驶。

（2）自导式。自由路径导引采用坐标定位原理，在车辆上预先设定运行线路的坐标信息。在车辆运行时，实时地测出实际的车辆位置坐标，再将两者进行比较后控制车辆的导向运行。

### 2.3.2.3 AGV 小车的结构与组成

AGV 小车通常由车体系统、车载控制系统、行走装置、安全与辅助系统四个子系统组成。AGV 小车的组成结构如图 2-33 所示。

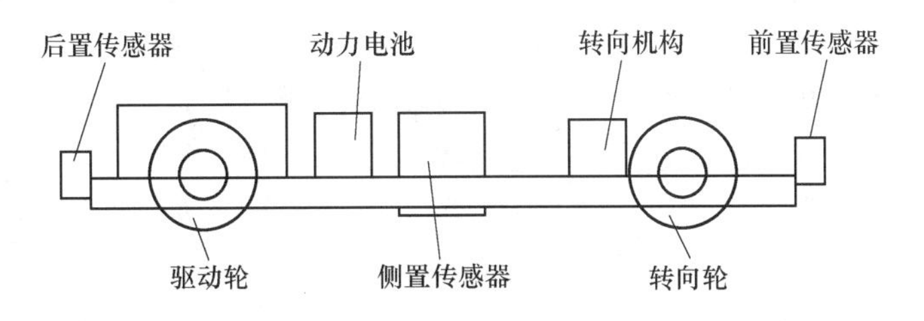


图 2-33　AGV 小车的组成

（1）车体系统。车体系统包括底盘、车架、壳体、驱动装置、转向机构和控制室等。该系统是 AGV 的躯体，具有电动车辆的基本特征。车架通常为钢结构件，要求具有一定的强度和刚度。驱动装置由车轮、减速器、制动器、电机及调速器等组成，是个伺服驱动的速度控制系统，驱动系统可由计算机或人工控制，可以驱动 AGV 运行并具有速度控制和制动能力。通过转向机构，AGV 可以实现向前、向后，或者纵向、横向、斜向及回转的全方位运动。

（2）车载控制系统。车载控制系统是 AGV 的核心，一般是由监控系统、导航系统、通信系统、电机驱动器及控制面板构成的。AGV 的运行、监测及各种智能化控制的实现，均须通过控制系统实现。

导航系统是 AGV 的一个重要的部分，最基本的技术要求就是从 A 点到 B 点可靠引导 AGV，避开已知的障碍物。采用不同的导向方式和导向系统有不同的组成。目前常见的导向方法有 10 种，分别是地链牵引导向、电磁感应导向、磁带导引、惯性导向、红外线导向、激光导向、光学导向、示教型导向、视觉导向和 GPS（全球定位系统）导向。

（3）行走装置。行走装置一般是由驱动轮、从动轮和转向机构组成。该装置有三轮、四轮、六轮及多轮等，三轮结构一般采用前轮转向和驱动，采用双轮驱动、四轮或六轮差速转向或独立转向方式。为了提高定位精度，驱动及转向电机都采用直流伺服电机。

（4）安全与辅助系统。为了避免 AGV 在运行过程中出现碰撞，保护人员和其他装置的安全，AGV 都具有障碍物探测和避撞、警音、警示、急停等安全措施。一般情况下，AGV 都采取多级硬件和软件的安全监控措施。

#### 2.3.2.4 传感器及原理

**灰色传感器**

本次任务需要利用传感器进行导向，接下来介绍能力源套件采用的灰度传感器。灰度传感器利用不同颜色的检测面对光进行不同程度的反射，光敏电阻对不同检测面返回的光（其阻值也不同）进行颜色深浅检测。在环境光干扰不是很严重的情况下，灰度传感器用于区别黑色与其他颜色，它还有比较宽的工作电压范围，在电源电压波动比较大的情况下仍能正常工作。灰度传感器输出的是连续的模拟信号，因而能很容易地通过 A/D 转换器或简单的比较器实现对物体反射率的判断，是一种实用的机器人巡线传感器。灰度传感器如图 2-34 所示。

图 2-34　灰度传感器

灰度传感器是模拟传感器，包含的一只发光二极管和一只光敏电阻安装在同一面上。灰度传感器利用不同颜色的检测面对光的反射程度不同，光敏电阻对不同检测面返回的光（其阻值也不同）进行颜色深浅检测。在有效的检测距离内，发光二极管发出白光，照射在检测面上，检测面反射部分光线，光敏电阻检测此光线的强度并将其转换为机器人可以识别的信号。

**调节方法**

灰度传感器上无信号指示灯，但是配有检测颜色返回模拟量大小调节器。欲使检测给定的颜色时，可以将发射/接收头置于给定颜色处，配合调节器即可调出合适的返回模拟量，其方法为：将调节器逆时针方向旋转，返回模拟量变大；将调节器顺时针方向旋转，返回模拟量变小。可以一直调节直到你需要的数值为止。若需要准确的模拟量，可以用程序在液晶屏幕上显示，配合调节器即可调出准确的模拟量。

注：用螺丝刀旋转调节器时，不要旋得太快，也不要旋得太用力，以防旋坏。在发现旋不动时，应马上停止。

**注意事项**

灰度传感器使用时需注意：

（1）检测面的材质不同也会引起其返回值的差异。

（2）外界光线的强弱对其影响非常大，会直接影响到检测效果，在对具体项目检测时注意包装传感器，避免外界光的干扰。

（3）根据它的工作原理，利用光敏探头根据检测面反射回来的光线强度，来确定其检测面的颜色深浅，因此测量的准确性与传感器到检测面的距离有直接关系。在机器人运动时机体的震荡同样会影响其测量精度。

### 2.3.3 任务实施

接下来任务实施环节用能力源创新课程套件仿真 AGV 小车并编程模拟运行，设计一个 AGV 小车。

### 2.3.3.1 方案设计

如图 2-35 所示该 AGV 小车采用两轮差动驱动，后轮主动，前轮导向；采用光学导向的形式，在小车的前部左右各安装两个灰度传感器。

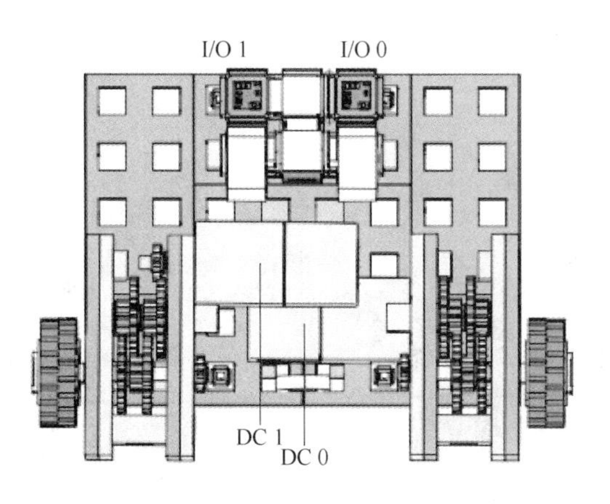


图 2-35 AGV 小车

### 2.3.3.2 材料准备

设计 AGV 小车时所需材料如图 2-36 所示。

| | | | |
|---|---|---|---|
| 正立方体 ×4 | 1号平板 ×6 (ϕ10mm) | 4号平板 ×1 (ϕ10mm) | 立方体连接器 ×1 |
| 短插销 ×6 (5mm) | 中L型连接器 ×10 | 中H型连接器 ×2 (ϕ10mm) | 中A型连接器 ×1 (ϕ10mm) |
| 12齿齿轮 ×4 (ϕ5mm) | 28齿齿轮 ×2 (ϕ5mm) | 小方管40 ×4 | 小方管55 ×2 |
| 小外方内圆管40 ×2 | 驱动轮毂 ×2 (ϕ5mm) | 驱动轮胎 ×2 | 滑动轴承 ×2 |
| 轴承 ×4 | 12/28组合齿轮 ×6 | 2轮组合体 ×1 | 调制地面灰度 ×2 |
| 电机 ×2 | 电机线 ×2 | Ccon102 ×1 | |

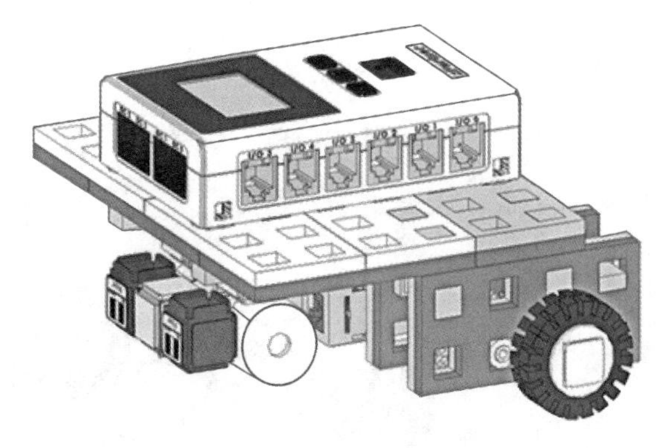

图 2-36 材料清单

### 2.3.3.3 动手搭建

接下来介绍小车搭建过程中的一些关键步骤，帮助同学们完成小车的硬件拼接，其搭建步骤分别如图 2-37～图 2-40 所示。

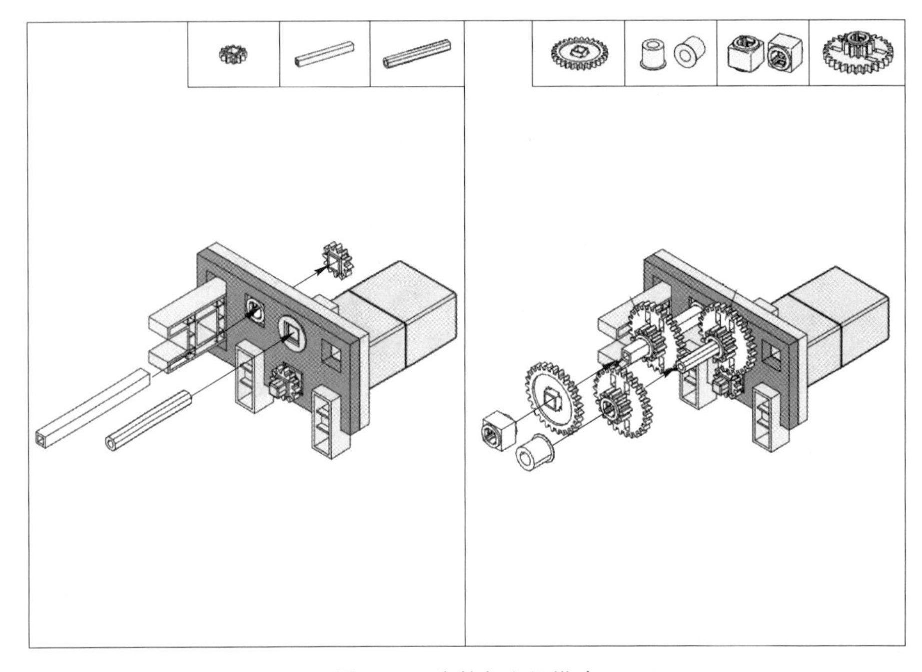

图 2-37 齿轮与电机搭建

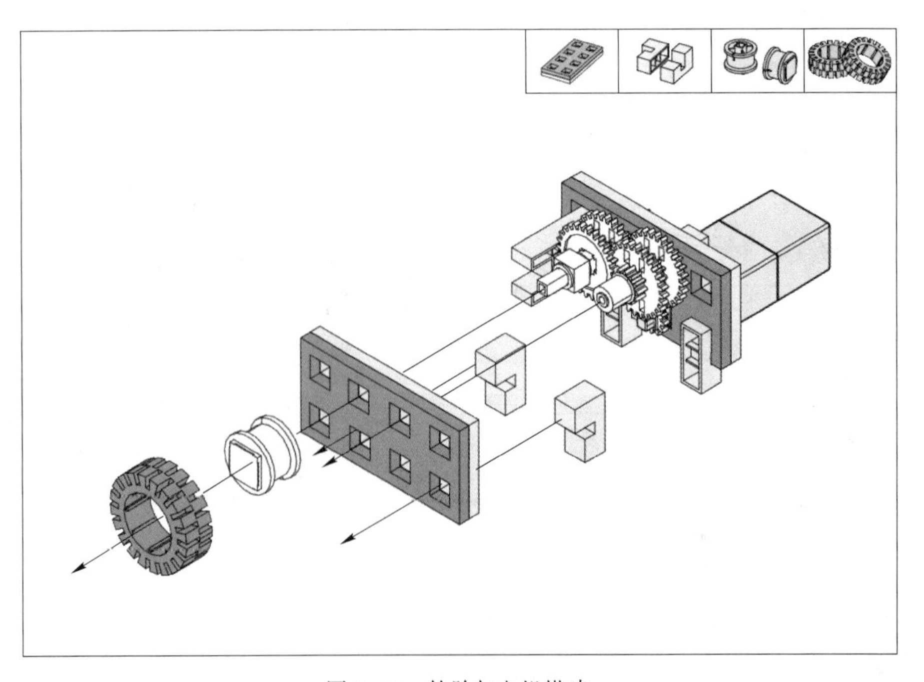

图 2-38 轮胎与电机搭建

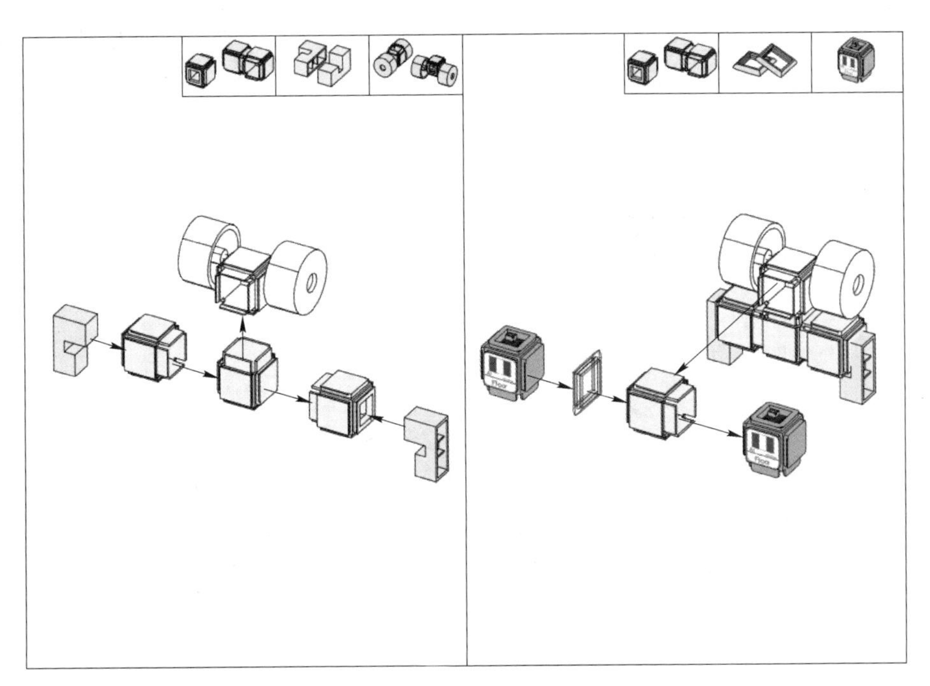

图 2-39　灰度传感器搭建

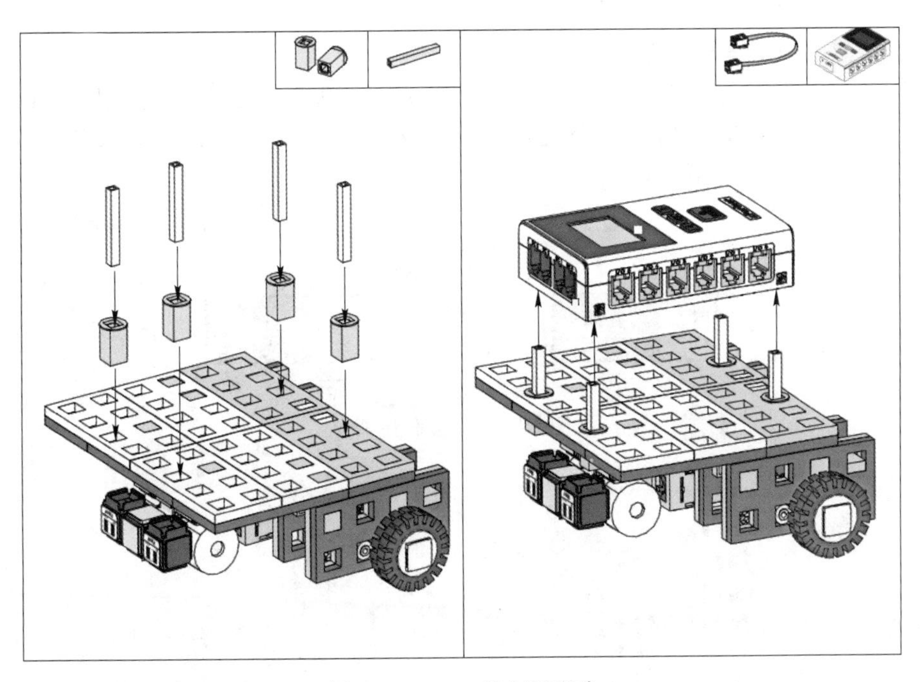

图 2-40　AGV 控制器搭建

#### 2.3.3.4　程序设计

**AGV 小车的控制要求**

AGV 小车控制要求为：自主导向的小车，能够在白板上沿着黑线条移动，实现导向功能。其程序设计流程如图 2-41 所示。

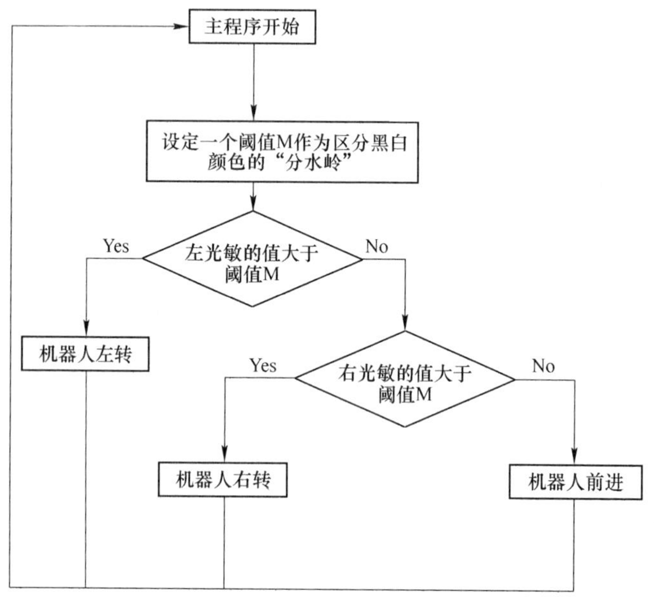


图 2-41　程序设计框图

**灰度传感器的用法与特性**

使用软件中的灰度检测模块，可以根据灰度值进行判断，完成黑白界限的判断，从而进行马达的转向和速度的设置，如图 2-42～图 2-44 所示。

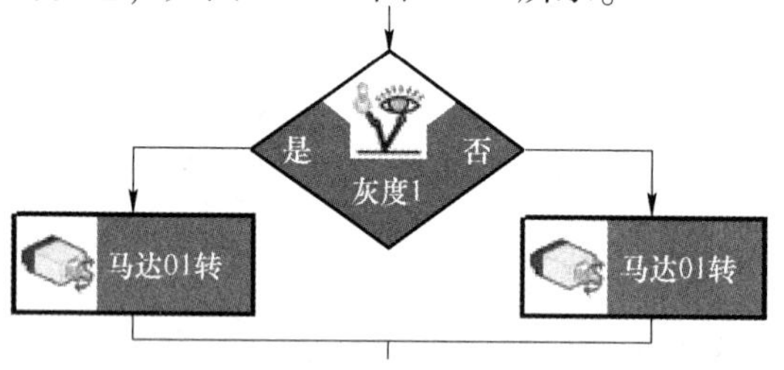


图 2-42　灰度传感器

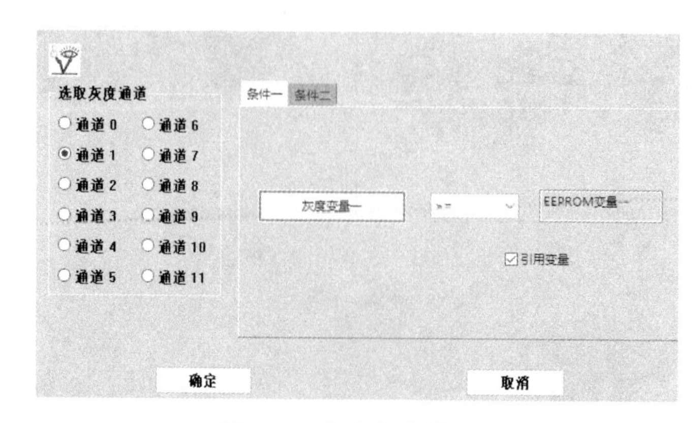


图 2-43　灰度传感器调试

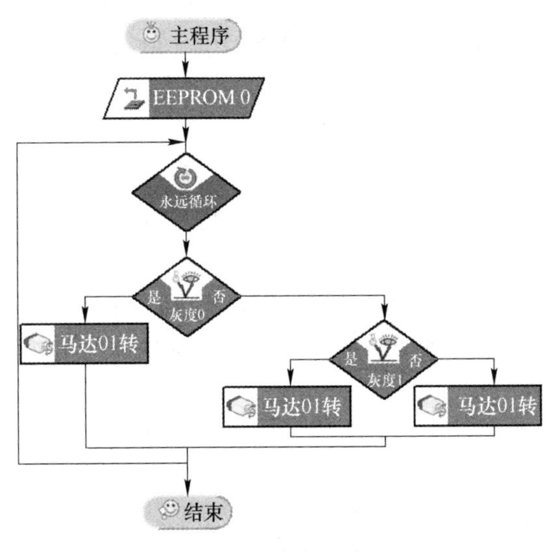

图 2-44　参考程序图

2.3.3.5　调试记录

测试电机的 DC 0 和 DC 1 两个端口、注意电机的正反转，将灰度传感器放在黑色和白色位置上，读取数值，从而求得左右灰度判断黑白色的临界值，分别赋予两个整型变量。最后，把数据填写到调试记录表中（见表 2-9）。

**表 2-9　调试记录表**

| 调试项目 | 调试过程记录 | 调试人员 |
|---|---|---|
| | | |
| | | |

## 2.3.4　任务评价

任务评价表见表 2-10。

**表 2-10　任务评价表**

| 序号 | 评价内容 | 成绩占比 | 自评 | 师评 |
|---|---|---|---|---|
| 1 | AGV 的导向原理 | 10 分 | | |
| 2 | 灰度传感器的原理 | 10 分 | | |
| 3 | AGV 小车的搭建 | 30 分 | | |
| 4 | AGV 小车的编程 | 30 分 | | |
| 5 | AGV 小车的调试 | 20 分 | | |

通过认识 AGV，了解 AGV 的应用、定义与分类；了解典型 AGV 的结构和常用部件，能说出日常见到的 AGV 的基本组成；分析 AGV 检测系统和控制系统，说出 AGV 的自动

导向原理，描述 AGV 的动力系统，熟知 AGV 的控制与通信方式和安全措施；利用能力源套件进行搭建与调试 AGV 小车控制系统，掌握灰度传感器的搭建和编程方法以及电机、齿轮等套件零件的使用及注意事项。

### 2.3.5 任务拓展

（1）尝试做一个不一样的自动引导小车（查找资料，看看更多有关自动引导小车的资料）。

（2）查阅资料，了解常见的 AGV 的型号和更多 AGV 的核心技术。

（3）通过 AGV 搭建，掌握工程工作方法，想想利用能力源创新课程套件还能搭建什么类型的 AGV。

（4）如果在 AGV 导向路径的前方有障碍物，AGV 怎样才能检测到，并且做出相应的反应。

（5）能不能给 AGV 小车增加搬运货物的功能？

（6）小车内部齿轮的作用是什么，传动比如何计算？

## 任务 2.4 搭建与调试工业机械手控制系统

### 2.4.1 任务引入

在工业产线中需要一个机器人将两种颜色的分拣瓶从产线上分拣，根据两种颜色，将它们放到不同的产线上。用能力源创新课程套件构建一个工业机器人，要求工业机器人有一个旋转自由度和两个直线自由度，可以将能力源创新课程套件中的分拣瓶从一个地方搬到另外一个地方。

### 2.4.2 相关知识

工业机械手（通常称为工业机器人，下文也称为工业机器人）是目前机器人领域中得到最广泛实际应用的自动化机械装置。在工业焊接、工业装配、工业搬运等领域都能见到它们的身影。尽管它们的形态各有不同，但它们有一个共同的特点，就是能够接收指令，精确地定位到三维（或二维）空间上的某一点进行作业。

#### 2.4.2.1 定义与分类

**工业机器人的定义**

美国机器人工业协会（U.S.RIA）提出的工业机器人的定义为：“工业机器人是用来进行搬运材料、零件、工具等可编程的多功能机械手，或通过不同程序的调用来完成各种工作任务的特种装置”。

1987 年国际标准化组织 ISO 对工业机器人的定义为：“工业机器人是一种具有自动操作和移动功能，能完成各种作业的可编程操作机”。

我国国家标准 GB/T 12643—2013 将工业机器人定义为：“工业机器人是一种能自动控制、可重复编程、多功能和多自由度的操作机，能搬运材料、工件或操持工具，用以完成各种作业”。

不管哪个定义，机器人具有四大特征，其分别是：

（1）仿生特征：模仿人的肢体动作；

（2）柔性特征：对作业具有广泛适应性；

（3）智能特征：具有对外界的感知能力；

（4）自动特征：自动完成作业任务。

**工业机器人的分类**

工业机器人按照不同的分类标准可以分为：

（1）按照机器人的运动形态可分为直角坐标型工业机器人、圆柱坐标型工业机器人、球坐标型工业机器人、多关节型工业机器人、平面关节型工业机器人和并联型工业机器人。这些机器人的结构如图 2-45 所示。

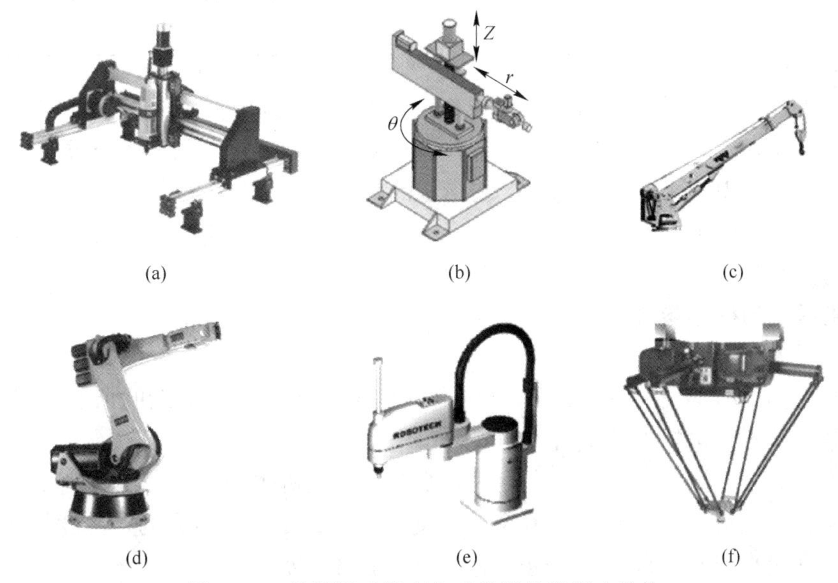


图 2-45　按照运动形态方式分类的机器人结构

（a）直角坐标型；（b）圆柱坐标型；（c）球坐标型；（d）多关节型；（e）平面关节型；（f）并联型

（2）按照输入信息的方式可分为操作机械手、固定程序工业机器人、可编程型工业机器人、程序控制工业机器人、示教型工业机器人和智能工业机器人（见表 2-11）。

**表 2-11　按照输入信息方式分类的机器人特点**

| 分　类 | 特　　点 |
|---|---|
| 操作机械手 | 一种由操作人员直接进行操作的具有几个自由度的机械手 |
| 固定程序工业机器人 | 按预先规定的顺序、条件和位置，逐步地重复执行给定作业任务的机械手 |
| 可编程型工业机器人 | 与固定程序机器人基本相同，但其工作次序等信息易于修改 |
| 程序控制型工业机器人 | 它的作业任务指令是由计算机程序向机器人提供，类似于机床控制 |
| 示教型工业机器人 | 能够按照记忆装置存储的信息来复现由人示教的动作，其示教动作可自动地重复执行 |
| 智能型工业机器人 | 采用传感器来感知工作环境或工作条件的变化，并借助自身的决策能力，完成相应的工作任务 |

（3）按照驱动方式可分为液压型工业机器人、电动型工业机器人和气压型工业机器人（见表2-12）。

**表2-12　按照驱动方式分类的机器人特点**

| 分　类 | 特　　点 |
| --- | --- |
| 液压型工业机器人 | 液压型工业机器人具有较大的抓举能力，可达上千牛顿，这类工业机器人结构紧凑、传动平稳、动作灵敏，但对于密封要求较高，且不宜在高温或者低温环境下使用 |
| 电动型工业机器人 | 目前用得最多的一类工业机器人，不仅因为电动机品种众多，也因为可以运用多种灵活控制的方法。电机直接驱动各轴，或者通过诸如减速器的装置来减速后驱动，结构十分紧凑、简单 |
| 气压型工业机器人 | 以压缩空气来驱动操作机，其优点是空气来源方便，动作迅速，结构简单造价低，无污染；缺点是由于空气具有可压缩性，导致工作速度的稳定性较差，这类工业机器人的抓举力较小，一般只有几十牛顿 |

（4）按照运动轨迹分类，可以分为点位型工业机器人和连续轨迹型工业机器人。点位控制是控制机器人从一个位姿到另一个位姿，其路径不限；连续轨迹控制是控制机器人的机械接口，按编程规定的位姿和速度，在指定的轨迹上运动。

通常见到的工业机械手属于程序控制型、连续轨迹、多关节工业机器人，末端多为气动或者电动控制装置。

**工业机器人的核心参数**

工业机器人的核心参数包括：

（1）自由度（Degree of Freedom）。自由度是指机器人确定运动时所具有的独立坐标轴运动的数目，不应包括末端执行器的开合自由度。

在工业机器人系统中，一个自由度就需要有一个电机驱动。在三维空间中描述一个物体的位置和姿态（简称位姿）需要6个自由度。但是，工业机器人的自由度是根据其用途而设计的，可能小于6个自由度，也可能大于6个自由度。

（2）机械原点（Mechanical Origin）。机械原点是指工业机器人各自由度共用的，在机械坐标系中的基准点。

（3）工作原点（Work Origin）。工作原点是指工业机器人工作空间的基准点。

（4）定位精度（Positioning Accuracy）。定位精度是指机器人手部实际到达位置与目标位置之间的差异。

（5）重复定位精度（Repetitive Positioning Accuracy）。重复定位精度是指机器人重复定位手部与同一目标位置的能力，以实际位置值的分散程度来表示。

（6）工作范围（Working Range）。工作范围是指机器人手臂末端或手腕中心所能达到的所有点的集合，一般指不安装末端操作器的工作区域。

（7）额定负载（Rated Load）。额定负载是指在规定性能范围内，在手腕机械接口处的最大负载允许值。

### 2.4.2.2　结构组成

从体系结构上看，工业机器人分为“三大部分六大系统”。“三大部分六大系统”

是一个统一的整体，三大部分是指用于实现各种动作的机械部分，用于感知内部和外部的信息的传感部分，以及用于控制机器人完成各种动作的控制部分；六大系统分别是驱动系统、机械结构系统、机器人—环境交互系统、感受系统、人机交互系统和控制系统。

工业机器人的驱动系统包括动力装置和传动机构，用以使执行机构产生相应的动作。

机械结构系统由机身、手臂、手腕、末端执行器四大件组成，如图 2-46 所示，有的机器人在底座安装有行走导轨。大多数工业机器人有 3~6 个运动自由度，其中手腕通常有 1~3 个运动自由度。

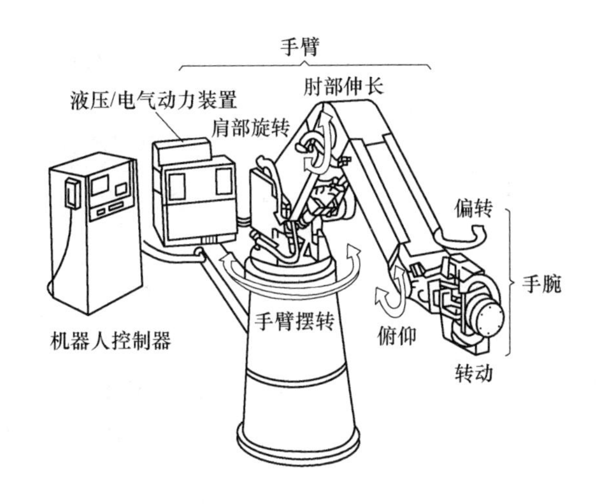


图 2-46　工业机器人结构

机器人—环境交互系统实现机器人与外部设备的联系和协调。

感受系统由内部传感器和外部传感器组成，用以检测其运动位置和工作状态，如位置、力、触角、视觉等传感器。

人机交互系统是人与机器人联系和协调的单元。

控制系统是按照输入的程序对驱动系统和执行机构发出指令信号，并进行控制。

### 2.4.2.3　核心技术

工业机器人核心技术主要包括传感技术、驱动技术、传动技术和编程技术。

**工业机器人的传感技术**

带有传感器的工业机器人可以更好地配合操作对象，进行适宜的操作。工业机器人经常使用的传感技术有传感器、触觉传感器、接近觉传感器和力传感器等。

视觉传感器主要用于零件或工件的位置补偿，零件的判别、确认等方面，视觉传感器通常安装在工业机器人的末端做局部视觉或安装在工业机器人系统外围做全局视觉，如图 2-47 所示。

触觉传感器和接近觉传感器一般固定在指端，用来补偿零件或工件的位置误差，防止碰撞等，如图 2-48 和图 2-49 所示。力传感器如图 2-50 所示。

图 2-47　视觉传感器

图 2-48　触觉传感器

图 2-49　接近觉传感器

图 2-50　力传感器

在本次的项目任务中，所用到的位置检测装置为磁敏传感器和旋转计数器，两者配合使用，通过磁敏传感器来检测设备复位回零状态，通过旋转计数器来检测设备在运行中的位置。

**工业机器人的驱动技术**

工业机器人有三种常见的驱动方式，分别是液压驱动、气压驱动和电动驱动。其中最常见的是电动驱动技术，电动驱动技术中的关键部件就是电动机，俗称电机。

电动机是机器人驱动系统中的执行元件，常采用的电动机为步进电机和直流/交流伺服电机。

交流伺服电机是目前在机器人上应用最多的电机。它内部的转子是永久磁铁，驱动器控制的 U/V/W 三相电形成电磁场，转子在此磁场的作用下转动，同时电机自带的编码器反馈信号给驱动器，驱动器根据反馈值与目标值进行比较，调整转子转动的角度。伺服电机的精度决定于编码器的精度（线数）。

**工业机器人的传动技术**

传动机构用来将原动机发出的机械能传递给关节或其他工作部分，以实现机器人各种必要的运动。工业机器人常用的传动有齿轮传动、螺旋传动和带、链传动。

除上述三种主要传动形式外，工业机器人中还采用液压传动、气压传动、连杆机构或凸轮机构。

**工业机器人的编程技术**

工业机器人的编程有三种不同的方式，包括：

（1）示教编程。示教编程是目前工业机器人广泛使用的编程方法。根据任务的需要，将机器人末端工具移动到所需的位置及姿态（简称位姿），然后把每一个位姿连同运行速度、各个参数等记录并存储下来，机器人便可按照示教的位姿再现。示教编程方式有手把手示教和示教盒示教两种方式。

目前大多数机器人还是采用示教方式编程。示教方式是一项成熟的技术，易于被熟悉工作任务的人员所掌握，而且用简单的设备和控制装置即可进行。示教过程进行的很快，示教过后，马上即可应用。如果需要，过程还可以重复多次，在某些系统中，还可以用于示教时不同的速度再现。

利用装在控制盒上的按钮可以驱动机器人按需要的顺序进行操作。在示教盒中，每一个关节都有一对按钮，分别控制该关节在两个方向上的运动，有时还提供附加的最大允许速度控制。虽然为了获得最高的运行效率，人们一直希望机器人能实现多关节合成运动，但在示教盒示教的方式下却难以同时移动多个关节。

（2）机器人语言编程。机器人语言编程提供了一种通用的人与机器人之间的通信手段。它是一种专用语言，用符号描述机器人的运动，与常人的计算机编程语言相似，如 ABB 机器人的 RAPID 语言，以及川崎机器人的 AS 语言。

（3）离线编程。在计算机中建立设备、环境及工件的三维模型，在这样一个虚拟的环境中对机器人进行编程。机器人离线编程（Off Line Progamming，OLP）系统是机器人语言编程的拓展，它充分利用了计算机图形学的成果，建立机器人及其工作环境的模型，再利用一些规划算法，通过对图形的控制和操作在离线的情况下进行编程。

### 2.4.3 任务实施

接下来任务实施环节用能力源创新课程套件仿真工业机器人并编程模拟运行，设计一个工业机器人模型（见图 2-51）。

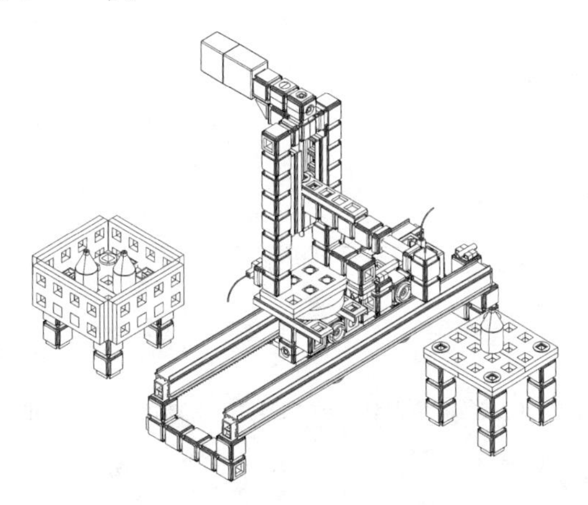

图 2-51 工业机械手模型

仿真的工业机器人要求可以升降、平移和旋转。工业机器人通过空间的运动将彩瓶从一个地方 A 搬运到指定的一个地方 B。

#### 2.4.3.1 方案设计

在图 2-51 中的驱动系统部分，工业机器人的平移是靠 0 号电机通过齿轮组减速，又通过齿轮与齿条的啮合，将旋转运动转为平移运动；旋转运动是通过 1 号电机齿轮组实现；升降是靠2号电机驱动丝杠实现。机器人抓取彩瓶是靠驱动电磁铁，通过磁力将带有铁块的彩瓶抓起来。

机械结构系统是靠能力源的各种结构件拼装实现，如正方体块、梁300mm、1号平板等；感受系统是靠旋转计数器来计算机器人各个姿态位置，以及控制各电机方向和速度；各自由度的复位位置的检测是靠磁敏传感器检测实现。

人机交互系统是控制器来实现，操作人员通过控制器显示屏当前状态，进行相应的操作，来实现人对机器的控制。

#### 2.4.3.2 材料准备

参考配套项目用料清单如图2-52所示。

图2-52 物料清单

#### 2.4.3.3 结构搭建

接下来介绍工业机器人搭建过程中的一些关键步骤，帮助同学们完成工业机器人的硬件拼接。其结构搭建模型分别如图2-53~图2-55所示。

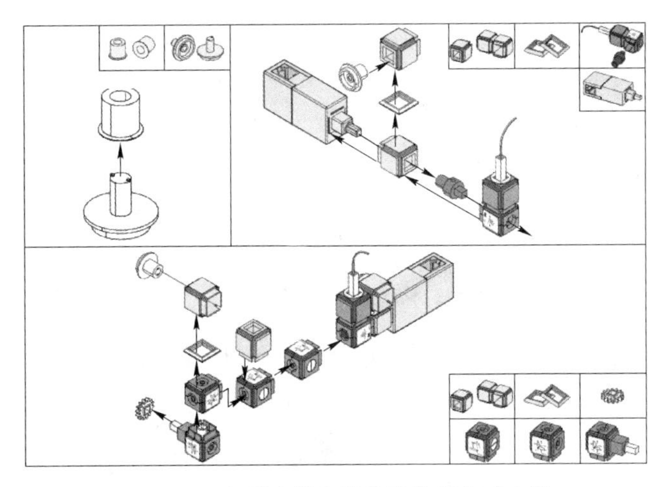

图 2-53　机器人搭配转数计数器和减速器

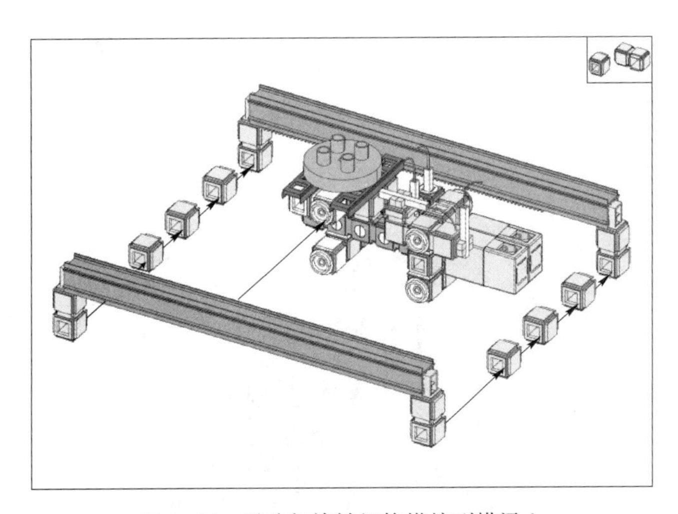

图 2-54　平移与旋转机构搭接到横梁上

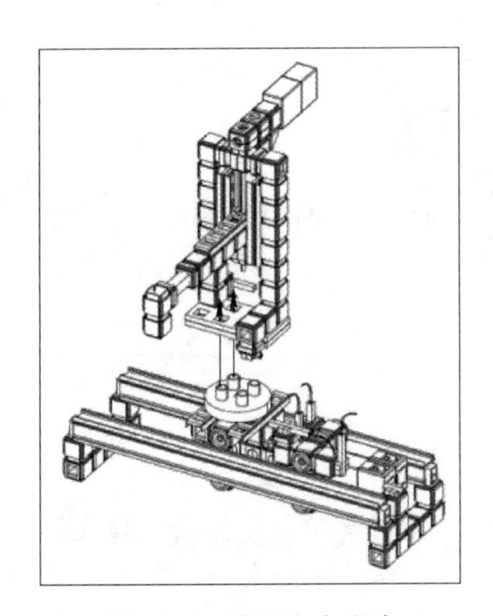

图 2-55　旋转平台搭建

工业机器人输入输出端口如图 2-56 和图 2-57 所示。

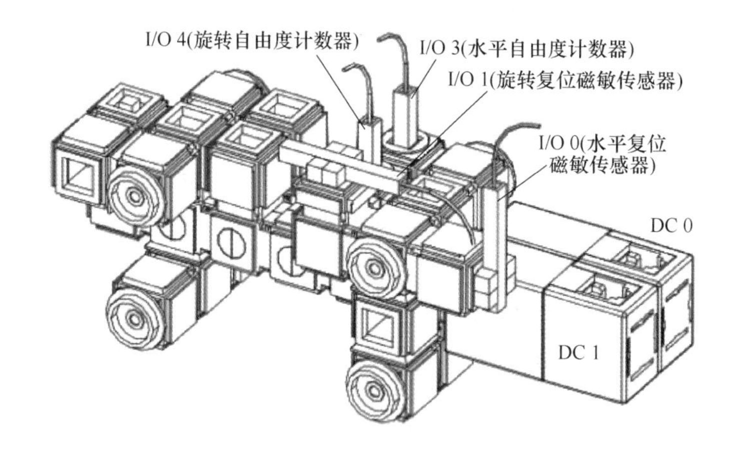


图 2-56　平移与旋转的信号接口

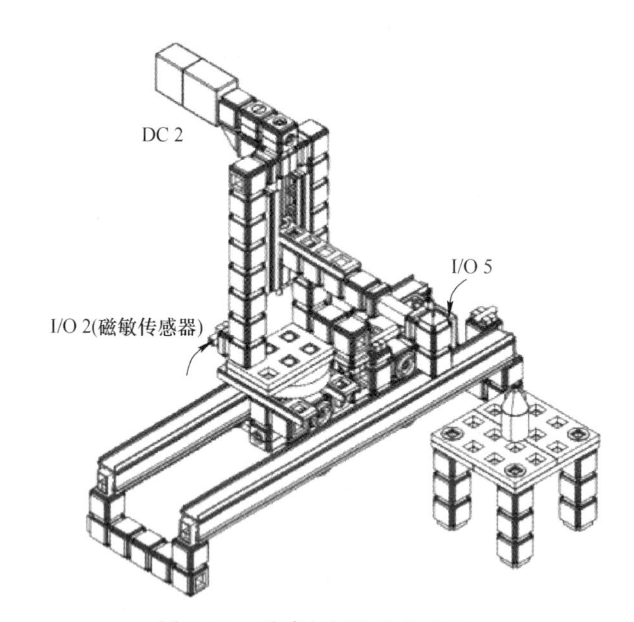


图 2-57　升降与抓取的信号接口

2.4.3.4　程序设计

**控制要求**

工业机器人的控制要求为：

程序运行后，升降、平移、旋转 3 个自由度都会首先运动到复位位置，此时电磁铁应刚好位于待搬运的彩瓶的上方。工作顺序依次为：电磁铁下降吸到彩瓶后抬起；通过旋转和平移将彩瓶搬运到目标处的上方；电磁铁下降并放下彩瓶。执行复位动作，进入下一循环，程序设计框图如图 2-58 所示。

**转数计数器的用法与特性**

能力源套件中旋转计数器是由一个磁敏传感器和一个带有磁铁的传动轴组成，如图 2-59 所示。

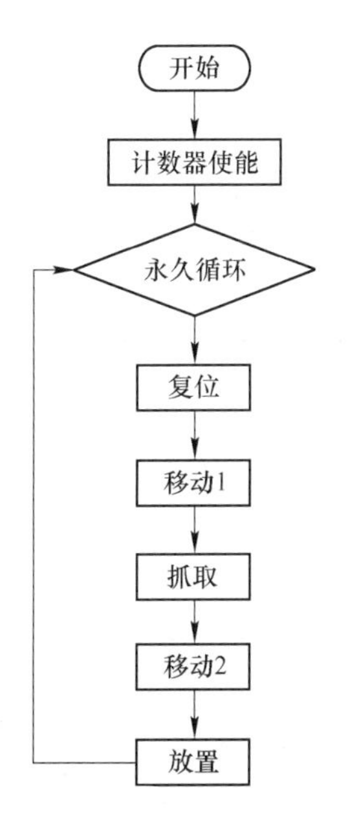


图 2-58　程序设计框图

图 2-59　旋转计数器结构

软件中的旋转计数器模块有四种功能，分别为计数器启动、清零、读取和终止。其中计数器的读取可进行逻辑判断。

根据图 2-60 设定条件，图 2-61 为当通道 4 计数数值大于 2000 时，马达 1 停止，否则马达 1 旋转。

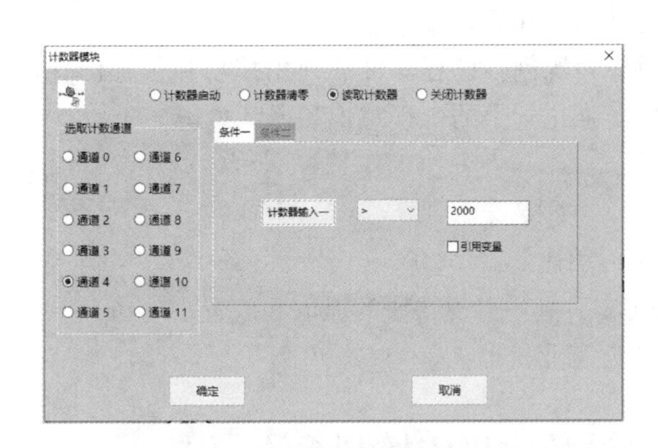


图 2-60　旋转计数器模块读取功能

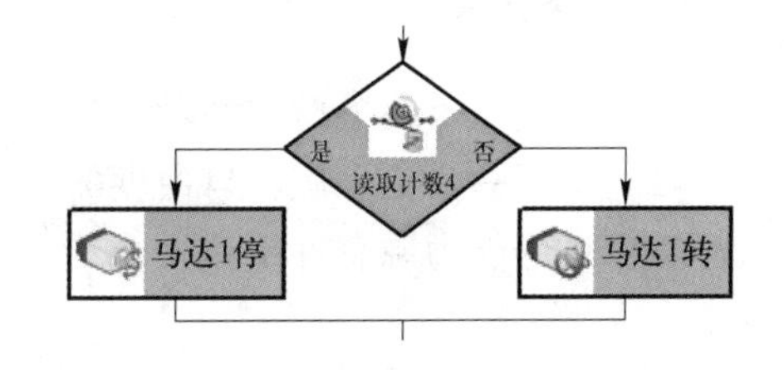


图 2-61　旋转计数器模块条件判断

在工业机器人中，系统可以根据旋转计数器来完成对马达的转向和速度的控制，并可以根据旋转计数器计算出各姿态的位置。

#### 2.4.3.5 调试记录

调试过程首先使用控制器上的“电机”界面测试，保证电机驱动的方向正确，从而保证每个自由度都可以顺利运动。

最后把数据填写到调试记录表中（见表 2-13）。

**表 2-13 调试记录表**

| 调试项目 | 调试过程记录 | 调试人员 |
| --- | --- | --- |
| | | |
| | | |

### 2.4.4 任务评价

任务评价表见表 2-14。

**表 2-14 任务评价表**

| 序号 | 评价内容 | 成绩占比 | 自评 | 师评 |
| --- | --- | --- | --- | --- |
| 1 | 工业机器人定义与分类 | 20 分 | | |
| 2 | 工业机器人结构组成 | 10 分 | | |
| 3 | 工业机器人核心技术 | 20 分 | | |
| 4 | 工业机器人项目搭建 | 10 分 | | |
| 5 | 工业机器人程序调试 | 40 分 | | |

工业机器人是一种具有自动操作和移动功能，能完成各种作业的可编程操作机，它具有仿生特征、柔性特征、智能特征和自动特征四大特征。工业机器人一般包括驱动系统、机械结构系统、机器人—环境交互系统、感受系统、人机交互系统和控制系统六大系统，其关键技术包括传感器技术、驱动技术、传动技术和编程技术。

通过构建工业机器人，进一步巩固了对能力源创新课程套件的使用技巧和工程仿真过程，有利于锻炼学生的创新思维和创新能力，强化学生的工程意识和团队意识。

### 2.4.5 任务拓展

（1）计算一下此模型中旋转部分的齿轮传动比。

（2）实现一个更多旋转自由度的工业机械手，并实现物块的抓取和放置。

（3）试用能力源套件组建一个平面关节型工业机器人。

## 任务 2.5 搭建与调试数控机床控制系统

### 2.5.1 任务引入

当有一个数控铣床需要改造时，先用能力源课程创新套件搭建一个仿真数控铣床，了解数控铣床的基本情况。仿真要求是：搭建的数控铣床要求具有 X、Y、Z 三个进给运动和一个主运动，并且对三个进给运动的位置进行检测。

## 2.5.2 相关知识

常见的数控机床主要有数控车床、数控铣床和加工中心。

数控车床是使用量最大、覆盖最广的一种数控机床，约占数控机床总数的 25%。数控车床主要用于进行车削加工，在车床上一般可以加工回转表面，如内外柱面、圆锥面、成形回转表面及螺纹面等，在数控车床上还可以加工高精度的曲面与断面螺纹。图 2-62 分别为卧式和立式数控车床，数控车床正在加工外圆柱面的实物图如图 2-63 所示，图 2-64 中所示的零件都是通过数控车床加工的。

图 2-62　卧式和立式数控车床

图 2-63　车床正在加工

图 2-64　数控车削加工的零件

数控铣床是使用计算机数字信号控制的铣床。它可以加工由直线和圆弧两种几何要素构成的平面轮廓，也可以直接用逼近法加工非圆曲线构成的平面轮廓，还可以加工立体曲面和空间曲线，如叶片、螺旋桨。典型数控铣床如图 2-65 所示，图 2-66 是采用数控铣床加工的零件，另外铣床还有孔加工的功能。

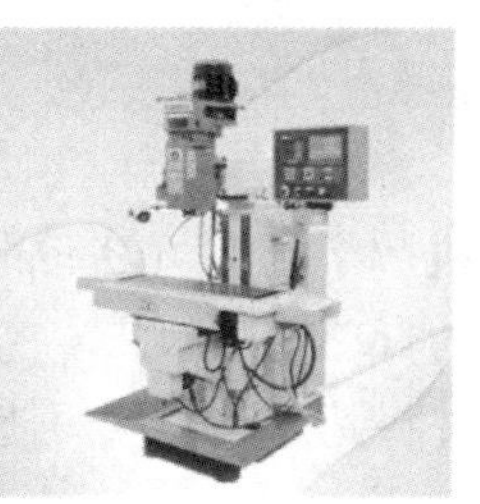

图 2-65　典型数控铣床

图 2-66　数控铣床加工的零件

加工中心是目前世界上产量最高、应用最广泛的数控机床之一。加工中心带有刀库和自动换刀装置，在加工时，工件经过一次装夹后，就能连续地对工件各加工表面自动完成铣、镗、钻、铰及攻丝等各种工序的加工。就中等加工难度的批量工件而言，其效率是普通设备的 5~10 倍，特别是它能完成许多普通设备不能完成的加工任务，对形状较复杂、精度要求高的单件加工或中小批量、多品种生产更为适用。数控加工中心如图 2-67 所示，图 2-68 是采用数控铣床加工的零件。

图 2-67　数控加工中心

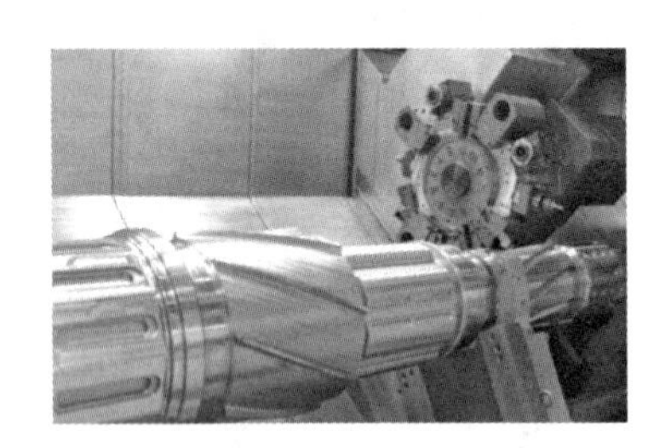

图 2-68　数控加工中心加工的零件

### 2.5.2.1　定义与分类

**数控机床的定义**

数控机床是计算机数字控制（Computer Numerical Control，CNC）机床的简称，是一种装有程序控制系统的自动化机床。该控制系统能够逻辑地处理具有控制编码或者其他符号指令规定的程序，并将其译码，从而使机床动作并加工零件。

**数控机床的分类**

数控机床可分为以下几类：

（1）按照工艺用途可分为三种类型：金属切削类、金属成型类和特种加工机床；

（2）按照驱动方式可分为点位控制、直线控制和轮廓控制机床。

点位控制机床的特点是机床的运动部件只能够实现从一个位置到另一个位置的精确运动，在运动和定位过程中不进行任何加工工序。典型的点位控制数控机床如数控钻床。

直线控制机床的特点是机床的运动部件不仅要实现一个坐标位置到另一个坐标位置的精确移动和定位，而且能实现平行于坐标轴的直线进给运动，或者控制两个坐标轴实现斜线进给运动。典型的直线控制数控机床如早期数控车床。

轮廓控制机床的特点是机床的运动部件能够实现两个坐标轴同时进行联动控制。它不仅要求控制机床运动部件的起点与终点坐标位置，还要求控制整个加工过程中每一点的速度和位移量，即要求控制运动轨小，将零件加工成在平面内的直线、曲线或在空间的曲面。典型的轮廓控制数控机床如加工中心。

（3）按照控制方式可以分为开环控制、半闭环控制和闭环控制机床（见图 2-69）。

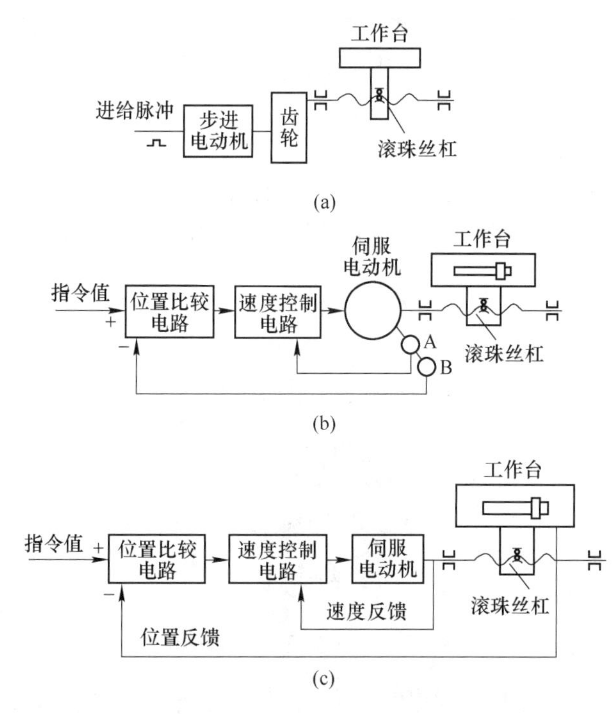


图 2-69　数控机床三种控制方式

（a）开环控制；（b）半闭环控制；（c）闭环控制

开环控制是指不带位置反馈装置的控制方式。

半闭环控制是指在开环控制伺服电动机轴上装有角位移检测装置，通过检测伺服电动机的转角间接地检测出运动部件的位移反馈给数控装置的比较器，与输入的指令进行比较，用差值控制运动部件。

闭环控制是指在机床最终的运动件的相应位置直接安装直线或回转式检测装置，将直接测量到的位移或角位移值反馈到数控装置的比较器中与输入指令位移量进行比较，用差值控制运动部件，使运动部件严格按实际需要的位移量运动。

### 2.5.2.2 结构组成

数控机床主要是由数控系统、驱动系统、辅助控制装置、机床本体等部分组成。

**数控系统**

数控系统是数控机床的核心和中心环节，包括硬件（印制电路板、显示器等）以及相应的软件。它用于输入数字化的零件程序，并完成输入信息的存储、数据的变换、插补运算以及实现各种控制功能。

**驱动系统**

驱动系统是数控机床执行机构的驱动部件，包括主驱动单元、进给驱动单元、主轴电机及进给电机等。它是在数控装置的控制下通过电气或电液伺服系统实现主轴和进给驱动。当几个进给联动时，可以完成定位、直线、平面曲线和空间曲线的加工。

**辅助控制装置**

辅助控制装置是指数控机床的一些必要的配套部件，用以保证数控机床的运行，如冷却、排屑、润滑、照明、监测等。它包括液压和气动装置、排屑装置、交换工作台、数控转台和数控分度头，还包括刀具及监控检测装置等。

**机床本体**

机床本体是数控机床的主体，由机床的基础大件（如床身、底座）和各运动部件（如工作台、床鞍、主轴等）所组成。立式数控铣床的结构如图 2-70 所示。

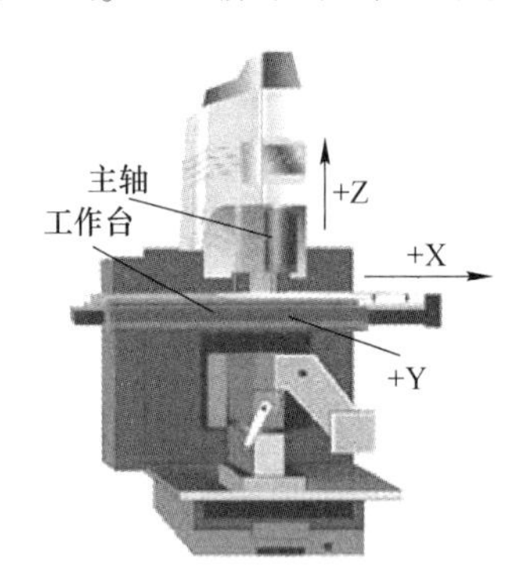


图 2-70 立式数控铣床的结构

### 2.5.2.3 核心技术

数控机床主要有四大核心技术，分别为机床传动技术、数控系统、伺服驱动系统和机床检测装置。

**数控机床的传动技术**

数控机床的驱动系统包含主轴驱动和进给驱动，相应的可以分为主运动和进给运动，主运动和进给运动都有传动部分。

（1）数控机床的主运动。数控机床的主运动是指生产切屑的传动运动，如数控车床上的主轴带动工件的旋转运动和立式数控铣床上主轴带动铣刀的旋转运动。数控机床的主运动是通过主运动电机拖动的，通常需要借助传动部件转换为机床的主运动，经常采用的传动方式包括：

1）调速电机。直接采用调速电机，大大简化了主轴箱体与主轴的结构，有效地提高了主轴部件的刚度。但主轴输出扭矩小，电机发热对主轴的精度影响大。

2）变速齿轮。变速齿轮是大中型数控机床采用较多的一种方式，通过少数几对齿轮减速，增大了输出扭矩，满足主轴对输出扭矩特性的要求。一部分小型数控机床也采用这种传动方式，以获得强力切屑时所需要的扭矩。

3）皮带传动。皮带传动主要应用在小型数控机床上，可以避免齿轮传动时引起的振动与噪声，但它只能使用满足扭矩特性要求的主轴。

（2）数控机床的进给运动。典型的数控机床通常采用闭环控制进给系统，由位置比较、放大元件、驱动单元、机械传动装置和检测反馈元件等几部分组成。进给系统接受控制系统下达的对每个运动坐标轴的速度指令，经速度与转矩调节输出驱动系统信号驱动进给电机转动，实现机床坐标轴运动，同时接受速度反馈信号实施速度闭环控制。机械传动装置是指将驱动源（进给电机）的旋转运动变为工作台各坐标轴的直线运动，进给传动装置有：

1）齿轮。齿轮传动是应用非常广泛的一种机械传动，各种机床的传动装置中几乎都有齿轮传动，通过齿轮传动将高转速低转矩的伺服电机的输出改变为低转速大转矩的执行件的输入。

2）同步齿形带。同步齿形带传动是一种新型的带传动，利用齿形带的齿形与带轮的轮齿依次啮合传递运动和动力，因而兼有带传动、齿轮传动及链传动的优点。且同步齿形带无相对滑动，平均传动比较准确，传动精度高，而且齿形带的强度高、厚度小、质量轻，故可用于高速传动。齿形带无须特别张紧，作用在轴和轴承上的载荷小、传动效率高，现已在数控机床中广泛应用。

3）滚珠丝杠螺母副。为了提高进给系统的灵敏度、定位精度和防止爬行，必须降低数控机床进给系统的摩擦，并减少静动摩擦系数之差。因此，行程不太长的直线运动机构常用滚珠丝杠副。

**数控机床的数控系统**

数控机床的数控系统是数控机床的核心部分，包括计算机系统、伺服驱动装置、位置检测装置、可编程控制器（PLC）和接口电路（见图 2-71）。

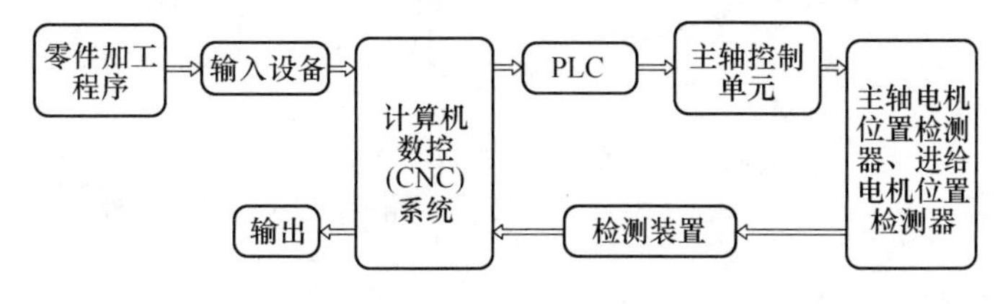


图 2-71　计算机数控系统组成

从自动化控制的角度来看，CNC 系统是一种位置控制系统，其本质上是以多执行部件的位置量为控制对象，并使其协调运动的自动控制系统，也是一种配有专用操作系统的计算机控制系统。从外部特征来看，数控系统是由硬件和软件两部分组成的。

计算机数控系统的硬件结构分为计算机基本系统、设备支持层和设备层。其中计算机基本系统是指计算机系统、显示设备、输入/输出设备等；设备支持层是指介于计算机和机床之间的人机交互编程系统、运动控制设备等。

从本质特征来看，数控系统软件是具有实时性和多任务性的专用操作系统，由数控机床管理软件和数控控制软件两部分组成，它是数控机床系统活的灵魂。数控系统的软件在硬件的支持下，合理的组织、管理整个系统的各项工作，实现各种数控功能，使数控机床按照操作者的要求，有条不紊地进行加工。数控系统的硬件和软件构成了数控系统平台。

**数控机床的伺服驱动系统**

数控机床伺服驱动系统主要有两种，一种是进给伺服系统，它控制机床各坐标轴的切削进给运动，以直线运动为主；另一种是主轴伺服系统，它控制主轴的切削运动，以旋转运动为主。

伺服驱动系统按控制原理可分分开环伺服系统、闭环伺服系统和半闭环伺服系统；按使用的执行元器件可分为步进电机驱动系统、直流伺服电机和交流伺服电机；按控制量性质可分为位置伺服系统和速度伺服系统。伺服系统常用的电机包括步进电机、直流伺服电机和交流伺服电机。

**数控机床的检测装置**

数控机床的检测装置以位置检测装置为主，下面主要讲解位置检测。

数控机床伺服系统中采用的位置检测装置分为直线型和旋转型两种。直线型位置检测装置用来检测运动部件的直线位移量，旋转型位置检测装置用来检测回转部件的转动位移量。在闭环系统中，位置检测的主要作用是检测位移量，并发出反馈信号和数控装置发出的指令信号相比较，若有偏差，经放大后控制执行部件，使其向消除偏差的方向运动直至偏差等于零为止。

常用的位置检测装置包括光栅位置检测装置和光电编码器（见图 2-72 和图 2-73）。

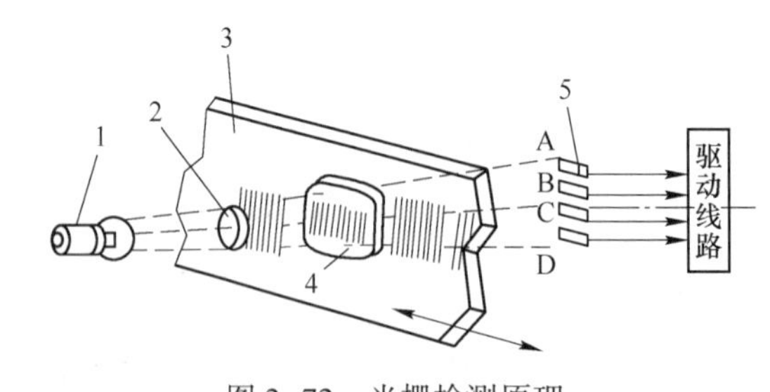


图 2-72 光栅检测原理

1—光源；2—透镜；3—标尺光栅；4—指示光栅；5—光电元件

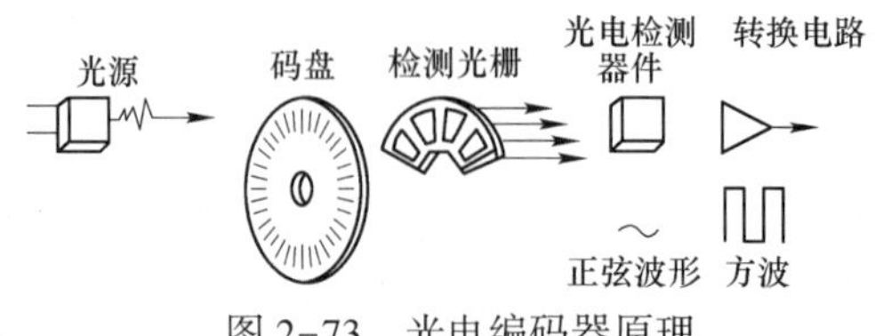


图 2-73 光电编码器原理

光栅作为数控机床的检测装置，已有几十年的历史，可用于检测长度、角度、速度、加速度、震动和爬行等，它是数控机床闭环系统用得较多的一种检测装置。光电编码器用来测量旋转运动的角位移，它是利用光电原理把机械角位移变成脉冲电信号，是一种使用广泛的角位移传感器。

在本次的项目任务中，所用到的位置检测装置为磁敏传感器和旋转计数器。两者配合使用，可通过磁敏传感器来检测设备复位回零状态，通过旋转计数器来检测设备在运行中的位置。

### 2.5.3 任务实施

接下来用能力源创新课程套件仿真数控铣床，并编程模拟运行。设计的数控铣床模型如图 2-74 所示。其中，数控铣床要求具有 X、Y、Z 三个进给运动，一个主运动，对三个进给运动的位置需要有检测。

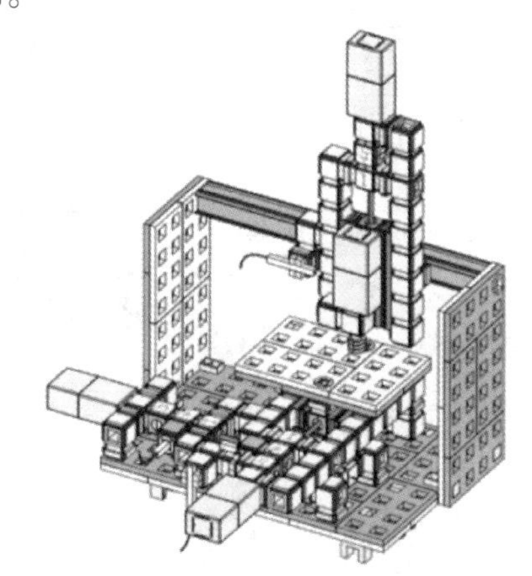

图 2-74　数控铣床模型

#### 2.5.3.1 方案设计

数控系统是数控机床的核心，用来输入数字化的零件程序。本模型中数控系统就是能力源控制器，通过操作控制器实现对 X、Y、Z 三轴的运动控制。

驱动系统是数控机床执行机构的驱动部件，包括主轴驱动和进给驱动。本模型采用电机驱动的方式，0 号电机驱动丝杠，将旋转运动转化成平移，实现数控铣床 X 轴方向的运动；1 号电机采用同样的方式，实现数控铣床 Y 方向的运动；2 号电机也采用同样的方式，实现数控铣床 Z 方向的运动。以上三个电机驱动实现数控机床的进给驱动。3 号电机实现主轴的旋转，同时实现数控的主轴驱动。

辅助控制系统是系统采用一些必要的配套部件，用于保证数控机床的运行。本模型采用 3 个磁敏传感器来实现进给系统 X、Y、Z 轴的复位位置检测；采用 2 个旋转计数器来实现 X、Y 轴的位置检测。

机床本体是数控机床的主体，由机床的基础大件（如床身、底垫）和各运动部件（如工作台、床鞍、主轴等）组成。本模型采用能力源各种结构件来组成铣床本体，如正立方体块、梁、4 号平板等。

### 2.5.3.2 材料准备

参考配套项目用料清单如图 2-75 所示。

图 2-75 物料清单

### 2.5.3.3 结构搭建

接下来介绍工业机器人搭建过程中的一些关键步骤，帮助同学们完成数控铣床的硬件拼接。其步骤模型分别如图 2-76~图 2-79 所示。

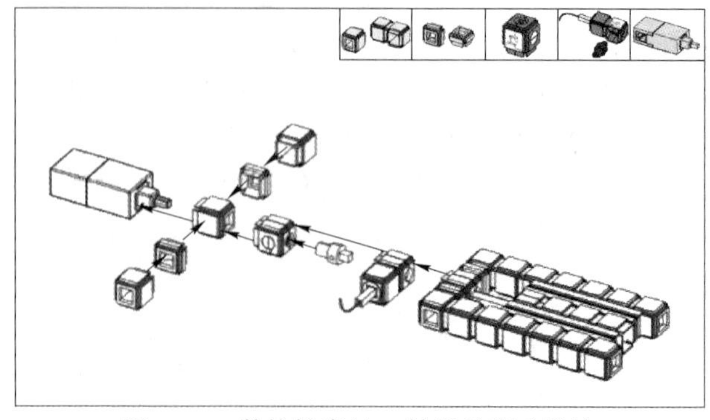

图 2-76 数控铣床的 X 轴进给驱动拼接

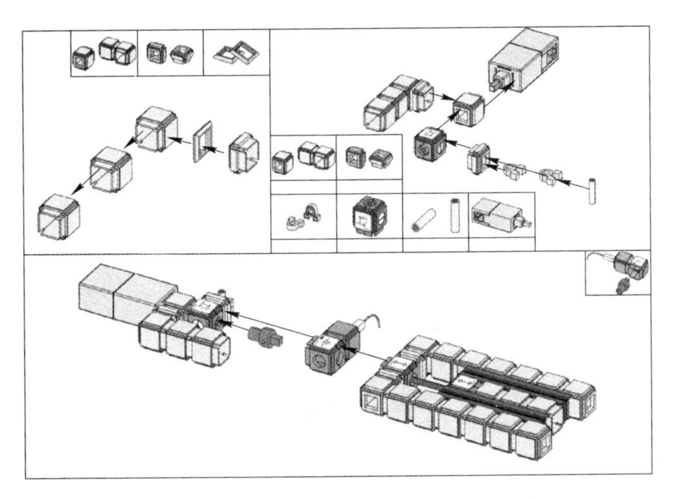

图 2-77 数控铣床的 Y 轴进给驱动拼接

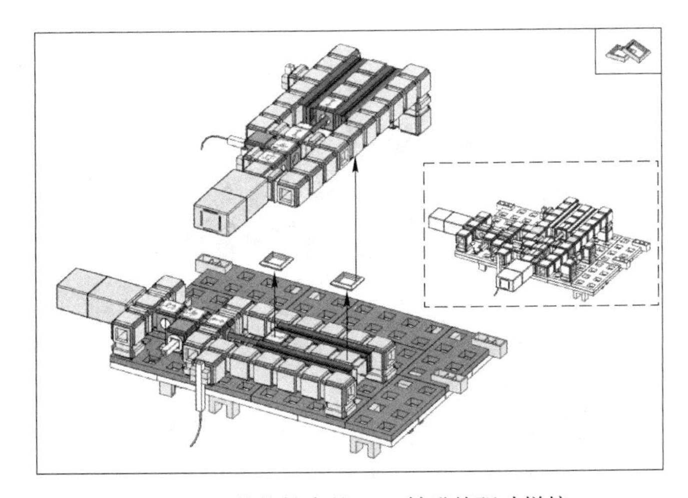

图 2-78 数控铣床的 X/Y 轴进给驱动拼接

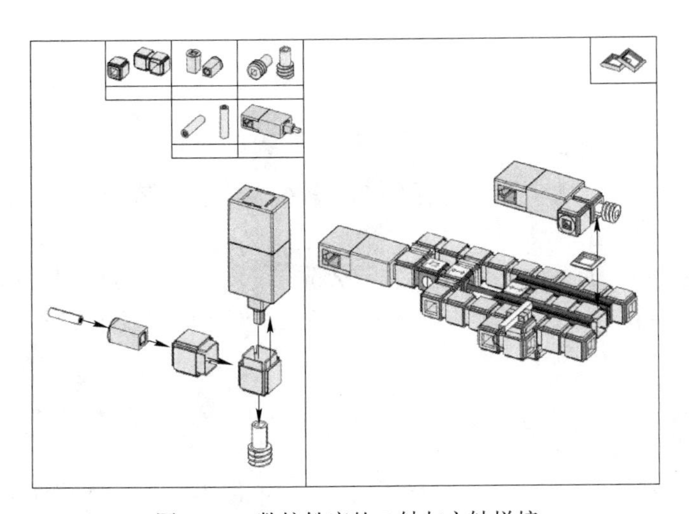

图 2-79 数控铣床的 Z 轴与主轴拼接

数控机床输入输出端口分别如图 2-80 和图 2-81 所示。

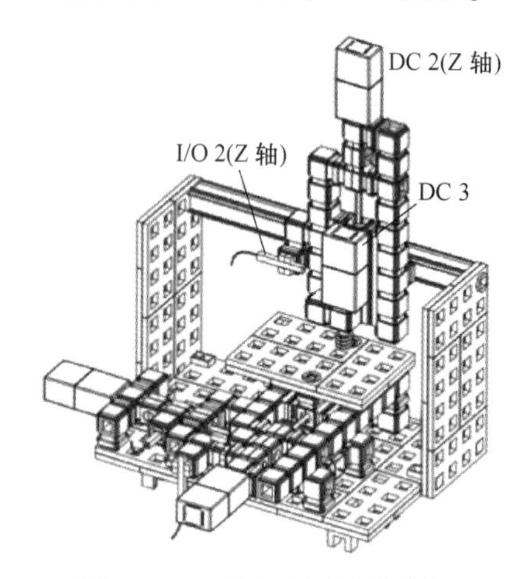


图 2-80　Z 轴与主轴的信号接口

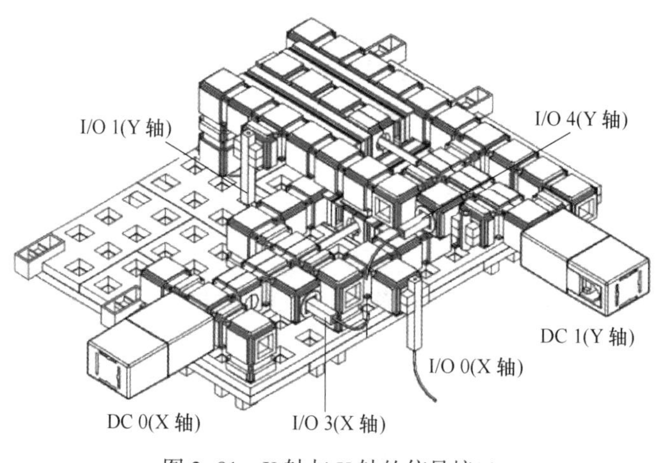


图 2-81　X 轴与 Y 轴的信号接口

#### 2.5.3.4　程序设计

数控铣床的控制要求为：

（1）把 3 个丝杠定义为直角坐标系的 3 个轴，从下往上依次为 X、Y、Z 轴，复位位置为原点，伸出方向为正。

（2）程序运行后，首先进行复位动作（Z 先抬起，X、Y 可以同时运动）。复位完成后鸣叫提示添加物料，然后进行第一个点的加工，本例程序仅做了 1 个点的加工动作，如图 2-82 所示。

#### 2.5.3.5　调试记录

调试过程首先使用控制器上的“电机”界面测试，保证电机驱动的方向正确，从而保证每个自由度都可以顺利运动。之后把数据填写到表中（见表 2-15）。

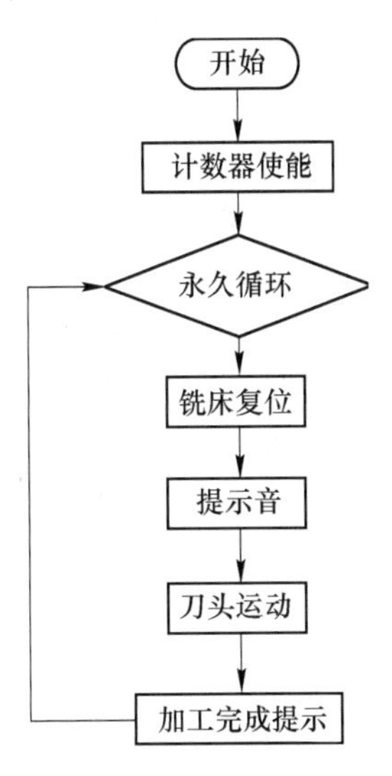


图 2-82　程序设计框图

**表 2-15　调试记录表**

| 调试项目 | 调试过程记录 | 调试人员 |
| --- | --- | --- |
| | | |
| | | |

### 2.5.4　任务评价

任务评价表见表 2-16。

**表 2-16　任务评价表**

| 序号 | 评价内容 | 成绩占比 | 自评 | 师评 |
| --- | --- | --- | --- | --- |
| 1 | 数控机床定义与分类 | 20 分 | | |
| 2 | 数控机床结构组成 | 10 分 | | |
| 3 | 数控机床核心技术 | 20 分 | | |
| 4 | 数控机床项目搭建 | 10 分 | | |
| 5 | 数控机床程序调试 | 40 分 | | |

数控机床主要由数控系统、伺服驱动系统、检测反馈装置、辅助控制装置、机床本体等部分组成。传动、数控系统、驱动技术、传感器技术和电气技术是数控机床的五大核心技术。数控机床中传动主要有丝杠传动、同步带传动和齿轮传动。数控系统是数控机床的中枢，它将接到的全部功能指令进行解码和运算，然后有序地发出各种需要的运动指令和各种机床功能的控制指令，直至运动和功能结束。主运动驱动和进给运动的驱动分别通过控制益对不同的电机实现控制。传感器主要实现对机床速度和位置的反馈。

通过采用能力源创新课程套件搭建数控机床系统，学习工程实践创新项目方案设计、材料准备、部件安装和程序编写与调试，进一步熟悉能力源控制器及其 VJC 开发系统的应用，掌握数控机床系统的工程实现主要技术环节。

### 2.5.5 任务拓展

（1）将刀头换成毛笔，通过移动合理驱动控制 X、Y、Z 轴的移动，在 X、Y 工作台上书写简单文字一、二、中、土等。

（2）思考如何控制 X、Y 轴（可以增加元器件），书写复杂的文字，例如“停止”。

（3）想一想通过增加怎样的器件及适当的算法来显示 X、Y、Z 的移动速度。

（4）想一想怎么把这台数控机床改造成数控雕刻机。

# 项目3　工业机器人的认识与操作

早在几千年前的神话故事中，类似机器人的概念就已经出现。但直到近代，机器人技术（尤其是工业机器人技术）才得到了飞速发展。机器人要代替人进行作业，首先要动起来，让机器人动起来的方法有多种，可以使用示教器进行手动控制，也可以编写 RAPID 程序自动控制，或者用外部信号来控制机器人。对于初学者来说，手动操作机器人是学习工业机器人的基础。

## 任务3.1　认识工业机器人

### 3.1.1　任务引入

工业机器人是典型的机电一体化设备，是面向工业领域的多关节机械手或多自由度的机器人。工业机器人的出现是人类利用机械推动社会发展的一个里程碑。通过本任务的学习，熟悉工业机器人的历史发展、应用以及工业机器人的典型结构、安全注意事项等。

### 3.1.2　相关知识

工业机器人（Industrial Robot，IR）是用于工业生产环境的机器人总称。1954 年美国戴尔沃最早提出了工业机器人的概念，并申请了专利；1959 年，第一台工业机器人在美国诞生，开创了机器人发展的新纪元。经过 60 年的发展，工业机器人在性能和用途等方面都有了很大的变化。现代工业机器人的结构越来越合理、控制越来越先进、功能越来越强大，正逐渐向着具有行走能力、多种感知能力和对作业环境较强的自适应能力的方向发展。

目前，工业机器人在工业生产各领域的应用也越来越广泛，其中，汽车制造业、电子电气业、金属制品及加工业是工业机器人主要应用领域。根据功能与用途，工业机器人通常可分为加工、装配、搬运、包装 4 大类。加工机器人是直接用于工业产品加工作业的工业机器人；装配机器人是将不同的零件或材料组合成组件或成品的工业机器人；搬运机器人是从事物体移动作业的工业机器人的总称；包装机器人是用于物品分类、成品包装、码垛的工业机器人。

目前，日本和欧盟是全球工业机器人的主要生产基地，主要企业有日本的 FANUC（发那科）、YASKAWA（安川）、瑞士和瑞典的 ABB、德国的 KUKA（库卡）等，这四个企业被誉为工业机器人的“四大家族”。本书就是以四大家族中的 ABB 工业机器人为载体开展教学的。

工业机器人最显著的特点有：

（1）可编程。生产自动化的进一步发展是柔性启动化。工业机器人可随其工作环境变化的需要再编程，因此它在小批量、多品种、具有均衡高效率的柔性制造过程中能发挥很好的功用，是柔性制造系统中的一个重要组成部分。

（2）拟人化。工业机器人在机械结构上有类似人的腰部、大臂、小臂、手腕、手爪等部分，在控制上有计算机。此外，智能化工业机器人还有许多类似人类的“生物传感器”，如皮肤型接触传感器、力传感器、负载传感器、视觉传感器、声觉传感器和语言功能等。传感器提高了工业机器人对周围环境的自适应能力。

（3）通用性。除了专门设计的专用的工业机器人外，一般工业机器人在执行不同的作业任务时具有较好的通用性。比如，更换工业机器人手部末端操作器（手爪、工具等）便可执行不同的作业任务。

### 3.1.3 任务实施

#### 3.1.3.1 工业机器人典型结构

**直角坐标工业机器人**

直角坐标机器人（见图3-1）一般做2~3个自由度运动，每个运动自由度之间的空间夹角为直角，可实现自动控制，可重复编程，所有的运动均按程序运行。直角坐标工业机器人可在恶劣的环境下工作，同时便于操作和维修。

**平面关节型工业机器人**

平面关节型机器人（见图3-2）又称为SCARA工业机器人，是圆柱坐标机器人的一种形式。SCARA机器人有三个旋转关节，其轴线相互平行，在平面内进行定位和定向；还有一个移动关节，用于完成末端件垂直于平面的运动。SCARA工业机器人精度高，动作范围较大，坐标计算简单，结构轻便，响应速度快，但负载较小，主要用于电子、分拣等领域。

图3-1　直角坐标工业机器人

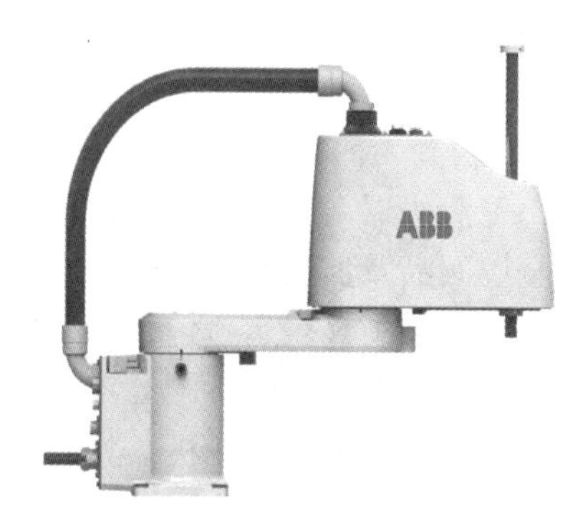


图3-2　平面关节型工业机器人

**并联工业机器人**

并联机器人（见图3-3）又称DELTA工业机器人，属于高速、轻载的工业机器人。一般通过示教编程或视觉系统捕捉目标物体，由三个并联的伺服轴确定夹具中心（TCP）的空间位置，实现目标物体的运输、加工等操作。Delta机器人主要用于食品、药品和电子产品等的加工和装配。

**串联工业机器人**

串联工业机器人（见图 3-4）拥有四个或四个以上旋转轴，其中六个轴是最普通的形式，类似于人类的手臂，有很高的自由度。该机器人可以自由编程，生产效率高，能代替人完成有害身体健康的复杂工作，应用于装货、卸货、喷漆、表面处理、测试、测量、弧焊、点焊、包装、装配、切屑机床、固定、特种装配操作、锻造、铸造等领域。

本书就是以串联工业机器人作为对象进行讲解的。

**协作工业机器人**

在传统的工业机器人逐渐取代单调、重复性高、危险性强的工作之时，协作机器人（见图 3-5）也将会慢慢渗入各个工业领域，与人共同工作。这将引领一个全新的机器人与人协同工作时代的来临。随着工业自动化的发展，人们发现需要协助型的工业机器人配合人来完成工作任务，这样比工业机器人的全自动化工作站具有更好的柔性和成本优势。

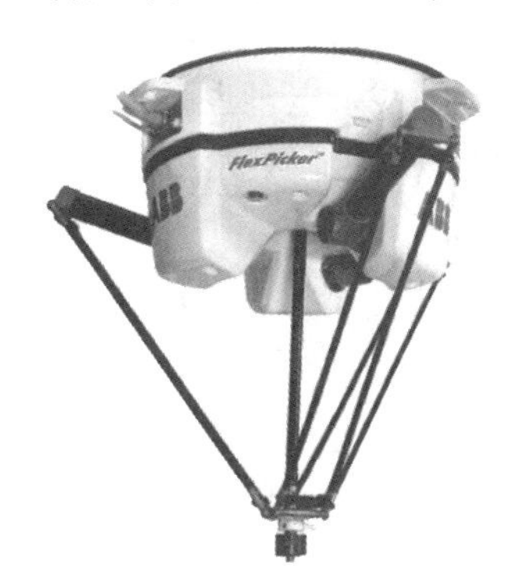

图 3-3　并联工业机器人

图 3-4　串联工业机器人

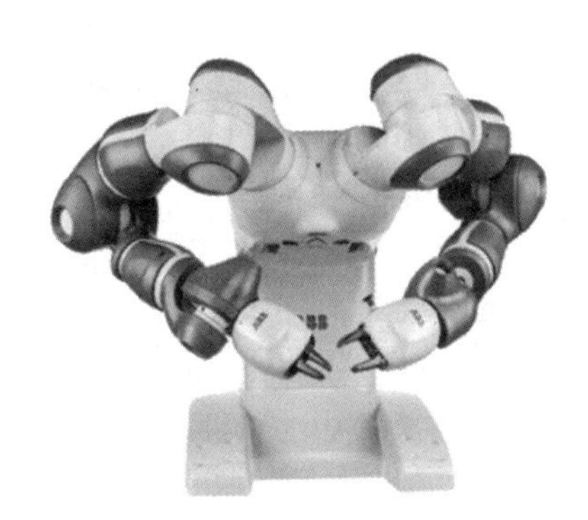
图 3-5　协作工业机器人

### 3.1.3.2　注意事项

工业机器人安全注意事项包括：

（1）关闭总电源！在进行机器人的安装、维修和保养时切记要将总电源关闭。带电作业可能会产生致命性后果。如不慎遭高压电击可能会导致心脏停搏、烧伤或其他严重伤害。

（2）与机器人保持足够安全距离！在调试与运行机器人时，它可能会执行一些意外的或不规范的运动。并且，所有的运动都会产生很大的力量，从而可能严重伤害个人或损坏机器人工作范围内的任何设备。因此时刻警惕与机器人保持足够的安全距离。

（3）静电放电危险！静电放电（ESD）是电势不同的两个物体间的静电传导，它可以通过直接接触传导，也可以通过感应电场传导。搬运部件或部件容器时，未接地的人员可能会传导大量的静电荷。这一放电过程可能会损坏敏感的电子设备。所以有此标识的情况下，要做好静电放电防护。

（4）紧急停止！紧急停止优先于任何其他机器人控制操作，它会断开机器人电机的驱动电源，停止所有运转部件，并切断由机器人系统控制且存在潜在危险的功能部件的电源。出现下列情况时请立即按下任意紧急停止按钮：

1）机器人运行中，工作区域内有工作人员。

2）机器人伤害了工作人员或损伤了机器设备。

（5）灭火！发生火灾时，请确保全体人员安全撤离后再行灭火，应首先处理受伤人员。

当电气设备（例如机器人或控制器）起火时使用二氧化碳灭火器，切勿使用水或泡沫。

（6）工作中的安全！机器人速度慢，但是很重，并且力度很大。运动中的停顿或停止都会产生危险。即使可以预测运动轨迹，但外部信号有可能改变操作，会在没有任何警告的情况下，产生料想不到的运动。

因此，当进入保护空间时，务必遵循以下安全条例：

1）如果在保护空间内有工作人员，请手动操作工业机器人系统。

2）当进入保护空间时，请准备好示教器，以便随时控制工业机器人。

3）注意旋转或运动过的工具，例如切削工具和锯。确保在接近工业机器人之前，这些工具已经停止运动。

4）注意工件和工业机器人系统的高温表面。工业机器人电动机长期运转后温度很高。

5）注意夹具并确保夹好工件。如果夹具打开，工件会脱落，并导致人员伤害或设备损坏。夹具非常有力，如果不按照正确方法操作，也会导致人员伤害。

6）注意液压、气压系统以及带电部件。即使断电，这些电路上的残余电量也很危险。

（7）示教器的安全！示教器是一种高品质的手持式终端，它配备了高灵敏度的一流电子设备。为避免操作不当引起的故障或损害，请在操作时遵循以下说明：

1）小心操作。不要摔打、抛掷或重击示教器。在不使用该设备时，将它挂到专门存放它的支架上，以防意外掉到地上。

2）示教器的使用和存放应避免被人踩踏电缆。

3）切勿使用锋利的物体操作触摸屏，应使用手指或触摸笔去操作示教器触摸屏。

4）定期清洁触摸屏。

5）切勿使用溶剂、洗涤剂或擦洗海绵清洁示教器，应使用软布蘸少量水或中性清洁剂清洁。

6）没有连接 USB 设备时务必盖上 USB 端口的保护盖。

（8）手动模式下的安全！在手动减速模式下，机器人只能减速（250mm/s 或更慢）操作（移动）。只要在安全保护空间之内工作，就应始终以手动速度进行操作。手动全速模式下，机器人以程序预设速度移动。手动全速模式应仅用于所有人员都位于安全保护空间之外时，而且操作人员必须经过特殊训练，深知潜在的危险。

（9）自动模式下的安全！自动模式用于在生产中运行机器人程序。在自动模式操作情况下，常规模式停止（GS）机制、自动模式停止（AS）机制和上级停止（SS）机制都将处于活动状态。

### 3.1.4 任务评价

任务评价表见表 3-1。

**表 3-1 任务评价表**

| 序号 | 评价内容 | 成绩占比 | 自评 | 师评 |
|---|---|---|---|---|
| 1 | 工业机器人的发展和应用 | 15 分 | | |
| 2 | 工业机器人的特点 | 15 分 | | |
| 3 | 工业机器人的典型结构 | 40 分 | | |
| 4 | 操作工业机器人的安全注意事项 | 30 分 | | |

（1）根据工业机器人的功能与用途，其主要产品大致可分为加工、装配、搬运和包装4大类。

（2）工业机器人主要企业有日本的FANUC（发那科）、YASKAWA（安川）、瑞士和瑞典的ABB以及德国的KUKA（库卡）等。它们被誉为工业机器人的“四大家族”。

（3）工业机器人最显著的特点有可编程、拟人化和通用性。

（4）工业机器人的典型结构有：直角坐标工业机器人、平面关节型工业机器人、并联工业机器人、串联工业机器人、协作型工业机器人等。

（5）工业机器人安全注意事项。

### 3.1.5 任务拓展

ABB在世界范围内安装了17.5万多台机器人。以下是ABB机器人部分型号的介绍。

#### 3.1.5.1 IRB 120

IRB 120（见图3-6）是迄今为止ABB制造的最小的六轴机器人，是ABB新型第四代机器人家族的最新成员。IRB 120具有敏捷、紧凑、轻量的特点，控制精度与路径精度俱优，主要应用在物流搬运、装配等工作。IRB 120仅25kg，荷重3kg（垂直腕为4kg），工作范围达580mm。如图3-7所示。

图3-6 IRB 120

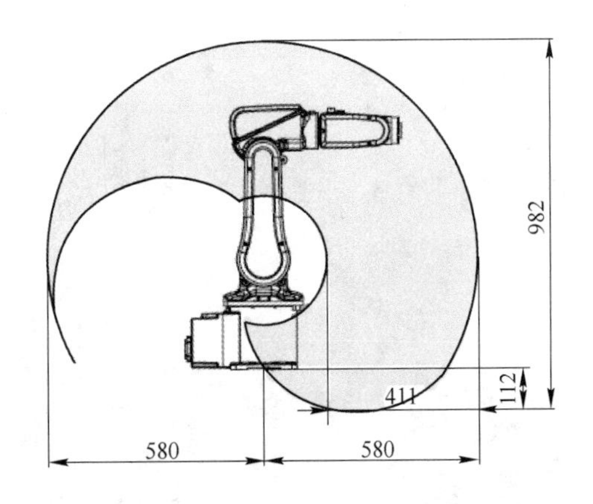


图3-7 IRB 120的工作范围

#### 3.1.5.2 IRB 260

IRB 260（见图3-8）机器人主要针对包装应用设计和优化，机身小巧，能集成于紧凑型包装机械中。工作范围（见图3-9）更靠近底座，最大限度缩小了占地面积。该机器人采用四轴运行设计，不但胜任各类包装作业，更是高产能和高柔性的保证。

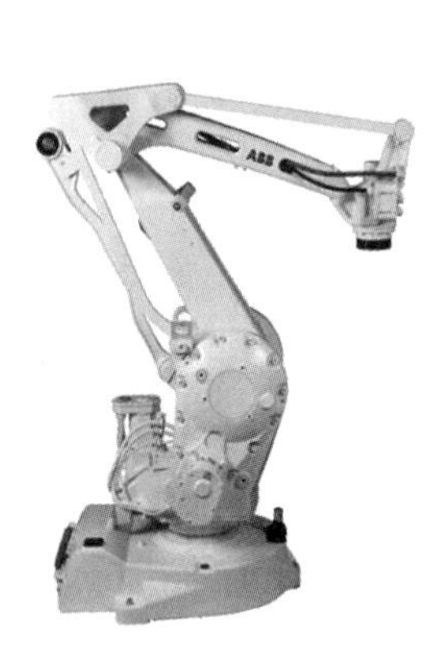

图 3-8　IRB 260

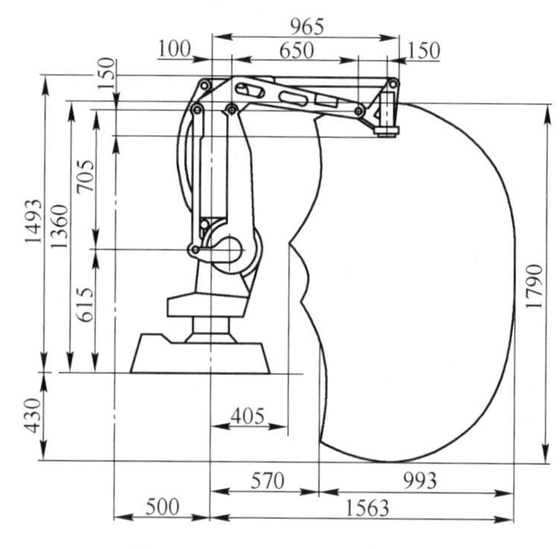


图 3-9　IRB 260 工作范围

### 3.1.5.3　IRB 360

近 15 年来，ABB 的 IRB 360（见图 3-10）拾料和包装技术一直处于领先地位。IRB 360 占地面积小（见图 3-11）、速度快、柔性强、负载大（荷重 8kg），采用可冲洗的卫生设计，出众的跟踪性能，集成视觉软件，步进式传送带同步集成控制。IRB 360 包括紧凑型、标准型、高载荷型、长臂型四个系列。

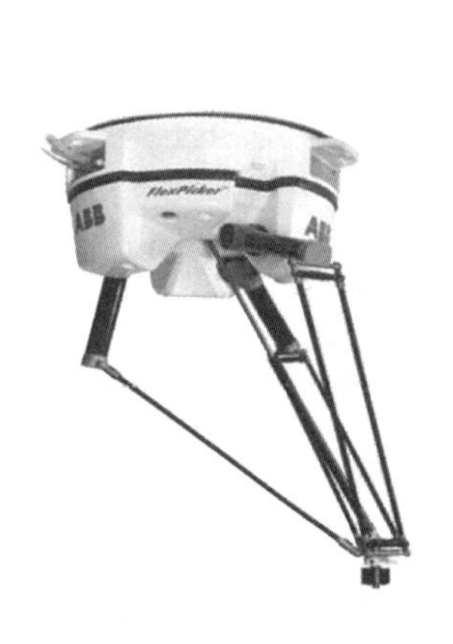

图 3-10　IRB 360

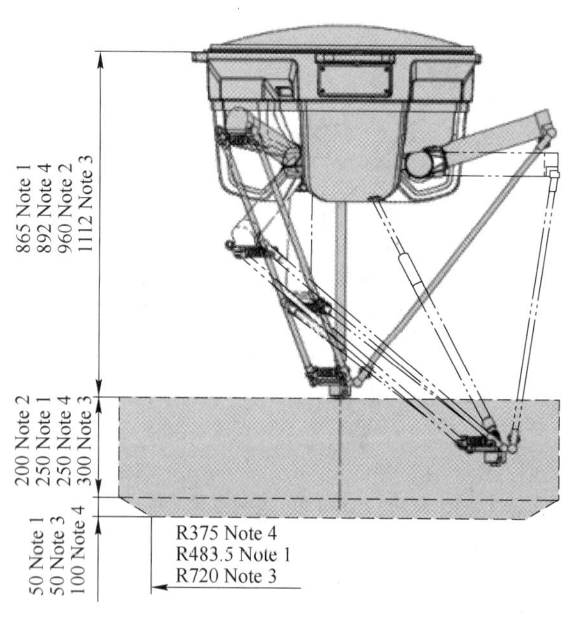


图 3-11　IRB 360 工作范围

#### 3.1.5.4 IRB 910SC

IRB 910SC 又称 SCARA（见图 3-2），是 ABB 生产的平面关节型工业机器人。它采用了铰接式机器人手臂，是一种能够在狭窄空间内操作的单臂机器人。SCARA 速度快，性价比高，是小零件装配，物料搬运和零件检测的理想选择。

#### 3.1.5.5 YuMi（IRB 14000）

YuMi 双臂工业机器人（见图 3-5）是多年的研究和发展的结果，使人类和机器人之间的协作变得更多。YuMi 为全新的自动化时代而设计，主要应用于小件搬运，小件装配。

YuMi 在未来有很大的发展前景，将改变我们组装自动化的思考方式。YuMi——你和我，共同探索无限可能。

## 任务 3.2 认识和使用示教器

### 3.2.1 任务引入

操作工业机器人就必须与 ABB 工业机器人的示教器（FlexPendant）打交道。通过本任务的学习，大家可以认识 ABB 工业机器人的示教器的组成与主界面，设定示教器的语言与系统时间，查看机器人常用信息与事件日志，以及完成工业机器人数据的备份与恢复。

### 3.2.2 相关知识

#### 3.2.2.1 示教器组成

示教器是一种手持式操作装置，由硬件和软件组成。其本身就是一套完整的计算机，用于处理与机器人系统操作相关的许多功能，如运行程序、微动控制操作器和修改机器人程序等。

示教器的主要组成如图 3-12 所示。在示教器上，绝大多数的操作都是在触摸屏上完成的，同时还保留了 12 个专用按钮（见图 3-13），各按钮名称及功能见表 3-2。

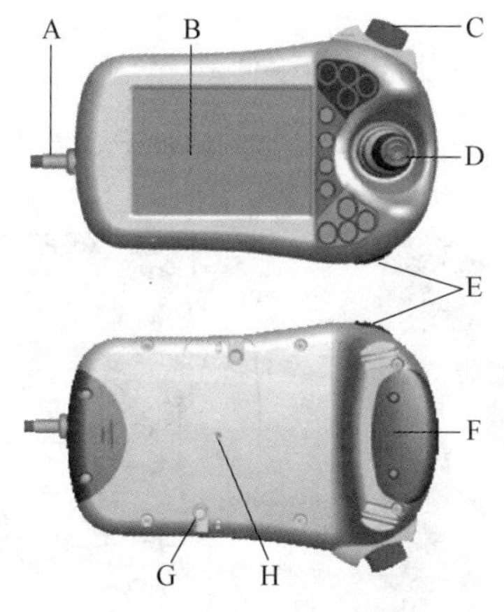


图 3-12 示教器组成

A—连接电缆；B—触摸屏；C—急停开关；D—操作杆；
E—USB 接口；F—使能器；G—触摸屏用笔；H—示教器复位按钮

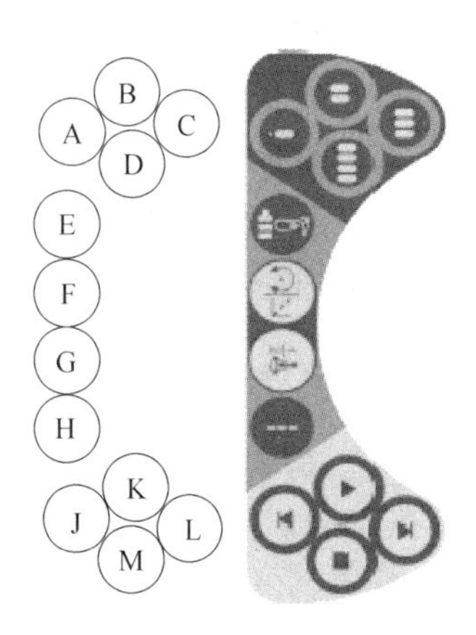


图 3-13　示教器上的按钮

**表 3-2　各按钮名称及功能**

| 代号 | 按键名称 | 功　　能 |
| --- | --- | --- |
| A～D | 自定义按键 | 用户可根据需要为这四个按键设置特定的功能；对这些按键进行编程后可简化程序编程或测试；它们也可用于启动 FlexPendant 上的菜单 |
| E | 机械单元选择按键 | 机器人轴/外轴的切换 |
| F | 线性运动/重定位运动切换按键 | 线性运动/重定位运动的切换 |
| G | 动作模式切换按键 | 关节 1-3 轴/ 4-6 轴的切换 |
| H | 增量开关按键 | 根据需要选择对应位移及角度的大小 |
| J | 步退执行按键 | 使程序后退至上一条指令 |
| K | START（启动）按键 | 开始执行程序 |
| L | 步进执行按键 | 使程序前进至下一条指令 |
| M | STOP（停止）按键 | 停止程序执行 |

操作示教器时，通常会手持该设备。右手便利者通常左手持设备，右手操作屏幕和按钮（见图 3-14）。左手便利者可以轻松通过将显示器旋转 180°，用右手持设备（见图 3-15）。

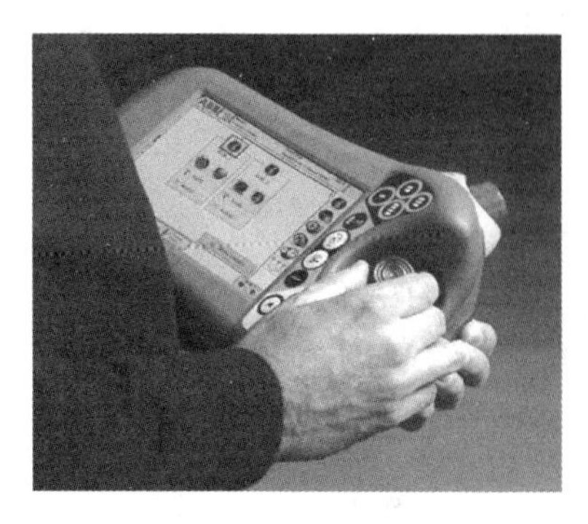

图 3-14　手持示教器

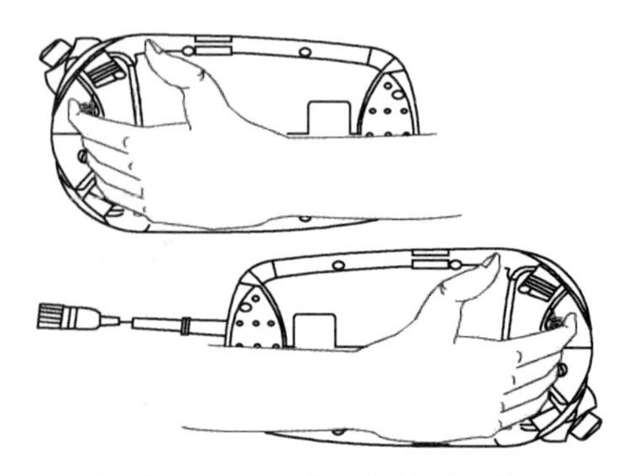

图 3-15　左/右手握持示教器

#### 3.2.2.2 认识示教器的主界面

机器人开机后的示教器初始界面如图 3-16 所示，各部分名称及功能见表 3-3。

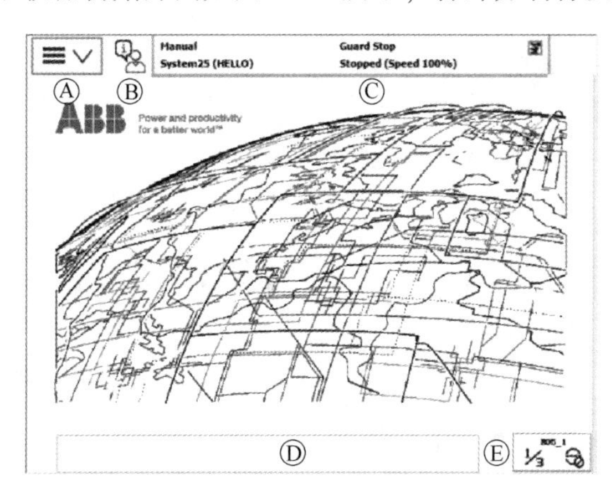


图 3-16　示教器初始界面

**表 3-3　各部分名称及功能**

| 代号 | 名称 | 功　能 |
|---|---|---|
| A | 主菜单 | 主菜单界面常用选项包括输入输出、手动操纵、程序编辑器、程序数据、校准和控制面板等 |
| B | 操作员窗口 | 操作员窗口显示来自机器人程序的消息；程序需要操作员做出某种响应，以便继续时出现此情况 |
| C | 状态栏 | 状态栏显示与系统状态有关的重要信息，如操作模式、电动机开启/关闭、程序状态等 |
| D | 任务栏 | 通过主菜单，可以打开多个视图，但一次只能操作一个；任务栏显示所有打开的视图，并可用于视图切换 |
| E | 快捷菜单 | 快捷菜单包含对微动控制和程序执行进行的设置 |

### 3.2.3 任务实施

#### 3.2.3.1 设定示教器的显示语言

示教器出厂时，默认的显示语言是英语。如有需要，可以通过以下操作将显示语言设定为其他语言。操作过程如下：

（1）单击左上角主菜单按钮，选择“Control Panel”（见图 3-17）。

（2）选择“Language”。

（3）选择需要的语言。例如选择“Chinese”，单击“OK”（见图 3-18）。

（4）点击“Yes”，系统重启，重启后，单击左上角按钮就能看到菜单已切换成中文界面。

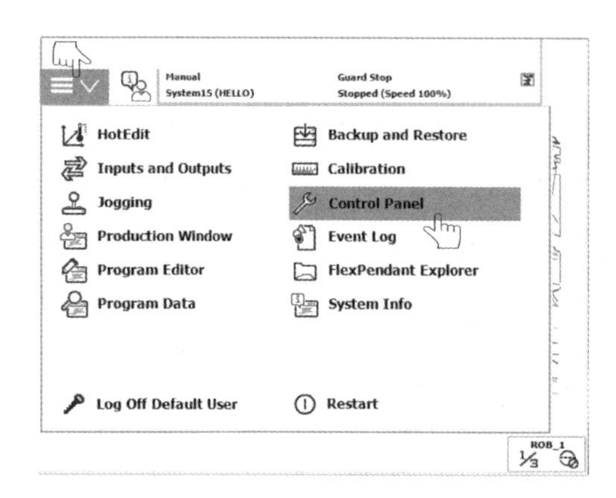


图 3-17　单击“Control Panel”

图 3-18　选择“Chinese”

### 3.2.3.2　设定工业机器人的系统时间

为了方便进行文件的管理和故障的查阅与管理，在进行各种操作之前要将机器人系统的时间设定为本地时区的时间。操作过程如下：

（1）单击左上角主菜单按钮，选择“控制面板”。

（2）选择“日期和时间”，在此画面就能对日期和时间进行设定（见图 3-19）。

（3）日期和时间修改完成后，单击“确定”。

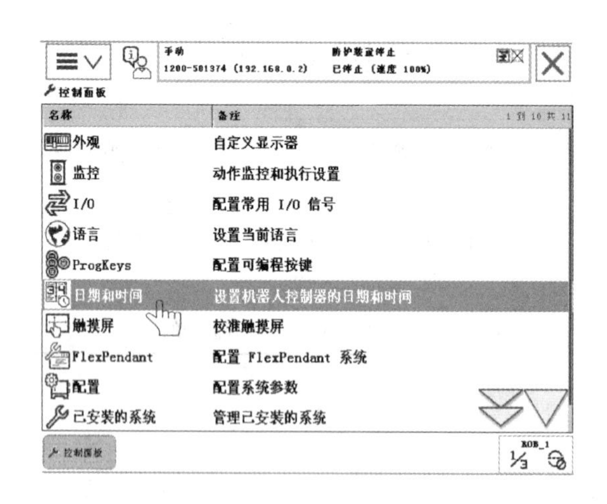


图 3-19　单击“时间和日期”

### 3.2.3.3　查看 ABB 工业机器人常用信息与事件日志

在示教器画面的状态栏上，可以查看 ABB 机器人常用信息，通过这些信息就可以了解到机器人当前所处的状态及一些存在的问题，如图 3-20 所示。

单击示教器界面上的状态栏，就可以查看机器人的事件日志。该界面显示出机器人运行的事件记录，包括时间日期等，为分析相关事件和问题提供准确信息。

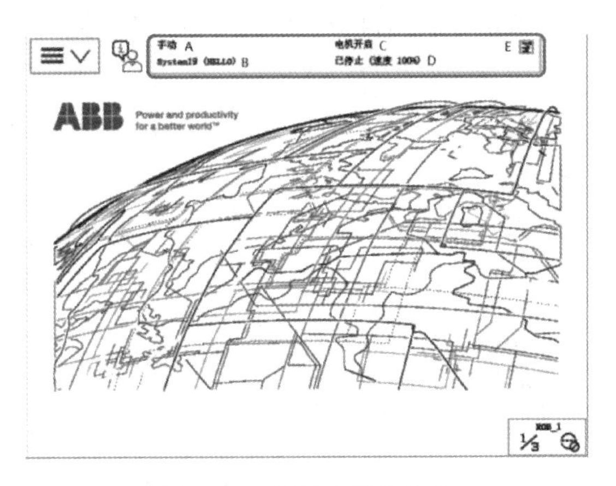


图 3-20　机器人常用信息

### 3.2.3.4　ABB 工业机器人数据的备份与恢复

定期对 ABB 机器人的数据进行备份，是保证 ABB 机器人正常工作的良好习惯。ABB 机器人数据备份的对象是所有正在系统内存运行的 RAPID 程序和系统参数。当机器人系统出现错乱或者重新安装新系统以后，可以通过备份快速地把机器人恢复到备份时的状态。

**对 ABB 工业机器人数据进行备份**

对 ABB 工业机器人数据进行备份的操作步骤为：

（1）单击左上角主菜单按钮，选择“备份与恢复”（见图 3-21）；

（2）单击“备份当前系统…”；

（3）单击“ABC…”，进行存放备份数据目录名称的设定；

（4）单击“…”按钮，选择备份存放的位置（机器人硬盘或 USB 存储设备），单击“备份”进行备份的操作（见图 3-22）；

（5）等待备份的完成。

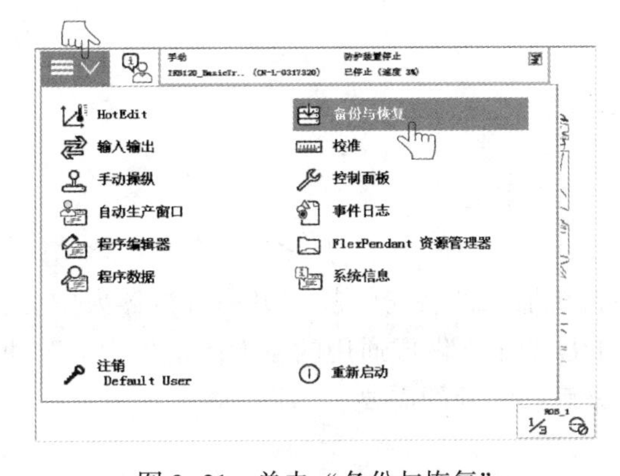


图 3-21　单击“备份与恢复”

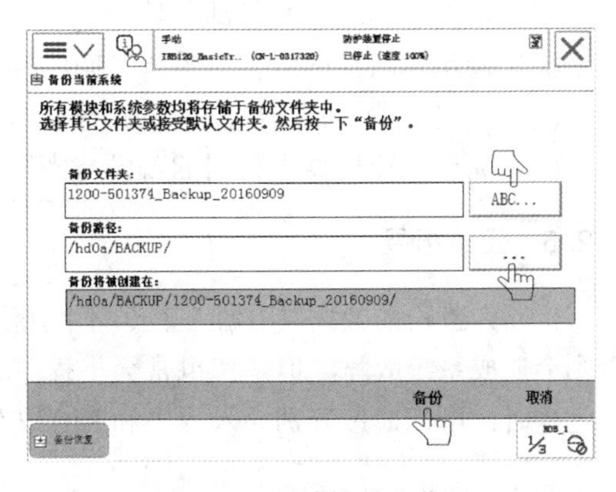


图 3-22　设定名称和位置

**对 ABB 工业机器人数据进行恢复**

对 ABB 工业机器人数据进行恢复的操作步骤为：

（1）单击“恢复系统…”（见图 3-23）；

（2）单击“…”，选择备份存放的目录，单击“恢复”（见图 3-24）；
（3）单击“是”；
（4）等待恢复的完成。

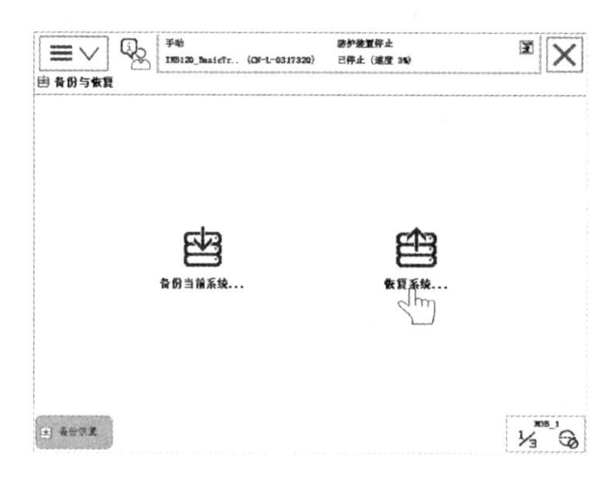

图 3-23 单击“恢复系统”

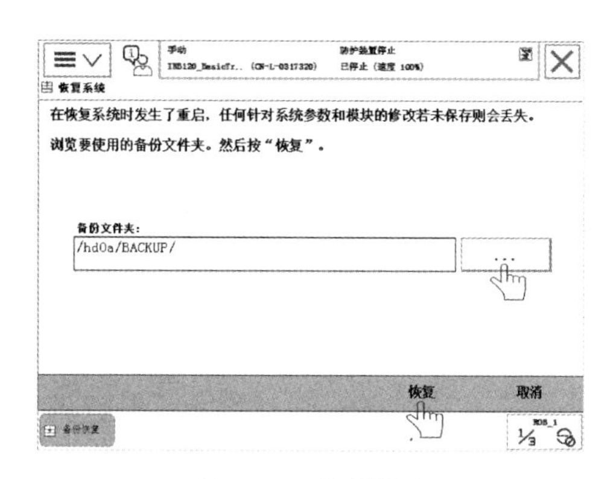

图 3-24 选择目录

### 3.2.4 任务评价

任务评价表见表 3-4。

**表 3-4 任务评价表**

| 序号 | 评价内容 | 成绩占比 | 自评 | 师评 |
| --- | --- | --- | --- | --- |
| 1 | 示教器的组成及硬件按钮 | 15 分 | | |
| 2 | 示教器主界面各部分名称及功能 | 15 分 | | |
| 3 | 在示教器上设置显示语言和系统时间 | 20 分 | | |
| 4 | 查看机器人常用信息与事件日志 | 20 分 | | |
| 5 | ABB 工业机器人数据的备份与恢复 | 30 分 | | |

（1）示教器的组成及各部分功能。
（2）设定示教器的语言与系统时间的操作。
（3）查看机器人常用信息与事件日志的操作。
（4）完成工业机器人数据的备份与恢复。

### 3.2.5 任务拓展

机器人备份的数据具有唯一性，不能将一台机器人的备份恢复到另一台机器人中去，否则会造成系统故障。但是，也常会将程序和 I/O 的定义做成通用的，方便批量生产时使用。这时，可以通过分别导入程序和 EIO 文件来解决实际的需要。

#### 3.2.5.1 单独导入程序

单独导入程序的操作步骤为：
（1）单击左上角主菜单按钮，选择“程序编辑器”（见图 3-25）；
（2）单击“模块”标签；

(3) 打开“文件”菜单，点击“加载模块”，从“备份目录/RAPID”路径下加载所需要的程序模块（见图 3-26）。

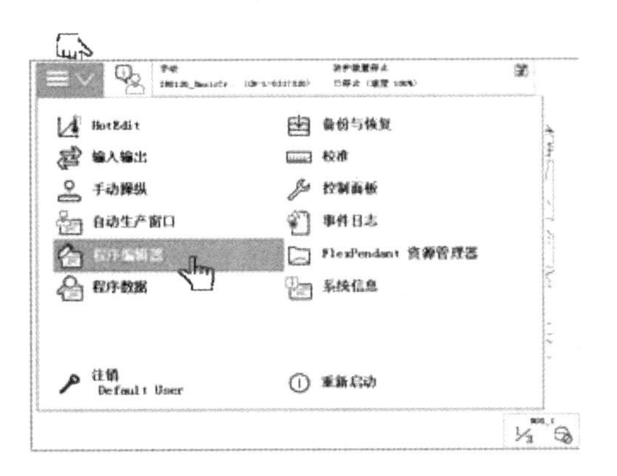


图 3-25 单击“程序编辑器”

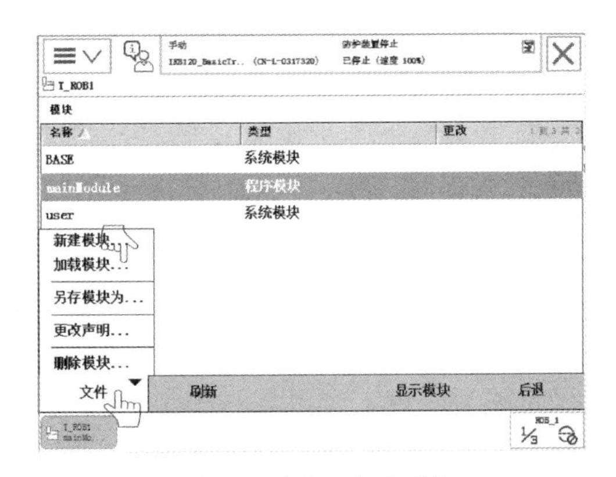


图 3-26 单击“加载模块”

### 3.2.5.2 单独导入 EIO 文件

单独导入 EIO 文件的操作步骤为：

(1) 单击左上角主菜单按钮，选择“控制面板”；

(2) 选择“配置”（见图 3-27）；

(3) 打开“文件”菜单，单击“加载参数”；

(4) 选择“删除现有参数后加载”，单击“加载”；

(5) 在“备份目录/SYSPAR”路径下找到“EIO.cfg”文件，单击“确定”（见图 3-28）；

(6) 单击“是”，重启后完成导入。

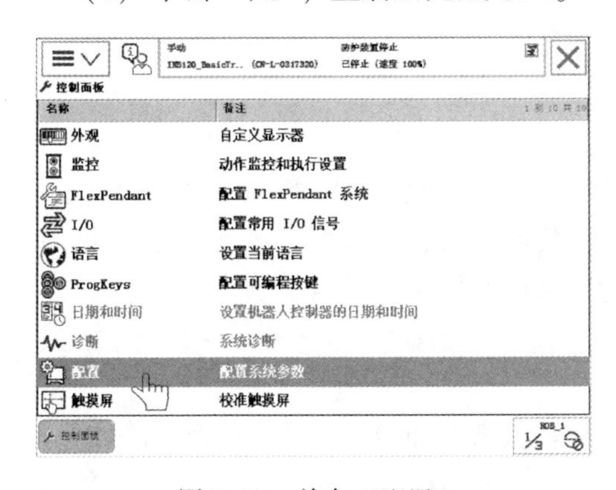


图 3-27 单击“配置”

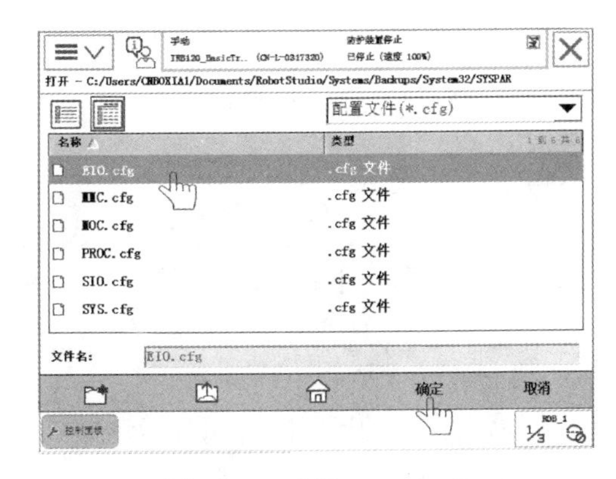


图 3-28 选择“EIO.cfg”

## 任务 3.3 手动操作 ABB 工业机器人

### 3.3.1 任务引入

ABB 工业机器人的运行模式有两种，分别为手动模式和自动模式。手动模式又分为手动减速模式和手动全速模式，手动减速模式下机器人的运行速度最高只能达到250mm/s。对初学者来说，手动操作机器人是学习工业机器人的基础。本任务主要学习单轴运动、线性运动、重定位运动的手动操作方法。

## 3.3.2 相关知识

### 3.3.2.1 使能器

使能器按钮位于示教器手动操作摇杆的右侧，如图3-29所示；操作者应用手的四个手指进行操作，如图3-30所示。使能器按钮分为两档，在手动状态下按第一档，工业机器人处于“电机开启”状态；按第二档（用力按到底），工业机器人就会处于“防护装置停止”状态。

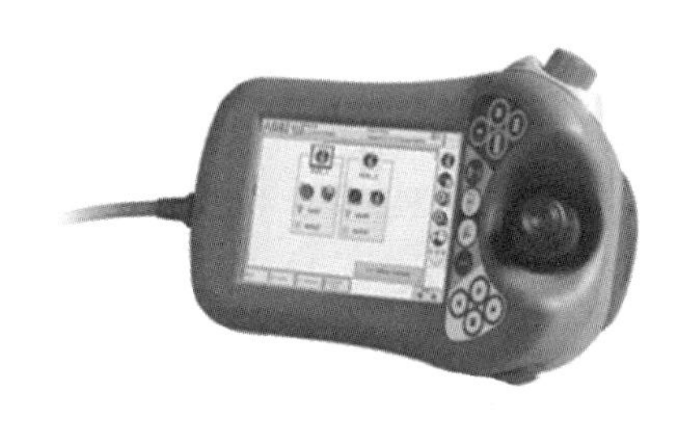

图3-29 使能器

图3-30 使能器操作

使能器是工业机器人为保证操作人员人身安全而设置的。只有在按下使能器按钮，并保持在“电机开启”的状态，才可对机器人进行手动的操作与程序的调试。当发生危险时，人会本能地将使能器按钮松开或按紧，机器人则会马上停下来，保证安全。

### 3.3.2.2 操纵杆的使用技巧

机器人的操纵杆可以比作汽车的节气门，操纵杆的操纵幅度与机器人的运动速度相关。操纵幅度较小，则机器人运动速度较慢；操纵幅度较大，则机器人运动速度较快。所以在操作时，尽量小幅度操纵，使机器人慢慢运动从而开始手动操纵学习。

### 3.3.2.3 增量模式的使用

机器人在手动运行时有两种运动模式，分别为默认模式和增量模式。选择方法如下：

（1）单击左上角主菜单按钮，选择“手动操纵”；

（2）选中“增量”（见图3-31）；

（3）根据需要选择增量模式，然后单击“确定”（见图3-32）。

在默认模式下，操纵杆的位移幅度越小，机器人的运动速度越慢；幅度越大，机器人的运动速度越快。

如果对使用操纵杆通过位移幅度来控制工业机器人运动的速度不熟练的话，那么可以使用“增量”模式来控制工业机器人的运动。在增量模式下，操纵杆每位移一次，机器人就移动一步（一个增量）；如果操纵杆持续1秒或数秒，机器人将持续移动（速率为10步

/秒)。增量移动幅度在小、中、大之间选择，也可以自定义增量运动幅度。

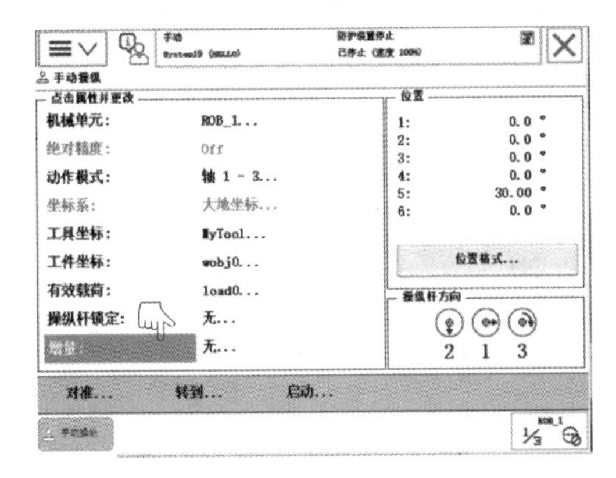

图 3-31　单击“增量”

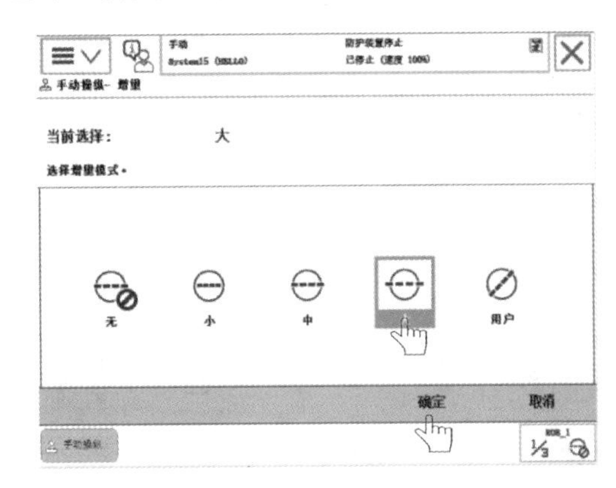


图 3-32　选择增量模式

### 3.3.3　任务实施

手动操纵工业机器人运动一共有三种模式，分别为单轴运动、线性运动和重定位运动。下面介绍如何手动操纵机器人进行这三种运动。

#### 3.3.3.1　单轴运动

一般 ABB 机器人是六个伺服电动机分别驱动机器人的六个关节轴（见图 3-33)，每次手动操作一个关节轴的运动，因此称为单轴运动。手动操纵单轴运动的方法为：

(1) 将控制柜上机器人状态钥匙切换到手动减速状态（小手标志）(图 3-34)。在状态栏中，确认机器人的状态已切换为“手动”。

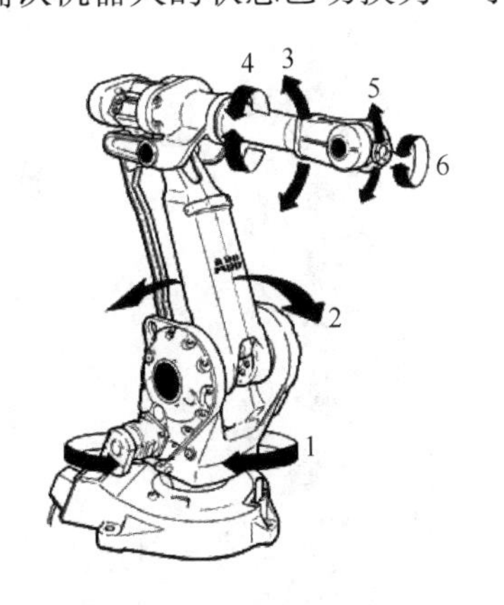


图 3-33　ABB 机器人
1~6—关节轴

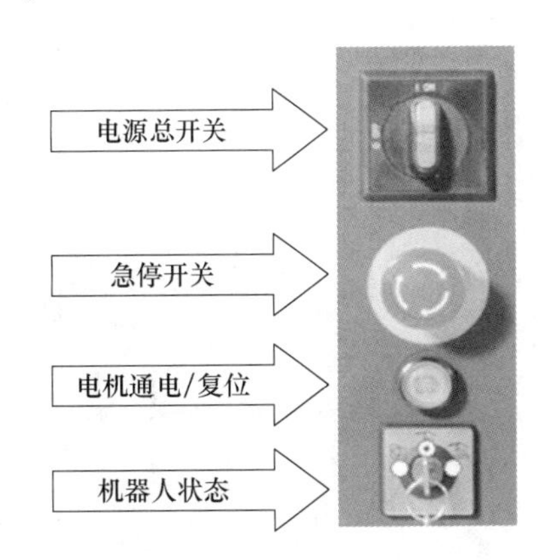


图 3-34　ABB 机器人控制柜

(2) 单击左上角主菜单按钮，选择“手动操纵”；单击“动作模式”；选中“轴 1-3”，然后单击“确定”，如图 3-35 所示（选中“轴 4-6”，就可以操纵轴 4-6)。

(3) 用手按下使能器，在状态栏中确认已进入“电机开启”状态，屏幕右下角显示“轴 1-3”的操纵杆方向，箭头代表正方向（见图 3-36)。此时，操作示教器上的操纵杆，完成单轴运动。

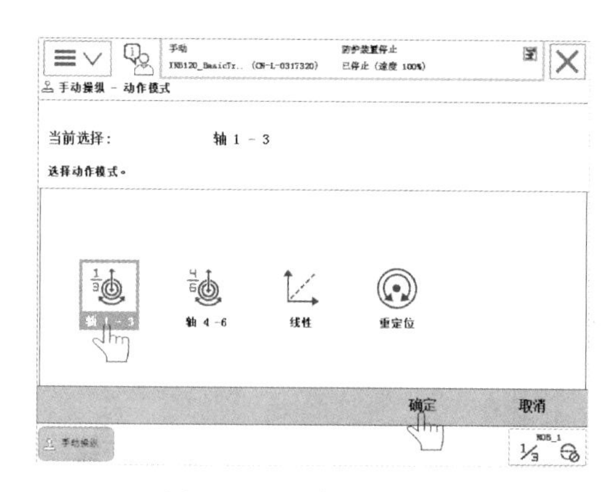


图 3-35　选中“轴 1-3”

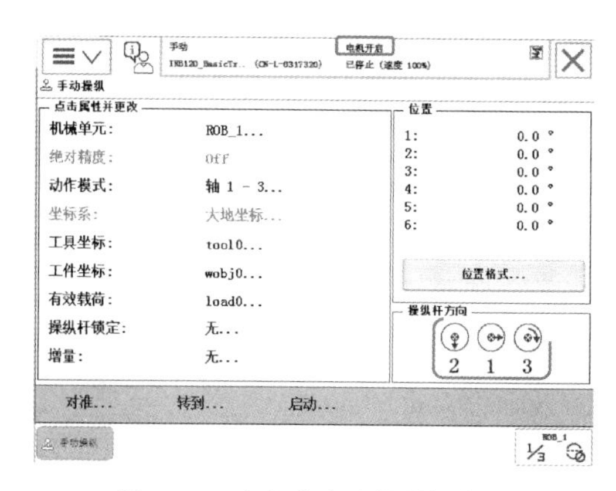


图 3-36　电机状态和操纵杆方向

### 3.3.3.2　线性运动

机器人的线性运动是指安装在机器人第六轴法兰盘上工具的 TCP 在空间中作线性运动。手动操纵线性运动的方法为：

（1）选择“手动操纵”；单击“动作模式”，选择“线性”，然后单击“确定”；单击“工具坐标”（机器人的线性运动要在“工具坐标”中指定对应的工具）；选中对应的工具“tool1”，然后单击“确定”（见图 3-37，工具数据的设定请参考任务 5.2）。

（2）用手按下使能器，在状态栏中确认已进入“电机开启”状态，屏幕右下角显示轴 X、Y、Z 的操纵杆方向，箭头代表正方向。此时，操作示教器上的操纵杆，工具的 TCP 点在空间中作线性运动（见图 3-38）。

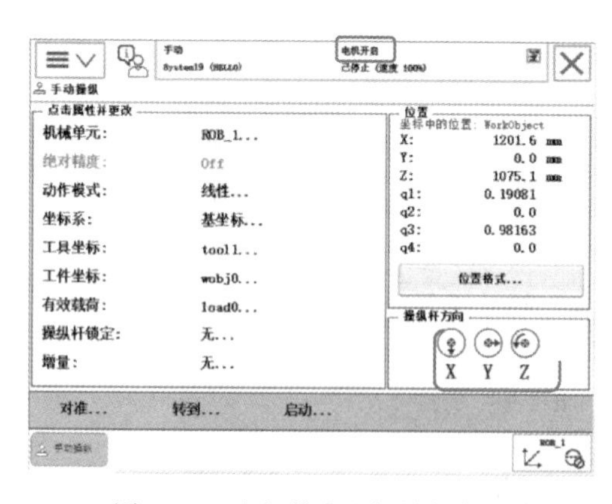


图 3-37　电机状态和操纵杆方向

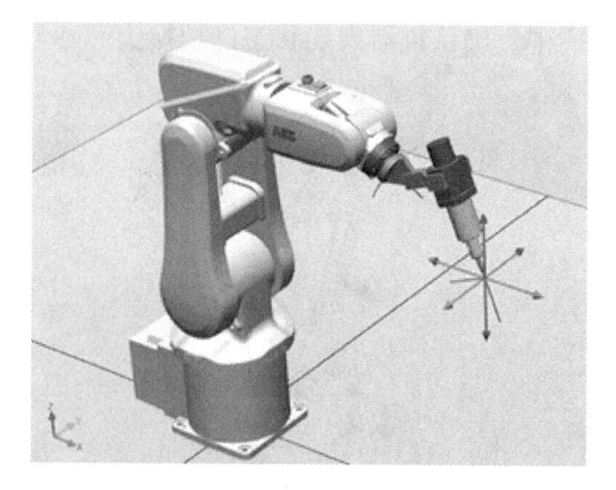

图 3-38　线性运动

### 3.3.3.3　重定位运动

机器人的重定位运动是指机器人第六轴法兰盘上的工具 TCP 点在空间中绕着坐标轴旋转的运动，也可以理解为机器人绕着工具 TCP 点作姿态调整的运动。手动操纵重定位运动的方法为：

（1）选择“手动操纵”；单击“动作模式”；选择“重定位”，然后单击“确定”；单击“坐标系”；选择“工具”，然后单击“确定”（见图 3-39）；单击“工具坐标”，选中对应的工具“tool1”，然后单击“确定”。

（2）用手按下使能器，在状态栏中确认已进入“电机开启”状态，屏幕右下角显示轴 X、Y、Z 的操纵杆方向，箭头代表正方向。此时，操作示教器上的操纵杆，机器人绕着工具 TCP 点作姿态调整的运动（见图 3-40）。

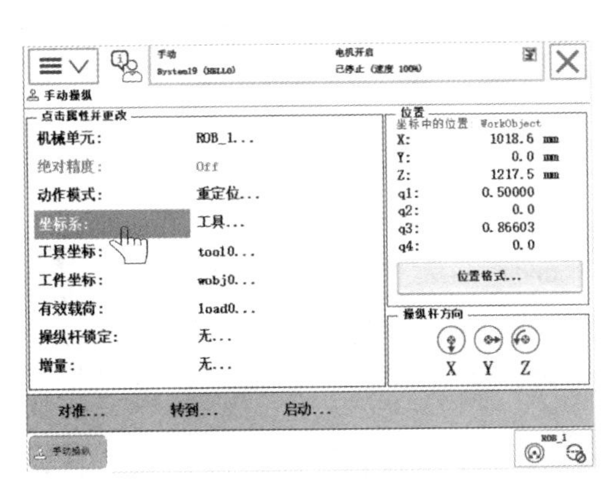


图 3-39　单击“坐标系”

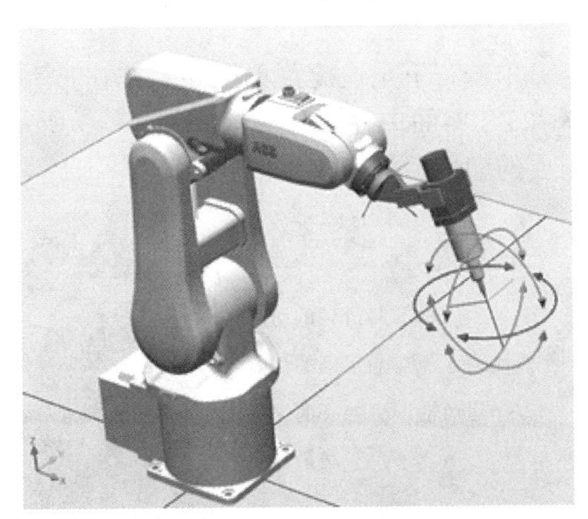

图 3-40　重定位运动

### 3.3.4　任务评价

任务评价表见表 3-5。

**表 3-5　任务评价表**

| 序号 | 评价内容 | 绩占比 | 自评 | 师评 |
| --- | --- | --- | --- | --- |
| 1 | 正确使用使能器 | 15 分 | | |
| 2 | 操纵杆的使用技巧 | 15 分 | | |
| 3 | 增量模式的使用 | 10 分 | | |
| 4 | 操纵工业机器人进行单轴运动 | 20 分 | | |
| 5 | 操纵工业机器人进行线性运动 | 20 分 | | |
| 6 | 操纵工业机器人进行重定位运动 | 20 分 | | |

（1）把使能器分为两档，是工业机器人为保证操作人员人身安全而设置的。

（2）操纵杆的操纵幅度与机器人的运动速度相关。

（3）机器人在手动运行时有两种运动模式，分别为默认模式和增量模式。

（4）手动操纵工业机器人运动一共有三种模式，分别为单轴运动、线性运动和重定位运动。

### 3.3.5　任务拓展

由于手动操纵机器人在程序编写和调试中应用广泛，如果每次都通过菜单操作，会比较烦琐。因此示教器提供了快捷按钮和快捷菜单简化操作步骤。

#### 3.3.5.1　手动操纵的快捷按钮

根据需要，ABB 工业机器人示教器上设置了 4 个手动操纵的快捷按钮，如图 3-41 所示。

#### 3.3.5.2　手动操纵的快捷菜单

单击右下角快捷菜单按钮；单击“手动操纵”按钮，单击“显示详情”按钮（见图 3-42），界面中各部分功能如图 3-43 所示；在快捷菜单上单击“增量模式”按钮，选择需要的增量，如图 3-44 所示。

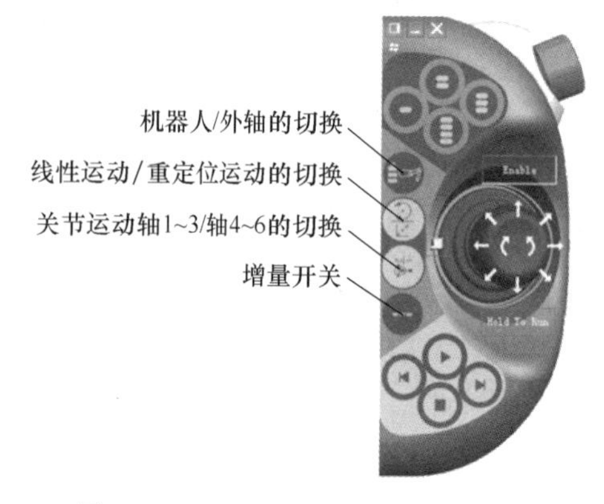


图 3-41　ABB 工业机器人示数器

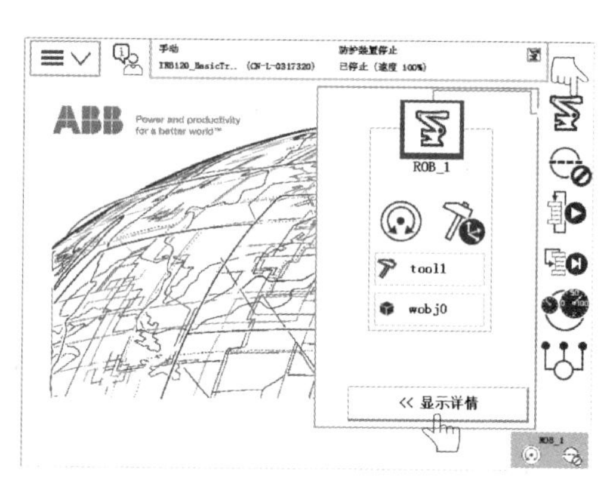


图 3-42　单击“手动操纵”

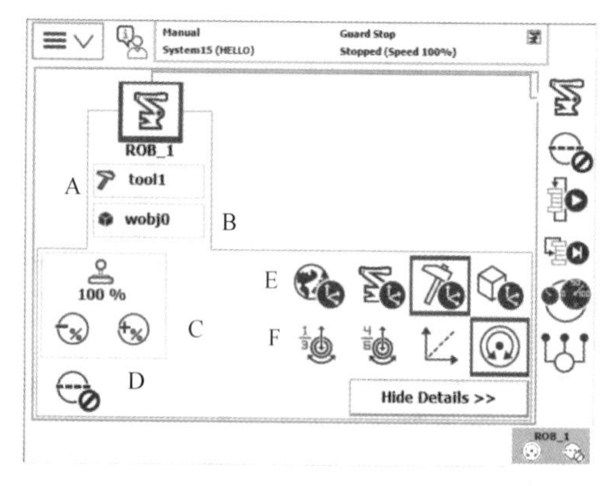


图 3-43　“显示详情”界面

A—选择当前使用的工具数据；B—选择当前使用的工件坐标；

C—操纵杆速率；D—增量开/关；E—坐标系选择；

F—动作模式选择

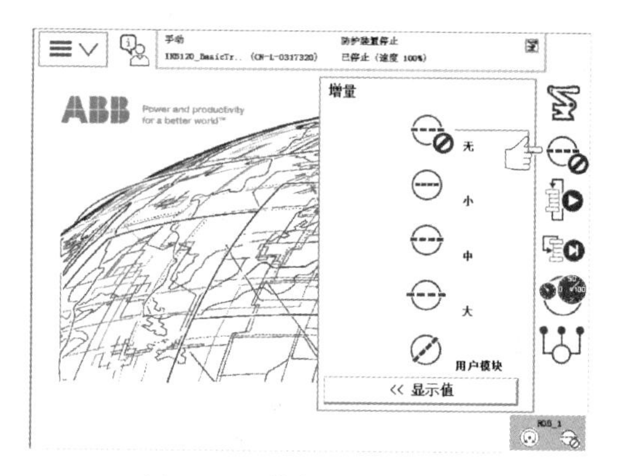


图 3-44　单击“增量模式”

## 任务 3.4　更新 ABB 工业机器人的转数计数器

### 3.4.1　任务引入

转数计数器的更新就是将机器人各个轴停到机械原点，把各关节轴上的同步标记对齐，然后在示教器上进行校准更新。通过本任务的学习，大家可以认识工业机器人的机械原点位置，手动操纵各关节轴运动到机械原点，并对转数计数器进行更新。

### 3.4.2 相关知识

ABB 工业机器人六个关节轴都有一个机械原点位置。以下情况下，需要对机械原点的位置进行转数计数器更新操作：

（1）更换伺服电机转数计数器电池后；

（2）当转数计数器发生故障，修复后；

（3）转数计数器与测量板之间断开过以后；

（4）断电后，机器人关节轴发生了位移；

（5）当系统报警提示“10036 转数计数器未更新”时。

### 3.4.3 任务实施

进行 ABB 机器人 IRB 1200 转数计数器更新的操作步骤为：

（1）用前面介绍的“单轴运动”操作方法，按 4-5-6-1-2-3 的顺序让各关节轴逐一回到机械原点位置，各轴的机械原点在机器人上都有显著标识，容易识别。

（2）六个轴都回到机械原点后的姿态如图 3-45 所示。再用示教器进行以下操作，对计数器进行更新：

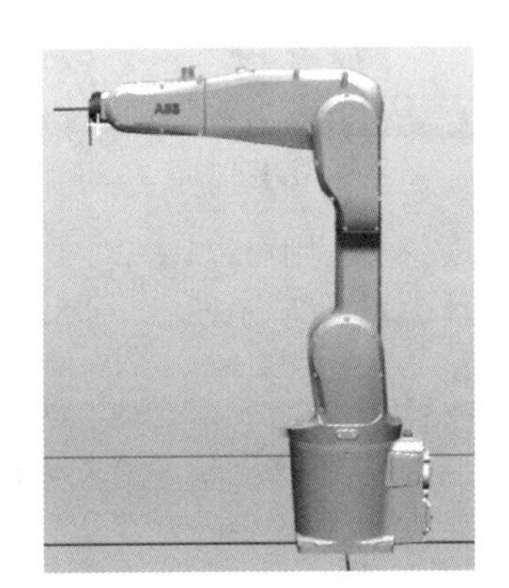

图 3-45　ABB 机器人 IRB 1200 转数计数器

1）单击左上角主菜单，选择“校准”（见图 3-46）；

2）单击“ROB_1”；

3）选择“校准参数”，单击“编辑电机校准偏移”（见图 3-47）；

4）单击“是”；

5）将机器人本机上电机校准偏移数据记录下来（见图 3-48），输入编辑电机校准偏移界面中，然后单击“确定”，如图 3-49 所示（如果示教器中显示的数值与机器人本体上的标签数值一致，则无须修改，单击“取消”即可）；

6）要使参数生效，必须重新启动系统，单击“是”。

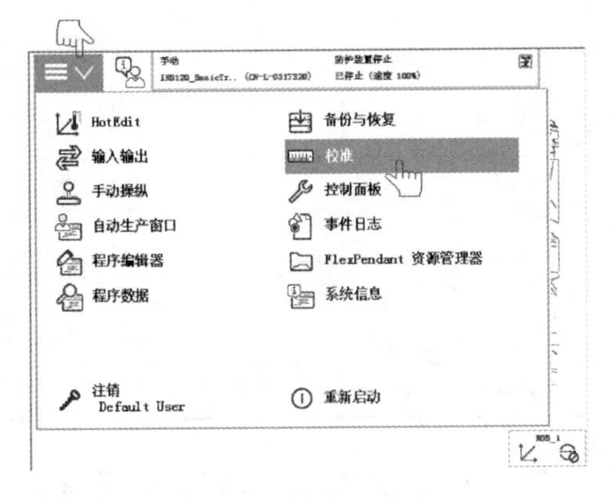


图 3-46　单击“校准”

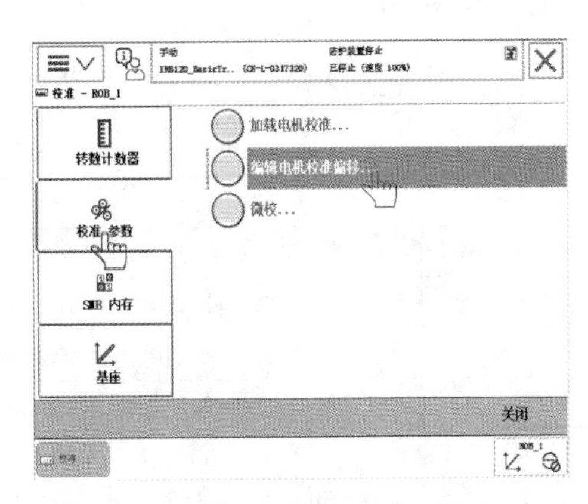


图 3-47　单击“编辑电机校准偏移…”

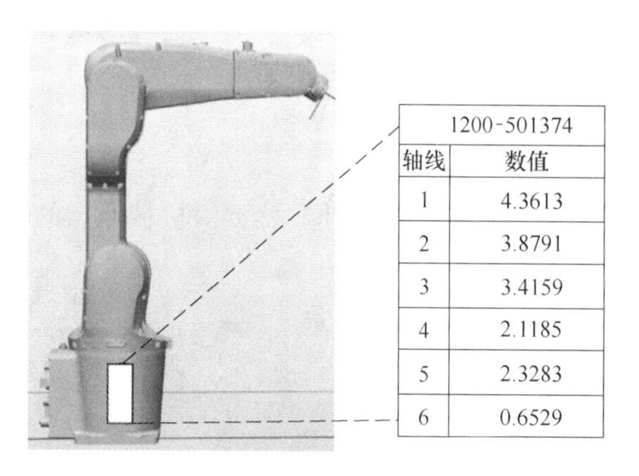

| 1200-501374 | |
|---|---|
| 轴线 | 数值 |
| 1 | 4.3613 |
| 2 | 3.8791 |
| 3 | 3.4159 |
| 4 | 2.1185 |
| 5 | 2.3283 |
| 6 | 0.6529 |

图 3-48　本机上电机校准偏移数据

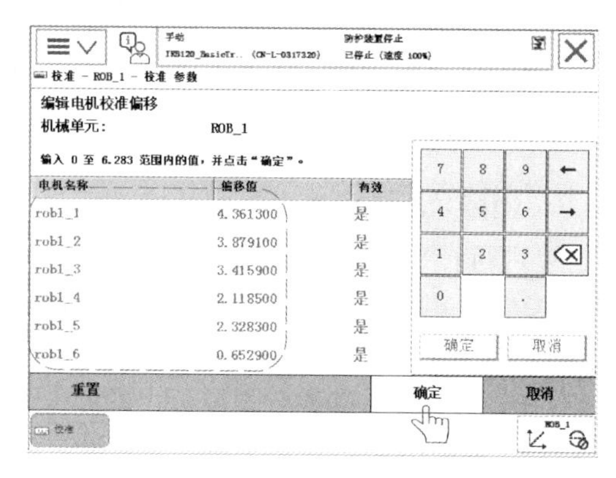


图 3-49　电机校准偏移界面

（3）重启后，再次选择“校准”，单击“ROB_1”，选择“转数计数器”，单击“更新转数计数器”（见图 3-50）；单击“是”；单击“确定”；单击“全选”（见图 3-51），然后单击“更新”（如果机器人由于安装位置的关系，无法让六个轴同时到达机械原点刻度位置，则可以逐一对关节轴进行转数计数器更新）；单击“更新”；操作完成后，转数计数器更新完成。

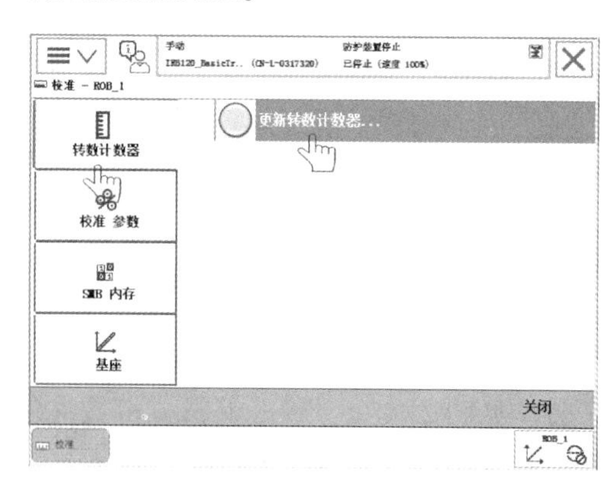


图 3-50　单击“更新转数计数器”

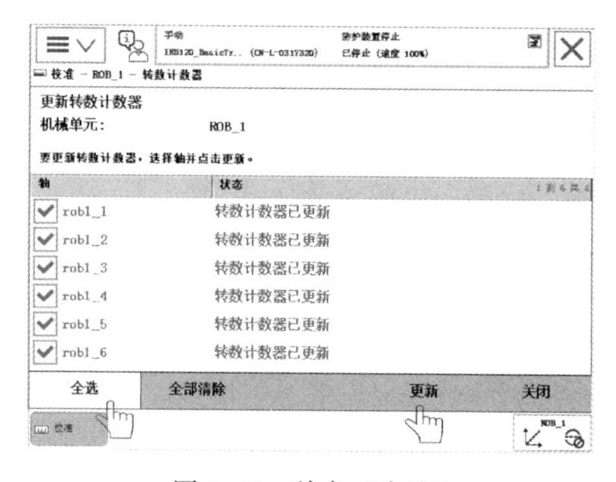


图 3-51　单击“全选”

### 3.4.4　任务评价

任务评价表见表 3-6。

**表 3-6　任务评价表**

| 序 号 | 评价内容 | 成绩占比 | 自评 | 师评 |
|---|---|---|---|---|
| 1 | 工业机器人各轴的机械原点位置 | 10 分 | | |
| 2 | 需进行转数计数器更新操作的情况 | 20 分 | | |
| 3 | 手动操纵使机器人各轴运动到机械原点 | 30 分 | | |
| 4 | 在示教器上对转数计数器进行更新 | 40 分 | | |

（1）转数计数器的更新是将机器人各个轴停到机械原点，把各关节轴上的同步标记对齐，然后在示教器进行校准更新。

（2）ABB工业机器人的六个关节轴都有一个机械原点位置，手动操纵使机器人各轴运动到机械原点。

（3）在示教器上对转数计数器进行更新。

### 3.4.5 任务拓展

#### 3.4.5.1 工业机器人本体电池的作用

本节所述型号的机器人，其零点信息数据存储在本体串行测量板上，而串行测量板在机器人系统接通外部主电源时，由主电源进行供电。当系统与主电源断开连接后，则需要串行测量板电池（本体电池）为其供电。

如果串行测量板断电，就会导致零点信息丢失，机器人各关节轴无法按照正确的基准进行运动。为了保持机器人机械零点位置数据的存储，需持续保持串行测量板的供电。当串行测量板的电池剩余后备电量（工业机器人电源关闭）不足2个月时，将显示电池低电量警告（38213电池电量低），此时需要更换新电池。否则电池电量耗尽，每次主电源断电后再次上电，都需要进行转数计数器更新的操作。

#### 3.4.5.2 工业机器人本体电池的使用寿命

通常，如果工业机器人电源每周关闭2天，则新电池的使用寿命为36个月；如果工业机器人电源每天关闭16小时，则新电池的使用寿命为18个月。对于较长时间的生产中断，通过电池关闭服务例行程序可延长电池的使用寿命（大约提高使用寿命3倍）。

# 项目 4　ABB 工业机器人的 I/O 通信

在实际生产过程中，为了更好地完成工作任务，工业机器人经常需要与周边设备进行通信，以获取环境和设备信息。工业机器人通信可借助 RS485、光纤等不同通信接口，并采用 Modbus 协议、TCP/IP 协议等多种通信协议来实现。本项目介绍了 ABB 工业机器人的通信种类、标准 I/O 通信板、DSQC651 板配置、通信总线的连接等，使学生们更好地认知 ABB 工业机器人的 I/O 通信及使用逻辑方法。

## 任务 4.1　ABB 工业机器人 I/O 通信种类

### 4.1.1　任务引入

ABB 工业机器人默认配置是基于 DeviceNet 通信总线协议的 I/O 通信板卡，通过此板卡进行主机控制系统与工业设备之间的通信。DeviceNet 协议是一种 CAN（控制器局域网，Controller Area Net）总线协议，这种传感器/执行器总线系统起源于美国，但在欧洲和亚洲地区正得到越来越广泛的应用。

### 4.1.2　相关知识

I/O 是 Input/Output 的缩写，即输入输出端口，机器人可通过 I/O 与外部设备进行交互，例如：

（1）数字量输入：各种开关信号反馈，如按钮开关、转换开关和接近开关等；传感器信号反馈，如光电传感器和光纤传感器；接触器、继电器触点信号反馈；触摸屏里的开关信号反馈。

（2）数字量输出：控制各种继电器线圈，如接触器，继电器，电磁阀；控制各种指示类信号，如指示灯，蜂鸣器。

ABB 机器人的标准 I/O 板的输入输出都是 PNP 类型。

#### 4.1.2.1　ABB 工业机器人 I/O 通信

ABB 机器人提供了丰富 I/O 通信接口，如 ABB 的标准通信，与 PLC 的现场总线通信，还有与 PC 机的数据通信（见图 4-1）。该通信可以轻松地实现与周边设备的通信。

ABB 的标准 I/O 板提供的常用信号处理有数字量输入、数字量输出、组输入、组输出、模拟量输入和模拟量输出。

ABB 机器人可以选配标准 ABB 的 PLC，省去了原来与外部 PLC 进行通信设置的麻烦，并且在机器人的示教器上就能实现与 PLC 的相关操作。

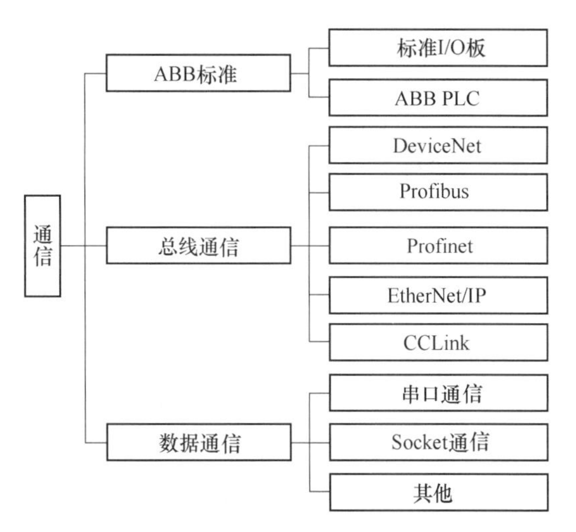


图 4-1　ABB 机器人 I/O 通信接口

这里以最常用的 ABB 标准 I/O 板 DSQC651 和 Profibus-DP 为例，进行详细的讲解。

#### 4.1.2.2　ABB 机器人通信介绍

主计算机单元，ABB 标准 I/O 板一般安装位置如图 4-2 所示；机器人总线及串口位置如图 4-3 所示；Profibus 总线及储存卡位置如图 4-4 所示。

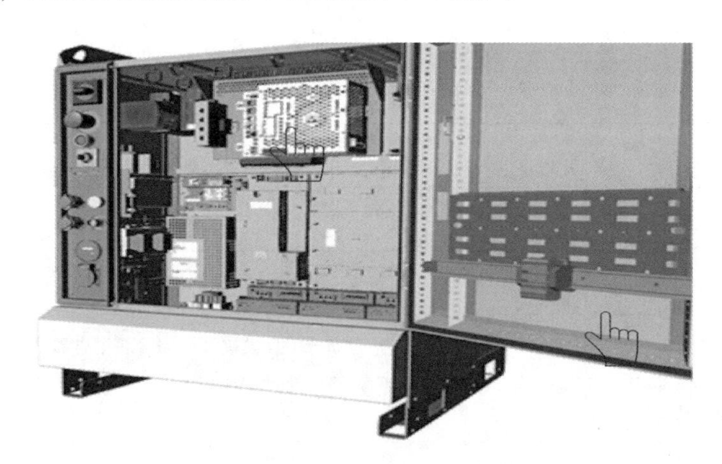

图 4-2　ABB 标准 I/O 板安装位置

说明：WAN 接口需要选择选项“PC INTERFACE”才可以使用。使用何种现场总线，要根据需要进行选配。如果使用 ABB 标准 I/O 板，就必须有 DeviceNet 的总线。

#### 4.1.2.3　ABB 标准 I/O 板卡

常用的 ABB 标准 I/O 板见表 4-1（具体规格参数以 ABB 官方最新公布为准）。

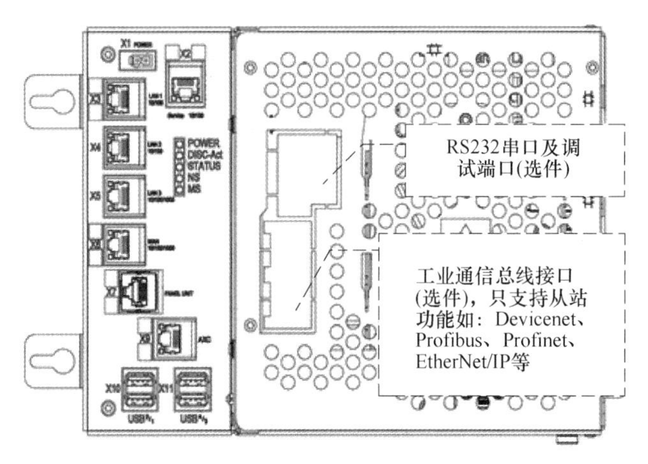


图 4-3　机器人总线及串口位置

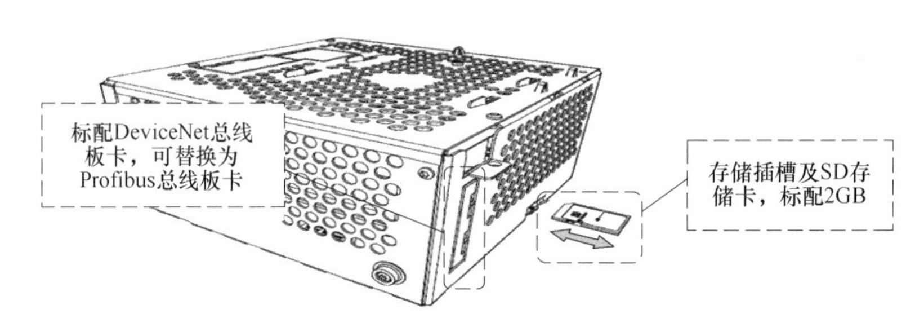


图 4-4　Profibus 总线及储存卡位置

**表 4-1　ABB 标准 I/O 板**

| 型　号 | 说　　明 |
| --- | --- |
| DSQC651 | 分布式 I/O 模块 di8 \ do8 \ ao2 |
| DSQC652 | 分布式 I/O 模块 di16 \ do16 |
| DSQC653 | 分布式 I/O 模块 di8 \ do8 带继电器 |
| DSQC355A | 分布式 I/O 模块 ai4 \ ao4 |
| DSQC377A | 输送链跟踪单元 |

DSQC651 板主要提供 8 个数字输入信号、8 个数字输出信号和 2 个模拟输出信号的处理。

**模块接口说明**

如图 4-5 所示，A 为数字输出信号指示灯，B 为 X1 数字输出接口，C 为 X6 模拟输出接口，D 为 X5 DeviceNet 接口，E 为模块状态指示灯，F 为 X3 数字输入接口，G 为数字输入信号指示灯。

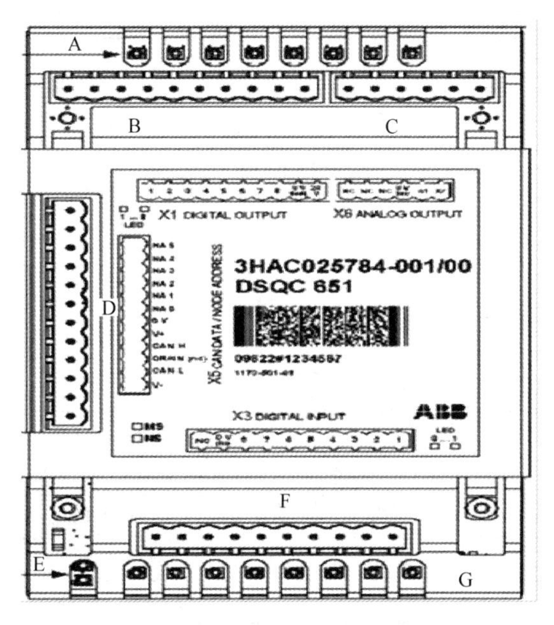


图 4-5　ABB 标准 DSQC651 板

**模块接口连接说明**

X1 端子的接口说明见表 4-2。

**表 4-2　X1 端子接口说明**

| X1 端子编号 | 使用定义 | 地址分配 |
| --- | --- | --- |
| 1 | OUTPUT CH1 | 32 |
| 2 | OUTPUT CH2 | 33 |
| 3 | OUTPUT CH3 | 34 |
| 4 | OUTPUT CH4 | 35 |
| 5 | OUTPUT CH5 | 36 |
| 6 | OUTPUT CH6 | 37 |
| 7 | OUTPUT CH7 | 38 |
| 8 | OUTPUT CH8 | 39 |
| 9 | 0V | — |
| 10 | 24V | — |

X3 端子的接口说明见表 4-3。

**表 4-3　X3 端子接口说明**

| X3 端子编号 | 使用定义 | 地址分配 |
| --- | --- | --- |
| 1 | INPUT CH1 | 0 |
| 2 | INPUT CH2 | 1 |
| 3 | INPUT CH3 | 2 |
| 4 | INPUT CH4 | 3 |
| 5 | INPUT CH5 | 4 |

续表 4-3

| X3 端子编号 | 使用定义 | 地址分配 |
| --- | --- | --- |
| 6 | INPUT CH6 | 5 |
| 7 | INPUT CH7 | 6 |
| 8 | INPUT CH8 | 7 |
| 9 | 0V | — |
| 10 | 未使用 | — |

X5 端子的接口说明见表 4-4。

**表 4-4　X5 端子接口说明**

| X5 端子编号 | 使 用 定 义 |
| --- | --- |
| 1 | 0V BLACK（黑色） |
| 2 | CAN 信号线 low BLUE（蓝色） |
| 3 | 屏蔽线 |
| 4 | CAN 信号线 high WHITE（白色） |
| 5 | 24V RED（红色） |
| 6 | GND 地址选择公共端 |
| 7 | 模块 ID bit 0（LSB） |
| 8 | 模块 ID bit 1（LSB） |
| 9 | 模块 ID bit 2（LSB） |
| 10 | 模块 ID bit 3（LSB） |
| 11 | 模块 ID bit 4（LSB） |
| 12 | 模块 ID bit 5（LSB） |

说明：ABB 标准 I/O 板是挂在 DeviceNet 网络上的，所以要设定模块在网络中的地址。端子 X5 的 6~12 的跳线就是用来决定模块的地址的，地址可用范围为 10~63。

如图 4-6 所示，将第 8 脚和第 10 脚的跳线剪去，2+8=10 就可以获得 10 的地址。

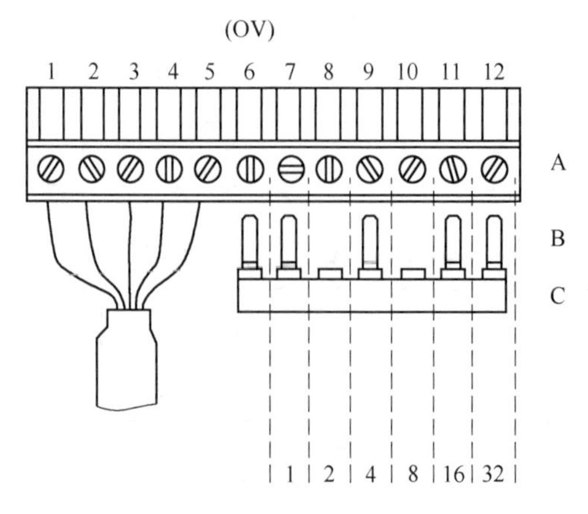


图 4-6　ABB 标准 I/O 板地址的跳线

X6 端子的接口说明见表 4-5。

**表 4-5　X6 端子接口说明**

| X6 端子编号 | 使用定义 | 地址分配 |
|---|---|---|
| 1 | 未使用 | — |
| 2 | 未使用 | — |
| 3 | 未使用 | — |
| 4 | 0V | — |
| 5 | 模拟输出 AO1 | 0-15 |
| 6 | 模拟输出 AO2 | 16-31 |

说明：模拟输出的范围为 0~+10V。

### 4.1.3　任务实施

ABB 标准 I/O 板 DSQC651 是最为常用的模块，下面以创建数字输入信号 di1、数字输出信号 do1、组输入信号 gi1 和组输出信号 go1 为例进行任务的实施。

#### 4.1.3.1　定义 DSQC651 板的总线连接

ABB 标准 I/O 板都是下挂在 DeviceNet 现场总线下的设备，通过 X5 端口与 DeviceNet 现场总线进行通信。

定义 DSQC651 板的总线连接的相关参数说明见表 4-6。

**表 4-6　定义 DSQC651 板的总线连接的相关参数**

| 参数名称 | 设定值 | 说　明 |
|---|---|---|
| Name | board10 | 设定 I/O 板在系统中的名字 |
| Network | DeviceNet | I/O 板连接的总线 |
| Address | 10 | 设定 I/O 板在总线中的地址 |

在系统中定义 DSQC651 板卡的操作步骤如下：

(1) 单击左上角主菜单按钮，选择“控制面板”，之后选择“配置”，双击“DeviceNet Device”；单击“添加”；单击“使用来自模板的值”对应的下拉箭头，选择“DSQC 651 Combi I/O Device”（见图 4-7）。

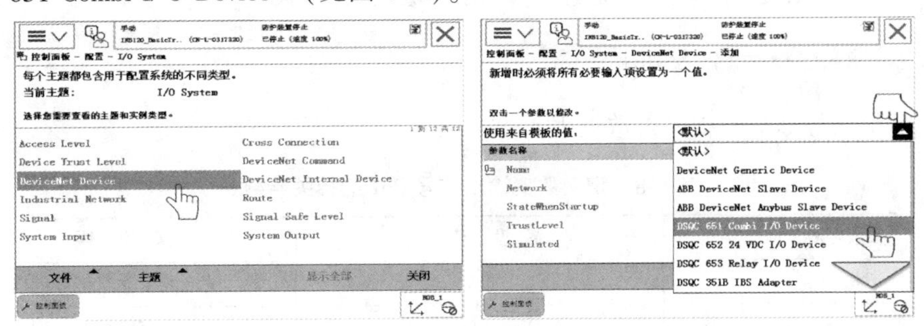


图 4-7　定义标准 DSQC651 板

（2）双击“Name”进行 DSQC651 板在系统中名字的设定（如果不修改，则名字是默认的“d651”），在系统中将 DSQC651 板的名字设定为“board10”（10 代表此模块在 DeviceNet 总线中的地址，方便识别），然后单击“确定”，将“Address”设定为 10，单击“确定”，然后单击“是”，这样 DSQC651 板的定义就完成了。

#### 4.1.3.2 定义数字输入信号 di1

数字输入信号 di1 的相关参数见表 4-7。

**表 4-7 定义数字输入信号 di1 的相关参数**

| 参数名称 | 设定值 | 说 明 |
|---|---|---|
| Name | di1 | 设定数字输入信号的名字 |
| Type of Signal | Digital Input | 设定信号的类型 |
| Assigned to Device | board10 | 设定信号所在的 I/O 模块 |
| Device Mapping | 0 | 设定信号所占用的地址 |

其操作如下：

（1）单击左上角主菜单按钮，选择“控制面板”；选择“配置”，双击“Signal”；单击“添加”，双击“Name”输入“di1”，然后单击“确定”；双击“Type of Signal”，选择“Digital Input”（见图 4-8）。

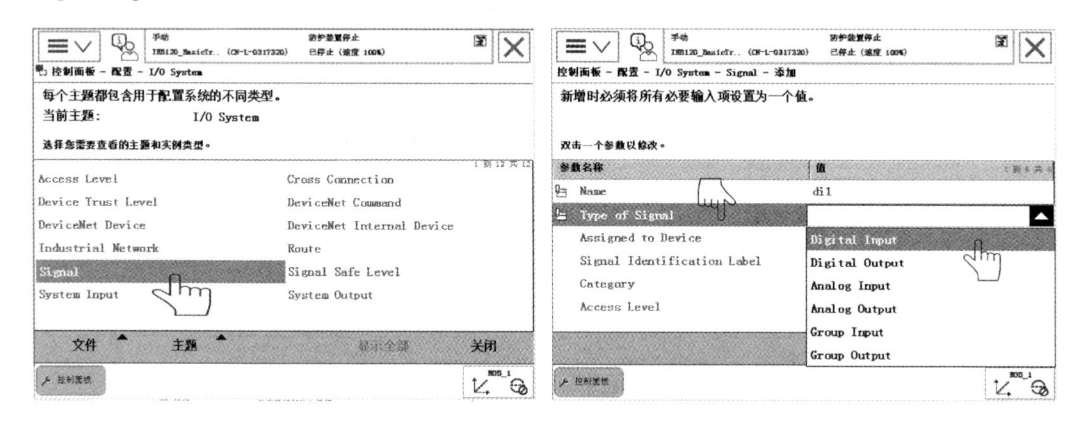


图 4-8 定义数字输入信号 di1

（2）双击“Assigned to Device”，选择“board10”；双击“Device Mapping”，输入“0”，然后单击“确定”；单击“确定”；单击“是”，完成设定。

#### 4.1.3.3 定义数字输出信号 do1

数字输出信号 do1 的相关参数见表 4-8。

**表 4-8 定义数字输出信号 do1 的相关参数**

| 参数名称 | 设定值 | 说 明 |
|---|---|---|
| Name | do1 | 设定数字输出信号的名字 |
| Type of Signal | Digital Output | 设定信号的类型 |

续表 4-8

| 参数名称 | 设定值 | 说　明 |
|---|---|---|
| Assigned of Device | board10 | 设定信号所在的 I/O 模块 |
| Device Mapping | 32 | 设定信号所占用的地址 |

其操作如下：

(1) 单击左上角主菜单按钮，选择“控制面板”，并选择“配置”，然后双击“Signal”；单击“添加”，双击“Name”，输入“do1”，然后单击“确定”；双击“Type of Signal”，选择“Digital Output”(见图 4-9)。

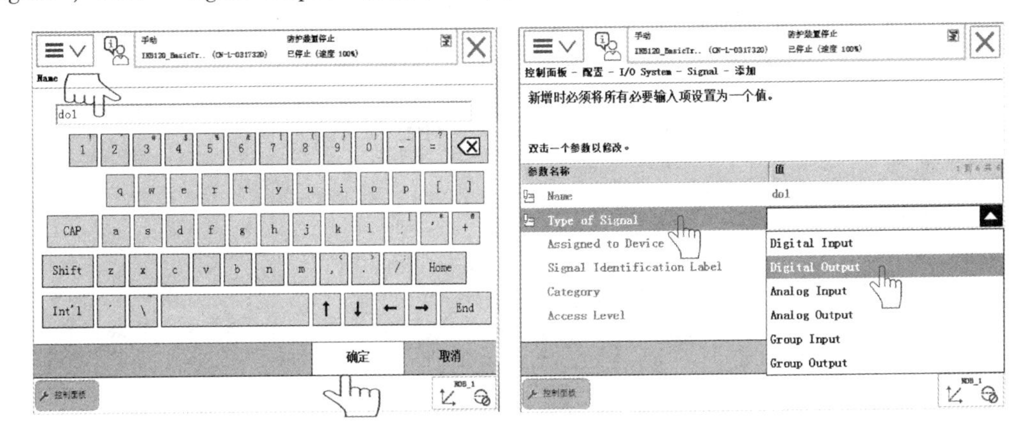


图 4-9　定义数字输出信号 do1

(2) 双击“Assigned to Device”，选择“board10”；双击“ Device Mapping”，输入“32”，然后单击“确定”；单击“确定”；单击“是”，完成设定。

#### 4.1.3.4　定义组输入信号 gi1

组输入信号 gi1 的相关参数及状态见表 4-9 和表 4-10。

**表 4-9　定义组输入信号 gi1 的相关参数**

| 参数名称 | 设定值 | 说　明 |
|---|---|---|
| Name | gi1 | 设定组输入信号的名字 |
| Type of Signal | Group Input | 设定信号的类型 |
| Assigned to Device | board10 | 设定信号所在的 I/O 模块 |
| Device Mapping | 1-4 | 设定信号所占用的地址 |

**表 4-10　定义组输入信号 gi1 的相关状态**

| 状态 | 地址 1 | 地址 2 | 地址 3 | 地址 4 | 十进制数 |
|---|---|---|---|---|---|
| | 1 | 2 | 4 | 8 | |
| 状态 1 | 0 | 1 | 0 | 1 | 2+8=10 |
| 状态 2 | 1 | 0 | 1 | 1 | 1+4+8=13 |

说明：组输入信号就是将几个数字输入信号组合起来使用，用于接受外围设备输入的BCD编码的十进制数。

此例中，gi1 占用地址 1~4 共 4 位，可以代表十进制数 0~15。如此类推，如果占用地址 5 位的话，可以代表十进制数 0~31。

其操作如下：

（1）单击左上角主菜单按钮，选择“控制面板”；选择“配置”，双击“Signal”；单击“添加”，双击“Name”，输入“gi1”，然后单击“确定”；双击“Type of Signal”，选择“Group Input”（见图 4-10）。

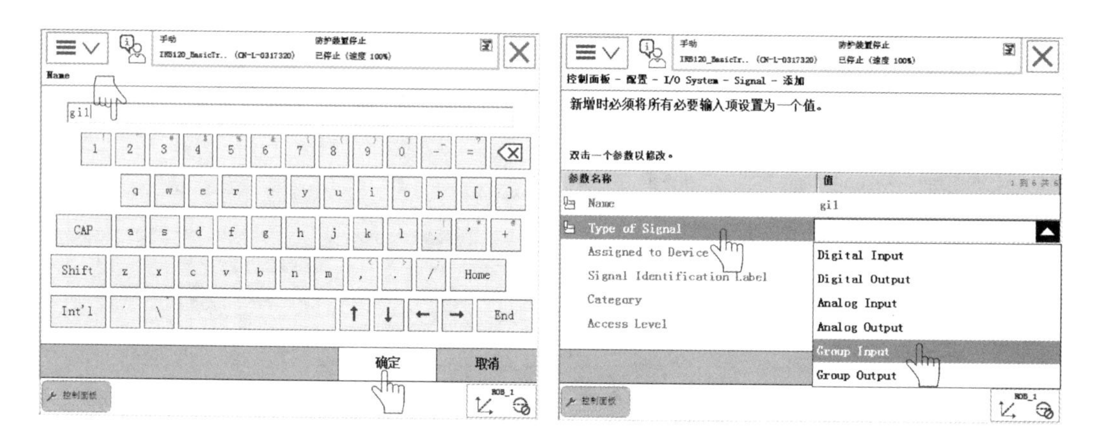


图 4-10　定义数字输入信号 gi1

（2）双击“Assigned to Device”，选择“board10”；双击“Device Mapping”，输入“1-4”，然后单击“确定”；单击“确定”，然后单击“是”，完成设定。

#### 4.1.3.5　定义组输出信号 go1

组输出信号 go1 的相关参数及状态见表 4-11 和表 4-12。

**表 4-11　定义组输入信号 go1 的相关参数**

| 参数名称 | 设定值 | 说　明 |
| --- | --- | --- |
| Name | go1 | 设定值输入信号的名字 |
| Type of Signal | Group output | 设定信号的类型 |
| Assigned to Device | board10 | 设定信号所在的 I/O 模块 |
| Device Mapping | 33-36 | 设定信号所占用的地址 |

**表 4-12　定义组输入信号 go1 的相关状态**

| 状态 | 地址 33 | 地址 34 | 地址 35 | 地址 36 | 十进制数 |
| --- | --- | --- | --- | --- | --- |
| | 1 | 2 | 4 | 8 | |
| 状态 1 | 0 | 1 | 0 | 1 | 2+8=10 |
| 状态 2 | 1 | 0 | 1 | 1 | 1+4+8=13 |

说明：组输出信号就是将几个数字输出信号组合起来使用，用于输出 BCD 编码的十进制数。

此例中，go1 占用地址 33~36 共 4 位，可以代表十进制数 0~15。如此类推，如果占用地址 5 位的话，可以代表十进制数 0~31。

其操作如下：

（1）单击左上角主菜单按钮，选择“控制面板”，选择“配置”；双击“Signal”，单击“添加”；双击“Name”，输入“go1”，然后单击“确定”；双击“Type of Signal”，选择“Group Output”（见图 4-11）。

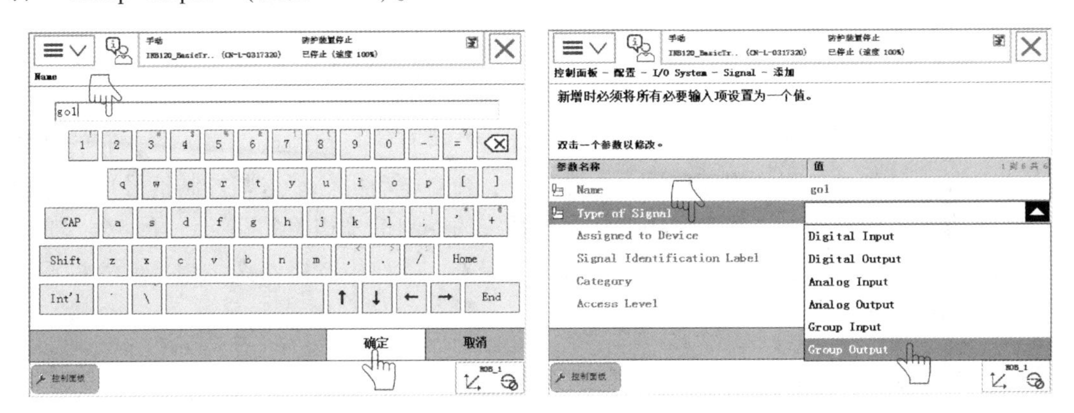


图 4-11　定义数字输出信号

（2）双击“Assigned to Device”，选择“board10”；双击“Device Mapping”，输入“33-36”，然后单击“确定”；单击“确定”；单击“是”，完成设定。

### 4.1.4　任务评价

任务评价表见表 4-13。

**表 4-13　任务评价表**

| 序号 | 评价内容 | 成绩占比 | 自评 | 师评 |
|---|---|---|---|---|
| 1 | ABB 工业机器人 I/O 通信种类 | 15 分 | | |
| 2 | ABB 工业机器人 DSQC652 板卡配置 | 35 分 | | |
| 3 | ABB 工业机器人 DI 与 DO 信号设置 | 15 分 | | |
| 4 | ABB 工业机器人 GI 与 GO 信号设置 | 15 分 | | |
| 5 | 操作工业机器人的安全注意事项 | 20 分 | | |

（1）ABB 工业机器人默认配置是基于 DeviceNet 通信总线协议的 I/O 通信板卡，通过此板卡进行主机控制系统与工业设备之间的通信。DeviceNet 协议是一种 CAN（Controller Area Net，控制器局域网）总线协议，这种传感器/执行器总线系统起源于美国，但在欧洲和亚洲地区正得到越来越广泛的应用。

（2）DSQC651 板卡可以提供对 8 位数字输入信号、8 位数字输出信号和 2 位模拟输出信号的处理。

（3）ABB 工业机器人还可以使用 DSQC667 模块通过 PROFIBUS 或者 DSQC688 模块通过 PROFINET 与 PLC 等外围控制设备进行快捷和大数据量进行通信。

### 4.1.5 任务拓展

（1）在示教器定义一块 DSQC652 的 I/O 板。

（2）请为 DSQC652 板上定义 di1、do2、gi3、go4 信号。

## 任务 4.2 系统输入/输出与 I/O 信号的关联

### 4.2.1 任务引入

工业机器人控制系统中有很多控制信号，这些信号有的与物理按键相关联，有的是系统控制动作，如紧急停止、程序启动、程序暂停、开启电动机等。这些控制信号可以分为输入信号和输出信号两种。将数字输入信号与系统的控制信号相关关联，就可以对系统进行控制。数字输出信号也可以与系统的信号关联起来，以作控制机器人之用，或者将系统的状态传输给外围设备。

### 4.2.2 相关知识

通过外部控制器 I/O 信号与工业机器人控制器 I/O 信号的连接，让机器人接收外部输入信号，并将机器人输出信号发送给外部控制器，让机器人与其他设备能够协同运行。

将数字输入信号与系统的控制信号关联起来，就可以对系统进行控制（例如电机开启、程序启动等）。

系统的状态信号也可以与数字输出信号关联起来，将系统的状态输出给外围设备，以作控制之用。

### 4.2.3 任务实施

#### 4.2.3.1 系统输入/输出与 I/O 信号的关联

建立系统输入“电机开启”与数字输入信号 di1 的关联的操作如下：

（1）单击左上角主菜单按钮，选择“控制面板”；选择“配置”；双击“System Input”；双击“添加”；双击“Signal Name”，选择“di1”，单击“确定”；双击“Action”，选择“Motors On”，单击“确定”（见图 4-12）。

（2）单击“确定”，单击“是”，完成设定。

建立系统输出“电机开启”状态与数字输出信号 do1 的关联的操作如下：

（1）单击左上角主菜单按钮，选择“控制面板”；选择“配置”，双击“System Output”；双击“添加”；双击“Signal Name”，选择“do1”，单击“确定”；双击“Status”，选择“Motors On State”，单击“确定”（见图 4-13）。

（2）单击“确定”，单击“是”，完成设定。

说明：关于系统输入/系统输出的定义详情，请查看 ABB 机器人随机光盘说明书。这

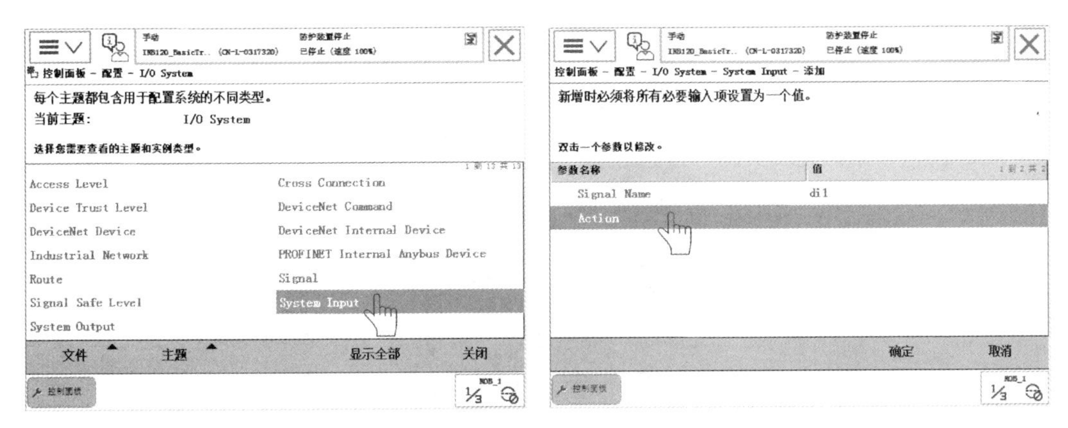


图 4-12　建立系统输入与数字输入信号 di1 关联

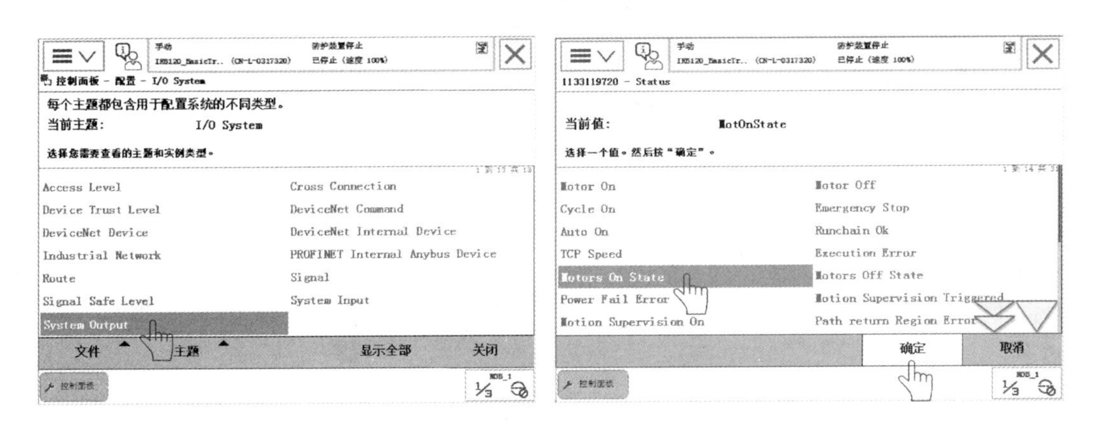


图 4-13　建立系统输出与数字输出信号 do1 关联

时就可以在示教器上的“输入输出”菜单查看相关信号的变化了。

#### 4.2.3.2　示教器可编程按键的使用

在示教器上的可编程按键可以为可编程按键分配想快捷控制的 I/O 信号，以方便对 I/O信号进行强制和仿真操作。示教器上可编程按键如图 4-14 所示。

例如，为可编程按键 1 配置数字输出信号 do1 的操作如下：

(1) 单击左上角主菜单按钮，选择“控制面板”；选择“配置可编程按键”，在“类型”中，选择“输出”(见图 4-15)。

(2) 选中“do1”，在“按下按键”中选择“按下/松开”，也可以根据实际需要选择按键的动作特性，单击“确定”，完成设定，现在就可以通过可编程按键<1>在手动状态下对 do1 进行强制的操作。

### 4.2.4　任务评价

任务评价表见表 4-14。

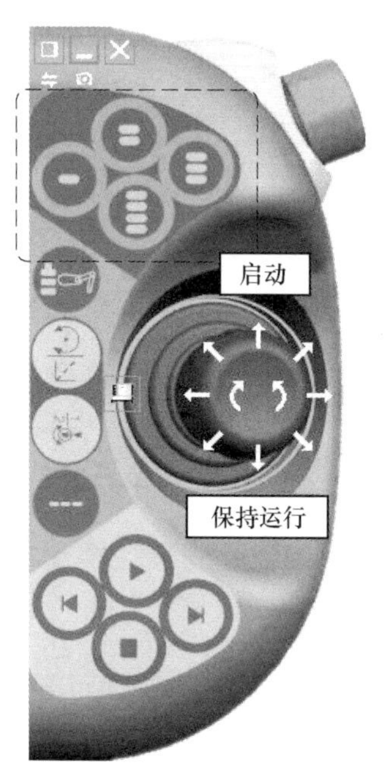


图 4-14　示教器上可编程按键

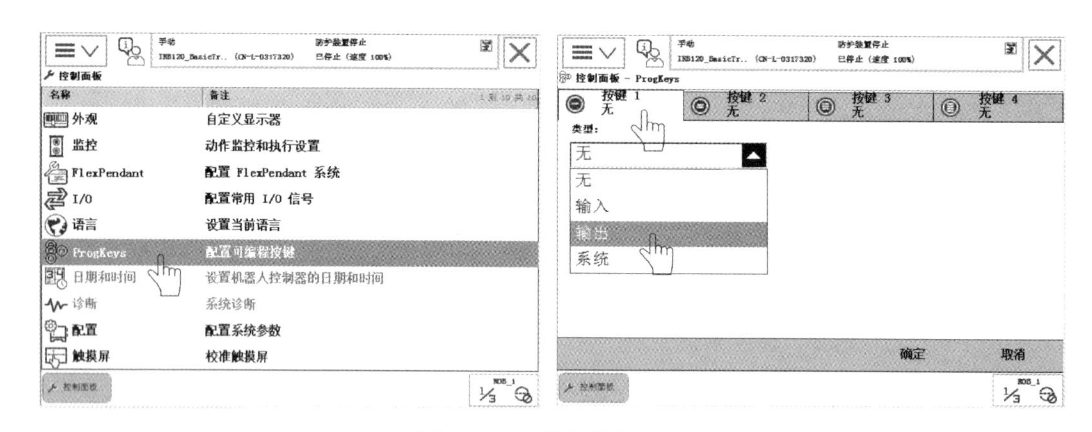


图 4-15　可编程按键配置

**表 4-14　任务评价表**

| 序号 | 评价内容 | 成绩占比 | 自评 | 师评 |
|---|---|---|---|---|
| 1 | ABB 工业机器人系统输入输出的设置 | 10 分 | | |
| 2 | ABB 工业机器人可编程快捷按键的使用 | 15 分 | | |
| 3 | ABB 工业机器人系统控制电机上电操作 | 25 分 | | |
| 4 | ABB 工业机器人系统控制灯的关联 | 35 分 | | |
| 5 | 操作工业机器人的安全注意事项 | 15 分 | | |

ABB 工业机器人控制系统中有很多的控制信号，这些信号有的与物理按键相关联，有

的是系统控制动作，如紧急停止、程序启动和程序暂停等。将数字输入信号与系统的控制信号关联起来，就可以对系统进行控制，例如开启电动机、启动程序等。

### 4.2.5 任务拓展

（1）请配置一个与电动机上电的关联系统输入信号。

（2）请配置一个灯与电动机状态的关联系统输出信号。

（3）请配置两个灯与电动机上电、断电的关联系统输入输出信号。

（4）请配置一个工业机器人手抓打开、关闭的可编程快捷按键。

（5）请配置一个工业机器人吸盘手抓吸气、放气的可编程快捷按键。

# 项目 5　ABB 工业机器人的程序数据

工业机器人数据是指可以被机器人程序处理的，具有特定含义的数字、字母、符号和模拟量的总称。工业机器人数据既可以是一种环境数据，也可以是单纯的值。对 ABB 工业机器人进行编程，需要使用特定的语言 RAPID 以及 ABB 机器人专用的编程环境。ABB 工业机器人的数据是在编程环境的程序模块或系统模块中建立的，并可以由同一个模块或其他模块中的指令引用。本项目介绍了 ABB 工业机器人认识和建立程序数据、三个关键程序数据的设定相关内容，能够更快了解 ABB 工业机器人编程时使用到的程序数据的类型和分类，如何创建程序数据，以及掌握最重要的三个关键程序数据的设定方法。

## 任务 5.1　认识和建立程序数据

### 5.1.1　任务引入

工业机器人是典型的机电一体化设备，在操作 ABB 工业机器人设备前必须要学会工业机器人的相关程序数据。程序数据是用于编辑机器人程序的指令、符号、运算等功能集合，使用户能够与机器人系统进行简单的人机交互。

### 5.1.2　相关知识

程序内声明的数据被称为程序数据。数据是信息的载体，它能够被计算机识别、存储和加工处理。它是计算机程序加工的原料，应用程序处理各种各样的数据。计算机科学中，数据是指计算机加工处理的对象，它可以是数值数据，也可以是非数值数据。数值数据是一些整数、实数或复数，主要用于工程计算、科学计算和商务处理等；非数值数据包括字符、文字、图形、图像、语音等。本任务可以了解 ABB 机器人编程会使用到的程序数据类型及分类，如何创建程序数据，以及最重要的三个关键程序数据（Tooldata，Wobjdata，and Loaddata）的设定方法。

学习程序数据的程序数据的类型分类与存储类型两个主题，可以使大家对程序数据有一个认识，并能根据实际的需要选择程序数据，根据不同的数据用途定义不同的程序数据。

### 5.1.3　任务实施

#### 5.1.3.1　程序数据的认知

程序数据是在程序模块或系统模块中设定值和定义一些环境数据。创建的程序数据由同一个模块或其他模块中的指令进行引用。如图 5-1 所示，虚线框中是一条常用的机器人关节运动的指令（MoveJ），并调用了 4 个程序数据。

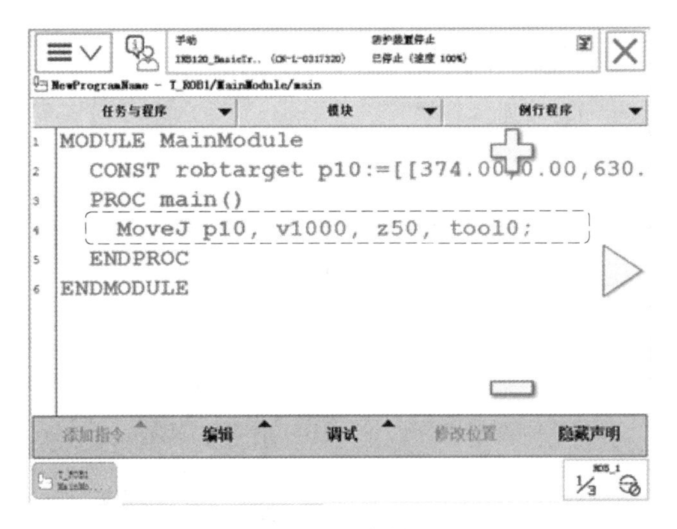


图 5-1　程序数据

图 5-1 中所使用的程序数据说明见表 5-1。

**表 5-1　程序数据说明**

| 程序数据 | 数据类型 | 说　明 |
|---|---|---|
| p10 | robtarget | 机器人运动目标位置数据 |
| v1000 | speeddata | 机器人运动速度数据 |
| z50 | zonedata | 机器人运动转弯数据 |
| tool0 | tooldata | 机器人工具数据 TCP |

### 5.1.3.2　程序数据的建立

程序数据的建立一般可以分为两种形式，一种是直接在示教器中的程序数据画面中建立程序数据；另一种是在建立程序指令时，同时自动生成对应的程序数据。

**建立 bool 类型程序数据**

建立 bool 类型程序数据的操作步骤为：

(1) 单击左上角主菜单按钮，选择“程序数据”，选择数据类型“bool”，然后单击“显示数据”(见图 5-2)。

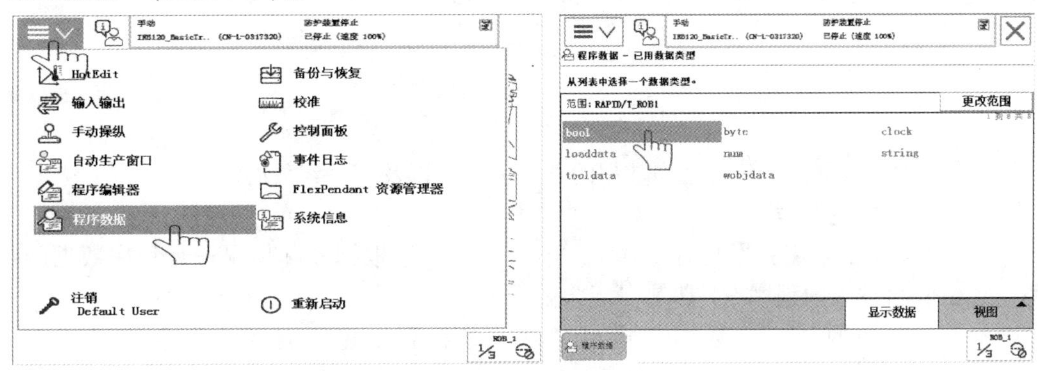


图 5-2　建立 bool 类型程序数据

（2）单击“新建...”，单击下拉按钮进行名称的设定，选择对应的参数，单击“确定”完成设定。

数据设定参数及说明见表 5-2。

**表 5-2　数据设定参数**

| 数据设定参数 | 说　明 |
| --- | --- |
| 名称 | 设定数据的名称 |
| 范围 | 设定数据可使用的范围 |
| 存储类型 | 设定数据的可存储类型 |
| 任务 | 设定数据所在的任务 |
| 模块 | 设定数据所在的模块 |
| 例行程序 | 设定数据所在的例行程序 |
| 维数 | 设定数据的维数 |
| 初始值 | 设定数据的初始值 |

**建立 num 类型程序数据**

建立 num 类型程序数据的操作步骤为：单击左上角主菜单按钮，选择“程序数据”，选择数据类型“num”；单击“显示数据”；单击“新建…”，单击此按钮进行名称的设定；单击下拉菜单选择对应的参数，单击“确定”完成设定（见图 5-3）。

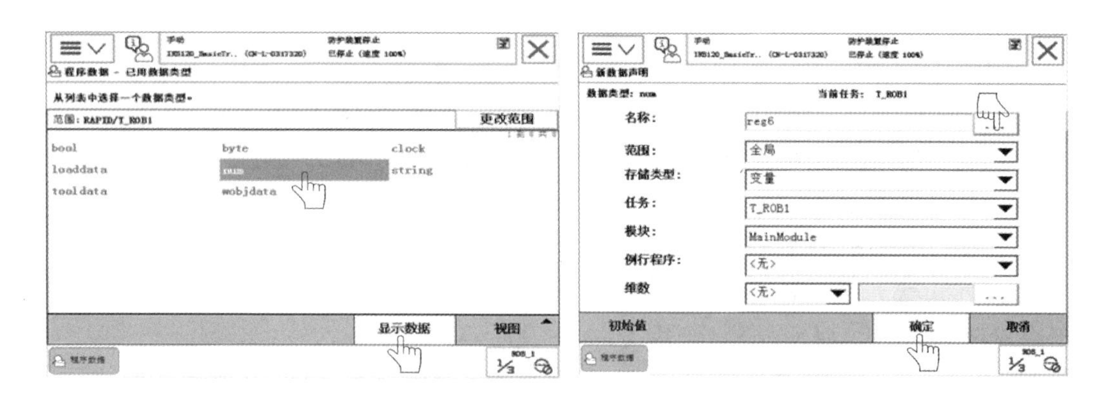


图 5-3　建立“num”类型程序数据

至此，大家就掌握了建立程序数据的基本方法，以及相关参数的定义与设定方法。

**程序数据的类型分类**

ABB 机器人的程序数据共有 100 个左右，并且可以根据实际情况进行程序数据的创建，为 ABB 机器人的程序设计带来了无限的可能。

在示教器中的“程序数据”窗口，就可查看和创建所需要的程序数据（见图 5-4）。

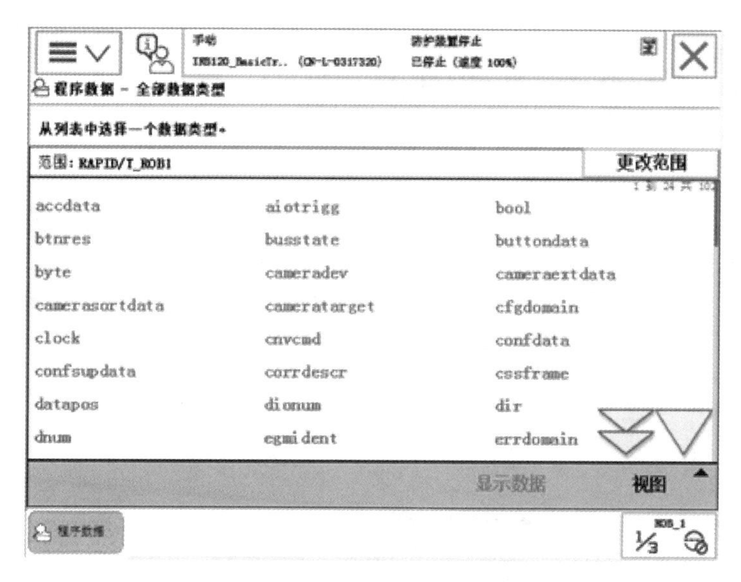


图 5-4　程序数据的分类

**程序数据的存储类型**

程序数据的存储类型包括：

（1）变量 VAR。变量型数据在程序执行的过程和停止时，会保持当前的值。但如果程序指针复位或者机器人控制器重启，数值会恢复为声明变量时赋予的初始值。

说明："VAR" 表示存储类型为变量；"num" 表示声明的数据是数字型数据（存储的内容为数字）。

如图 5-5 所示，"VAR num length:=0;" 是指名称为 "length" 的变量型数值数据；"VAR string name:="Tom";" 是指名称为 "name" 的变量型字符数据；"VAR bool finished:=FALSE;" 是指名称为 "finished" 的变量型布尔量数据。

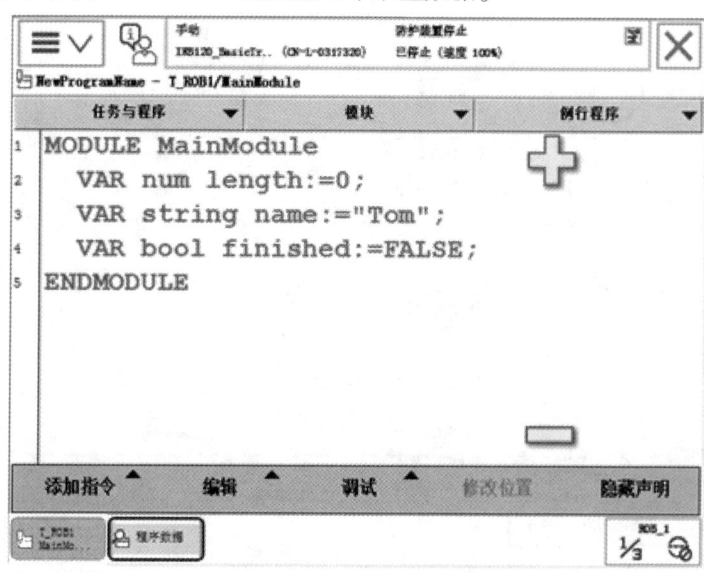


图 5-5　程序编辑窗口（变量存储类型）

说明：在声明数据时，可以定义变量数据的初始值。如：length 的初始值为 0；name 的初始值为 Tom；finished 初始值为 FALSE。

在机器人执行的 RAPID 的程序中也可以对变量存储类型程序数据进行赋值的操作，如图 5-6 所示。

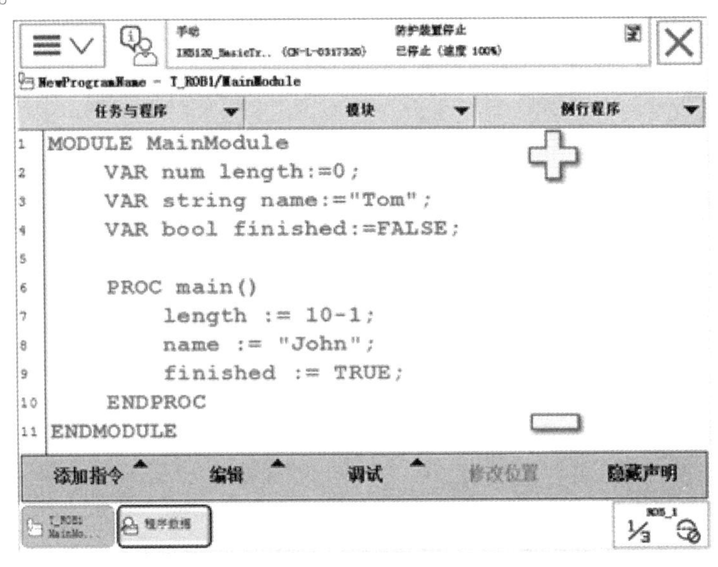


图 5-6　程序编辑窗口（变量赋值）

说明：在程序中执行变量型程序数据的赋值，以及指针复位或者机器人控制器重启后，都将恢复为初始值。

（2）可变量 PERS。无论程序的指针如何变化、机器人控制器是否重启，可变量型的数据都会保持最后赋予的值。

如图 5-7 所示，“PERS num numb:=1;”是指名称为“numb”的数值数据；“PERS string text:="Hello";”是指名称为“text”的字符数据。

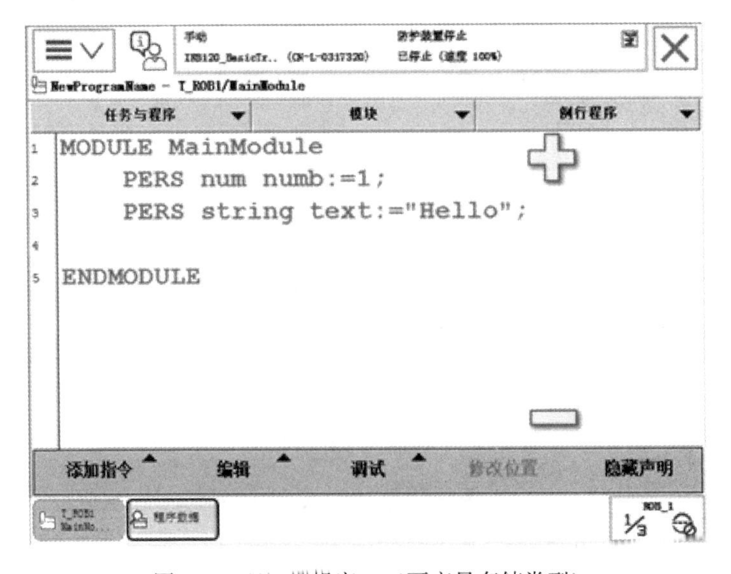


图 5-7　程序编辑窗口（可变量存储类型）

说明：PERS 表示存储类型为可变量。

在机器人执行的 RAPID 的程序中也可以对可变量存储类型程序数据进行赋值的操作，如图 5-8 所示。

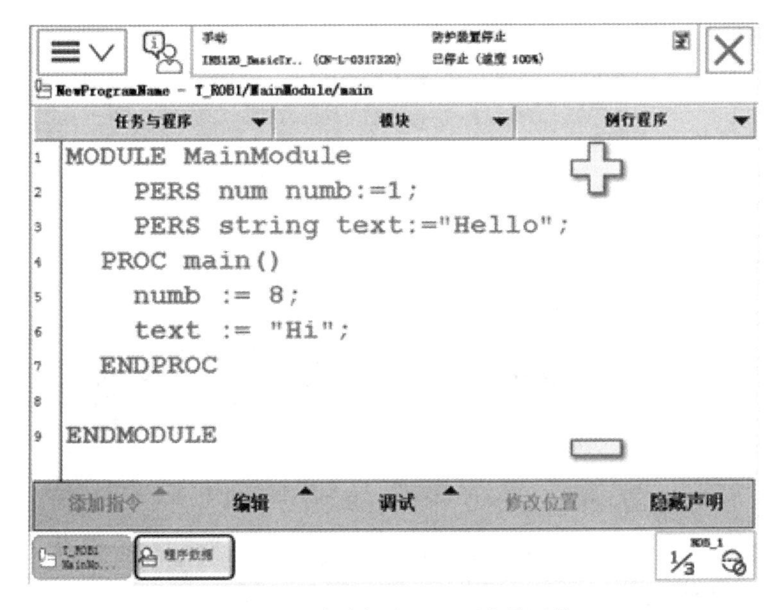


图 5-8　程序编辑窗口（可变量赋值）

在程序执行以后，赋值的结果会一直保持到下一次对其进行重新赋值，如图 5-9 所示。

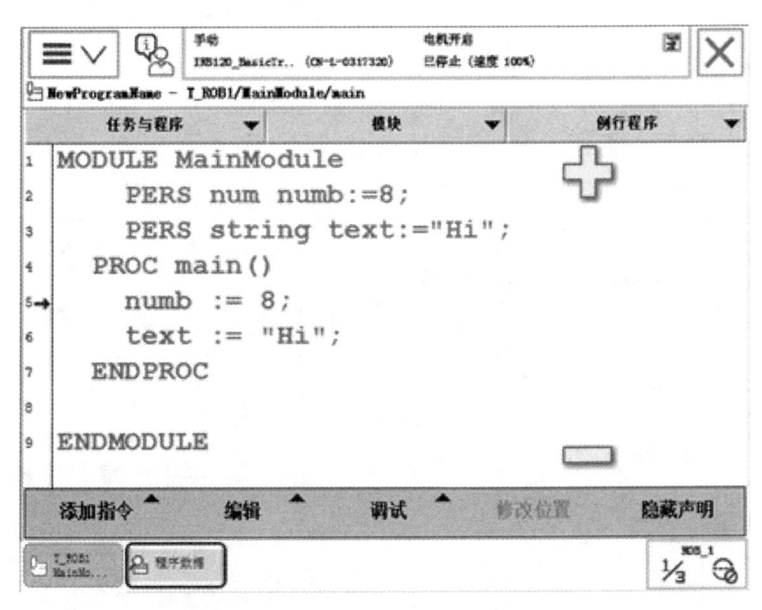


图 5-9　程序编辑窗口（可变量赋值）

（3）常量 CONST。常量的特点是在定义时已赋予了数值，并不能在程序中进行修改，只能手动修改。

如图 5-10 所示，“CONST num gravity：= 9. 81;” 是指名称为 “gravity” 的数值数据。

“CONST string greating：="Hello"；” 是指名称为 “greating” 的字符数据。

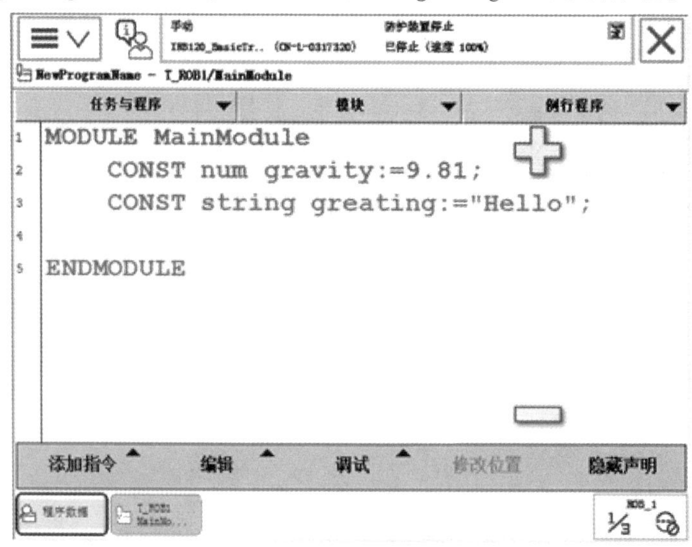


图 5-10　程序编辑窗口（常量存储类型）

说明：存储类型为常量的程序数据，不允许在程序中进行赋值的操作。

**常用程序数据说明**

常用的程序数据包括：

（1）数值数据 num。num 用于存储数值数据，比如计数器。

num 数据类型的值可以为：整数（如-5）；小数（如 3.45）；也可以指数的形式写入，比如 2E3 （ = 2 * 10^3 = 2000），2.5E-2 （ = 0.025） 等。整数数值始终将 - 8388607 ~ +8388608的整数作为准确的整数储存。小数数值仅为近似数字，因此不得用于等于或不等于对比。若为使用小数的除法和运算，则结果亦将为小数。

数值数据 num 示例：将整数 3 赋值给名称为 count1 的数值数据（见图 5-11）。

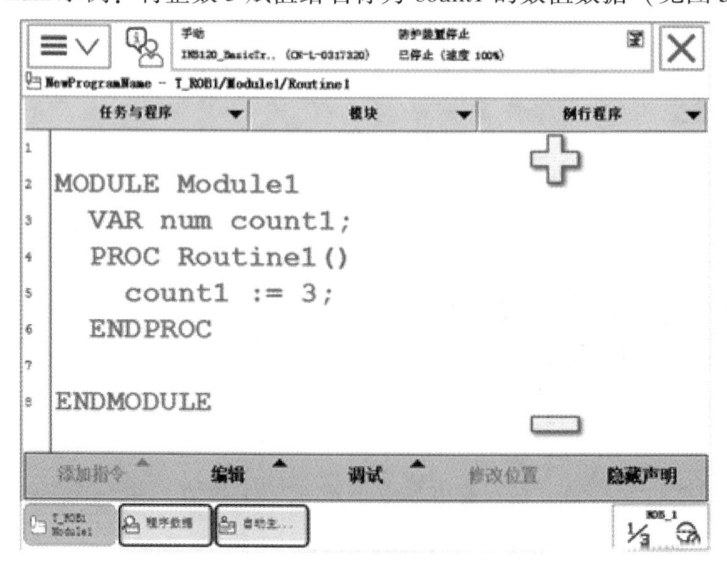


图 5-11　数值数据

（2）逻辑值数据 bool。bool 用于存储逻辑值（真/假）数据，即 bool 型数据值可以为 TRUE 或 FALSE。

逻辑值数据 bool 示例：首先判断 count1 中的数值是否大于 100，如果大于 100，则向 highvalue 赋值 TRUE，否则赋值 FALSE（见图 5-12）。

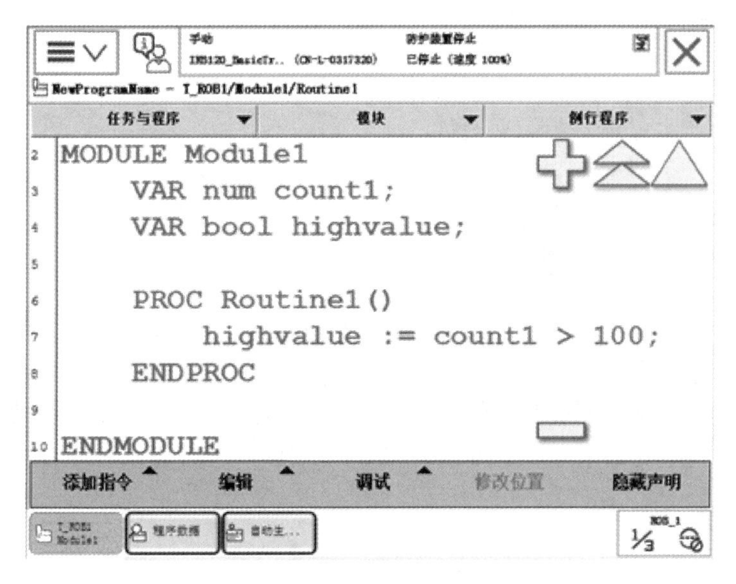


图 5-12　逻辑值数据

（3）字符串数据 string。string 用于存储字符串数据。字符串是由一串前后附有引号（" "）的字符（最多 80 个）组成，例如，" This is a character string" 。

如果字符串中包括反斜线（\），则必须写两个反斜线符号，例如，" This string contains a \\ character" 。

字符串数据 string 示例：将 start welding pipe 1 赋值给 text，运行程序后，在示教器中的操作员窗口将会显示 start welding pipe 1 这段字符串（见图 5-13）。

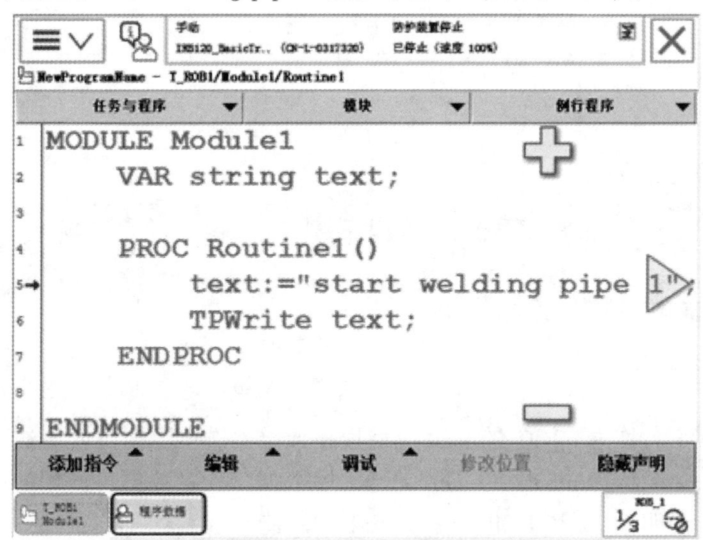


图 5-13　字符串数据

## 5.1.4 任务评价

任务评价表见表 5-3。

**表 5-3 任务评价表**

| 序号 | 评价内容 | 成绩占比 | 自评 | 师评 |
|---|---|---|---|---|
| 1 | 工业机器人的程序数据 | 10 分 | | |
| 2 | 工业机器人的程序数据构成 | 20 分 | | |
| 3 | 工业机器人的程序数据建立 | 25 分 | | |
| 4 | 工业机器人程序数据的类型及说明 | 25 分 | | |
| 5 | 操作工业机器人的安全注意事项 | 20 分 | | |

（1）RAPID 语言中，工具、位置、负载等不同的信息都以数据形式保存。数据由用户创建和声明，并可任意命名。

（2）数据建立时需要明确数据的存储方式，即数据存储类型，以分配存储空间。常用的数据存储类型包括变量、可变量和常量三种，分别用 VAR、PERS 和 CONST 表示。

## 5.1.5 任务拓展

（1）请分别指出图 5-14 中的程序数据。

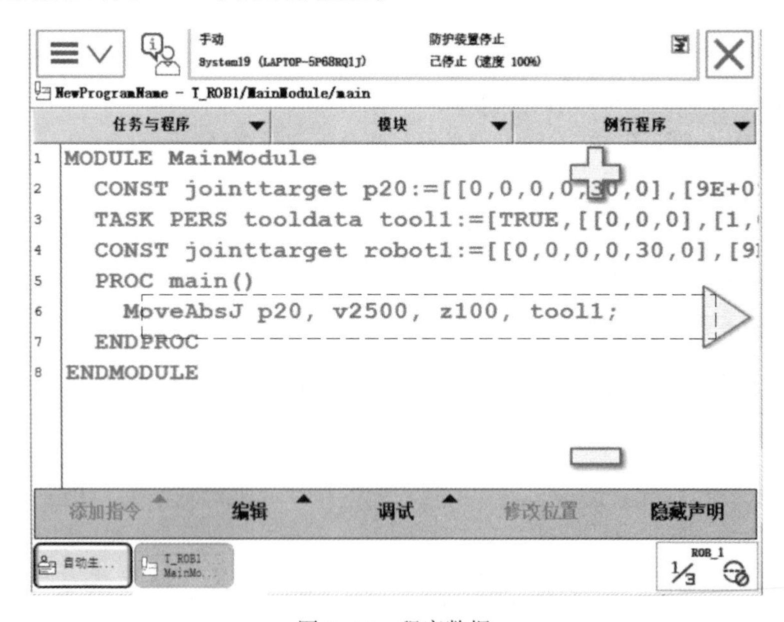


图 5-14 程序数据

（2）建立一个名称为 robot1 的 bool 类型程序数据，设定数据可使用的范围为本地。

（3）建立一个名称为 robot2 的 num 类型程序数据，设定数据可使用的范围为任务。

（4）请简要说明 VAR 与 PERS 与 CONST 的功能及区别。

（5）建立一个将“Hello my friend”赋值给 text 的字符串程序数据。

# 任务 5.2　三个关键程序数据的设定

## 5.2.1　任务引入

工业机器人在工作之前一定要根据工作内容进行程序编程，编辑工业机器人应用程序。搭建编程环境的第一步就是要建立三个必要的基础数据（工具数据、工件坐标和负荷数据。

## 5.2.2　相关知识

### 5.2.2.1　工具数据

工具数据 tooldata 是用于描述安装在机器人第六轴上的工具的 TCP，质量，重心等参数数据。不同的机器人应用就可能配置不同的工具，比如弧焊的机器人使用弧焊枪作为工具；用于搬运板材的机器人使用吸盘式的夹具作为工具（见图 5-15）。

默认工具（tool0）的工具中心点（Tool Center Ponit）位于机器人安装法兰的中心，如图 5-16 所示，图中的 A 点就是原始的 TCP 点。

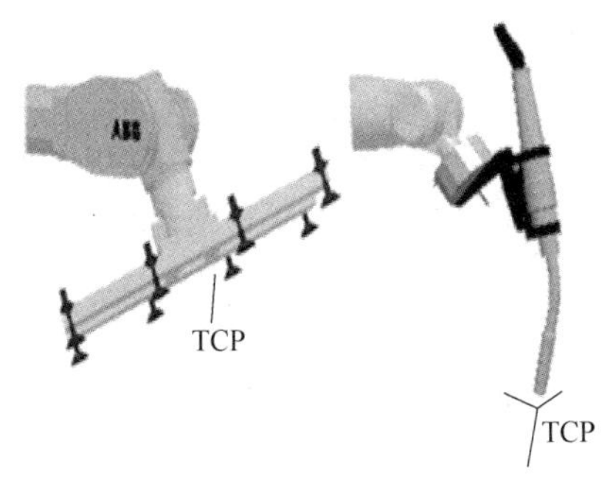


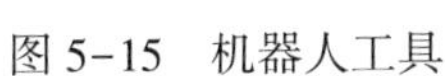

图 5-15　机器人工具

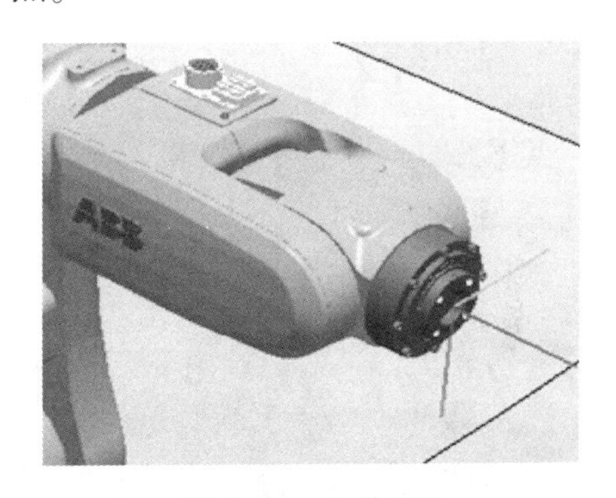

图 5-16　机器 TCP

数据组成 tooldata 用于描述工具（如焊枪或夹具）的特征。此类特征包括工具中心点（TCP）的位置和方位，以及工具负载的物理特征。

说明：如果是使用固定工具，则定义的工具坐标系是相对于世界坐标系。

工具数据 tooldata 示例：

```
PERS tooldata gripper:=[TRUE,[[97.4,0,223.1],[0.924,0,0.383,0]],[5,[23,0,75],[1,0,0,0],0,0,0]];
```

工具数据 gripper 定义内容为：

（1）机械臂正夹持着工具。

（2）TCP 所在点沿着工具坐标系 X 方向偏移 97.4mm，沿工具坐标系 Z 方向偏移 223.1 mm。

（3）工具的 X 方向和 Z 方向相对于腕坐标系 Y 方向旋转 45°。

（4）工具质量为 5kg。

（5）重心所在点沿着腕坐标系 X 方向偏移 23mm，沿腕坐标系 Z 方向偏移 75mm。可将负载视为一个点质量，即不带转矩惯量。

#### 5.2.2.2 工件坐标数据 Wobjdata

工件坐标系对应工件定义工件相对于大地坐标系（或其他坐标系）的位置。机器人可以拥有若干工件坐标系，或表示不同工件，或表示同一工件在不同位置的若干副本。

对机器人进行编程时就是在工件坐标系中创建目标和路径。其优点包括：

（1）当重新定位工作站中的工件时，只需更改工件坐标系的位置，所有路径将即刻随之更新。

（2）因为整个工件可连同其路径一起移动，所以允许操作以外轴或传送导轨移动的工件。

说明：A 是机器人的大地坐标，这是为了方便编程为第一个工件建立一个工件坐标 B，并在这个工件坐标 B 进行轨迹编程。

如果台子上还有一个一样的工件需要走一样的轨迹，只需要建立一个工件坐标 C，将工件坐标 B 中的轨迹复制一份，然后将工件坐标从 B 更新为 C，无需再对一样的工件重复轨迹编程（见图 5-17）。

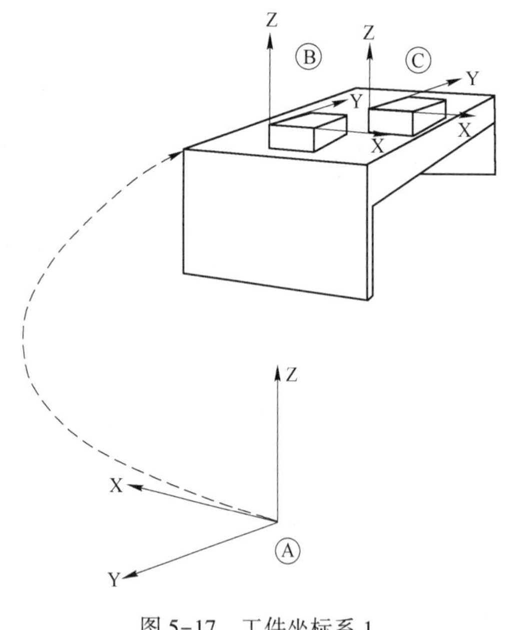


图 5-17　工件坐标系 1

说明：在工件坐标 B 中对 A 对象进行轨迹编程。如果工件坐标的位置变化成工件坐标 D 后，只需在机器人系统重新定义工件坐标 D，则机器人的轨迹就自动更新到 C，不需要再次轨迹编程。因 A 相对于 B，C 相对于 D 的关系是一样，并没有因为整体偏移而发生变化（见图 5-18）。

如果在运动指令中指定了工件，则目标点位置将基于该工件坐标系。其优势包括：

（1）便捷地手动输入位置数据，比如离线编程可从图纸获得位置数值。

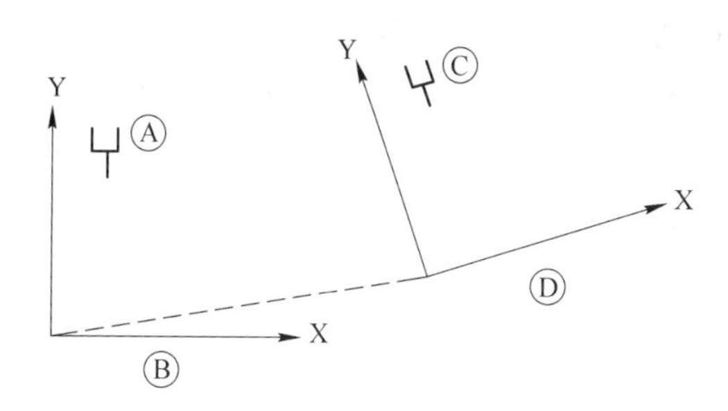


图 5-18　工件坐标系 2

（2）轨迹程序可以根据变化，快速重新使用。比如，如果移动了工作台，则仅需重新定义工作台工件坐标系即可。

（3）可根据变化对工件坐标系进行补偿。比如利用传感器来获得偏差数据来定位工件。

工件数据 wobjdata 示例如下：

```
PERSwobjdatawobj1:=[FALSE,TRUE,"",[[300,600,200],[1,0,0,0]],[[0,200,30],[1,0,0,0]]];
```

工件数据 wobj1 定义内容为：

（1）机械臂未夹持着工件。

（2）使用固定的用户坐标系。

（3）用户坐标系不旋转，且在大地坐标系中用户坐标系的原点为 $x=300$mm，$y=600$mm，$z=200$mm。

（4）目标坐标系不旋转，且在用户坐标系中目标坐标系的原点为：$x=0$mm，$y=200$mm，$z=30$mm。

### 5.2.2.3　有效载荷数据 Loaddata

对于搬运应用的机器人，应该正确设定夹具的质量、重心 Tooldata，以及搬运对象的质量和重心数据 Loaddata（见图 5-19）。

Loaddata 用于设置机器人轴 6 上安装法兰的负载载荷数据。

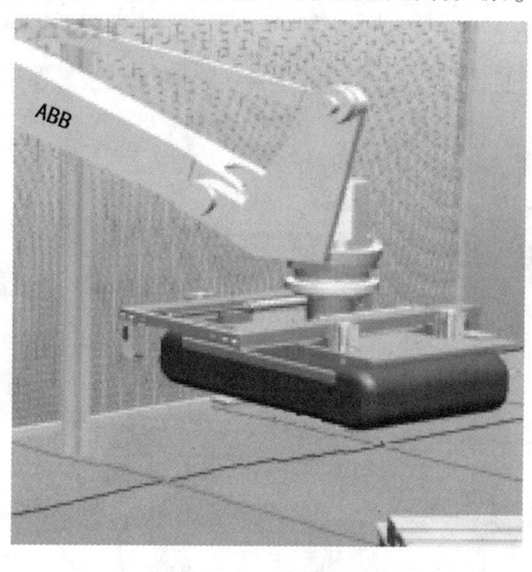


图 5-19　有效载荷

载荷数据常常定义机器人的有效负载或抓取物的负载（通过指令 Grip Load 或 Mech Unit Load 来设置），即机器人夹具所夹持的负载。同时将 Loaddata 作为 Tooldata 的组成部分，以描述工具负载。

载荷数据 Loaddata 示例：

```
PERS loaddata piece1:=[5,[50,0,50],[1,0,0,0],0,0,0];
```

载荷数据 piece1 定义内容如下：

（1）质量 5kg；

（2）重心为 $x=50$mm，$y=0$mm 和 $z=50$mm，相对于工具坐标系；

（3）有效负载为一个点质量。

### 5.2.3 任务实施

#### 5.2.3.1 工具坐标 Tooldata 数据的设定

工具中心点的设定原理如下：

（1）首先在机器人工作范围内找一个非常精确的固定点作为参考点。

（2）然后在工具上确定一个参考点（最好是工具的中心点）。

（3）通过之前学习到的手动操纵机器人的方法，去移动工具上的参考点（最少四种不同的机器人姿态尽可能与固定点刚好碰上）。为了获得更准确的 TCP，下面例子中使用六点法进行操作，第四点是用工具的参考点垂直于固定点，第五点是工具参考点从固定点向将要设定为 TCP 的 X 方向移动，第六点是工具参考点从固定点向将要设定为 TCP 的 Z 方向移动。

（4）机器人可以通过这四个位置点的位置数据计算求得 TCP 的数据，然后 TCP 的数据就保存在 Tooldata 这个程序数据中被程序进行调用。

其操作方法如下：

（1）单击左上角主菜单按钮，选择“手动操纵”；选择“工具坐标”；点击“新建”；对工具数据属性进行设定后，点击“确定”；选中 tool1 后，点击“编辑”菜单中的“定义”选项，选择“TCP 和 Z，X”方法设定 TCP，选择合适的手动操纵模式，按下使能键，使用摇杆使工具参考点去靠上固定点，作为第一个点（见图 5-20）。

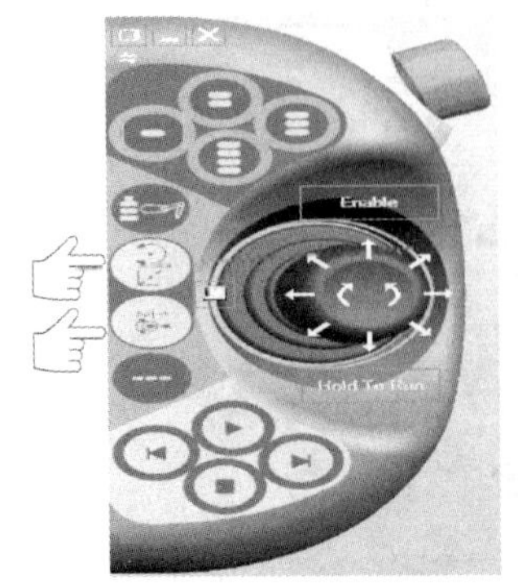

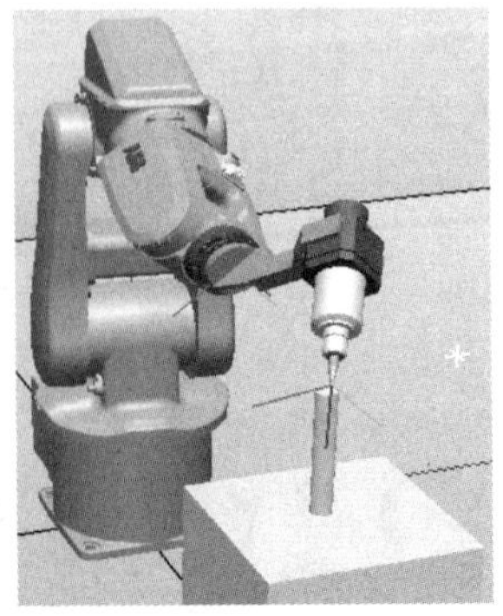

图5-20 建立工具数据“点1”

（2）选中“点 1”，点击“修改位置”，将“点 1”位置记录下来，工具参考点以此姿态靠上固定点（见图 5–21）。

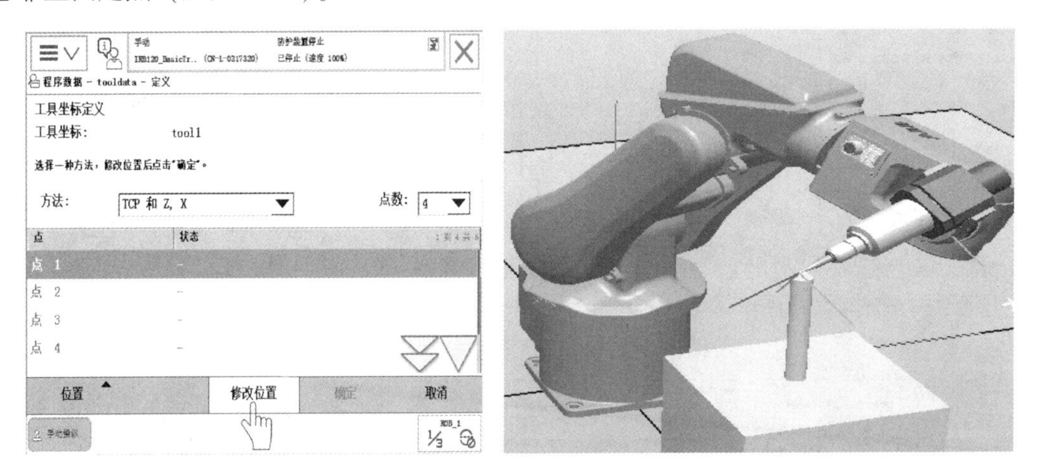


图 5–21　建立工具数据“点 2”

（3）选中“点 2”，点击“修改位置”，将“点 2”位置记录下来，工具参考点在变化姿态靠上固定点；选中“点 3”，点击“修改位置”，将“点 3”位置记录下来，工具参考点以此姿态靠上固定点（见图 5–22）。

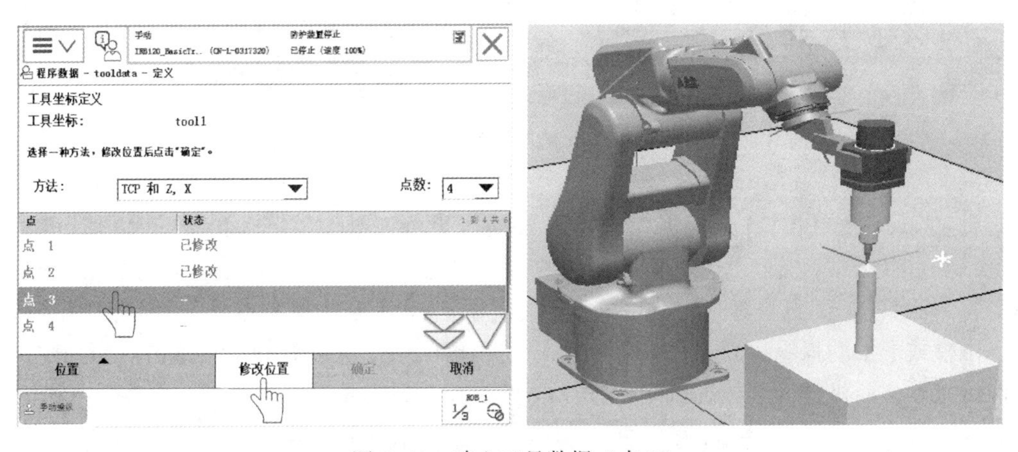


图 5–22　建立工具数据“点 3”

（4）选中“点 4”，点击“修改位置”，将“点 4”位置记录下来，工具参考点以此姿态靠上固定点（见图 5–23）。

（5）选中“延伸器点 X”，点击“修改位置”，将“延伸器点 X”位置记录下来，单击“是”，完成设定（见图 5–24）。

（6）选中延伸器点 Z，点击“修改位置”，将延伸器点 Z 位置记录下来，点击“确定”，完成设定，对误差进行确认（当然是越小越好了，但也要以实际验证效果为准）。接着设置 tool1 的质量和重心，选中 tool1，然后打开编辑菜单选择“更改值”，页面显示的内容就是 TCP 定义时生成的数据；点击箭头向下翻页，在此页面中，根据实际情况设定工

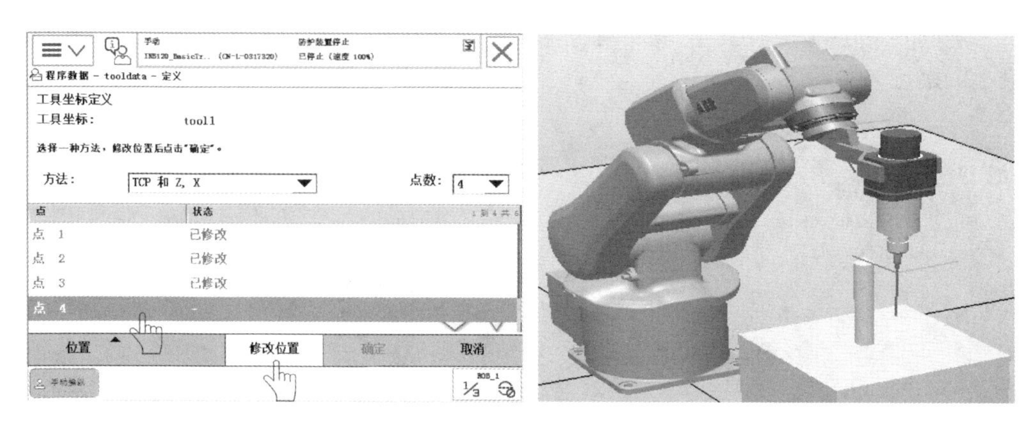


图 5-23 建立工具数据“点 4”

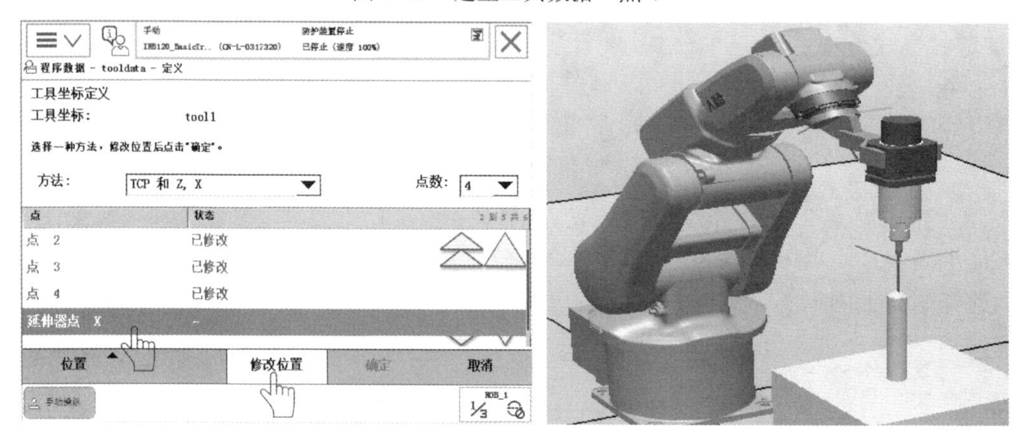


图 5-24 建立工具数据“延伸器点 X”

具的质量 mass（单位：kg）和重心位置数据（此重心是基于 tool0 的偏移值，单位：mm），然后点击“确定”；选中 tool1，点击“确定”，动作模式选定为“重定位”，坐标系选定为“工具”，工具坐标选定为“tool1”。使用摇杆将工具参考点靠上固定点，然后在重定位模式下手动操纵机器人，如果 TCP 设定精确的话，可以看到工具参考点与固定点始终保持接触，而机器人会根据重定位操作改变姿态（见图 5-25）。

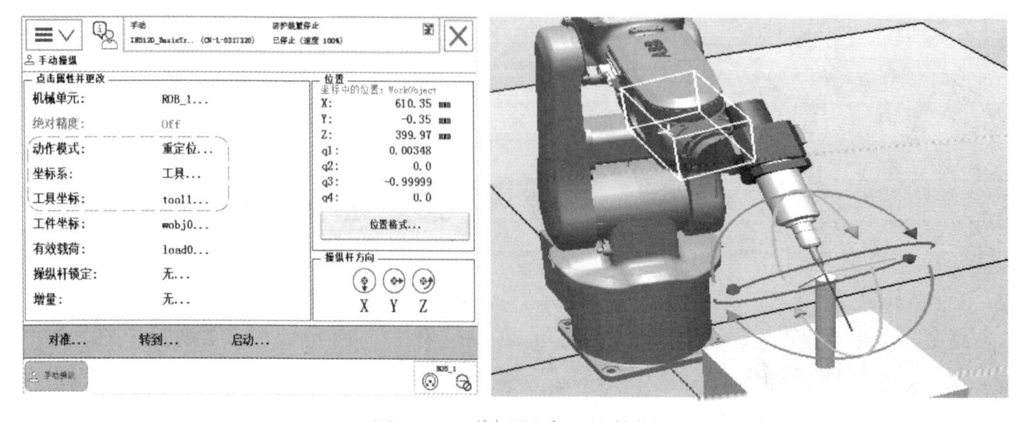


图 5-25 验证新建工具数据

#### 5.2.3.2 工件坐标 Wobjdata 数据的设定

在对象的平面上，只需要定义三个点，就可以建立一个工件坐标（见图 5-26），即：

（1）X1、X2 确定工件坐标 X 正方向；

（2）Y1 确定工件坐标 Y 正方向；

（3）工件坐标系的原点是 Y1 在工件坐标 X 上的投影。

工件坐标符合右手定则如图 5-27 所示。

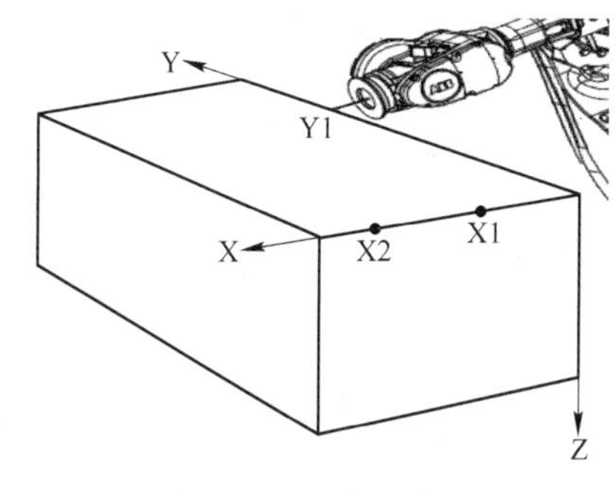


图 5-26 工件坐标系

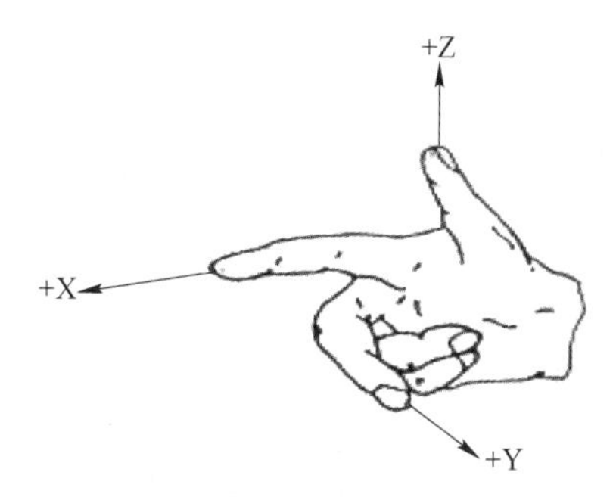


图 5-27 右手定则

建立工件坐标的操作步骤为：

（1）单击左上角主菜单按钮，选择“手动操纵”，选择“工件坐标”；点击“新建”，对工件数据属性进行设定后，点击“确定”；选中 wobj1 后，点击“编辑”菜单中的“定义”选项，用户方法选择“3 点”；手动操作机器人的工具参考点靠近定义工件坐标系的 X1 点，选中“用户点 X1”，点击“修改位置”，将点 X1 位置记录下来（见图 5-28）。

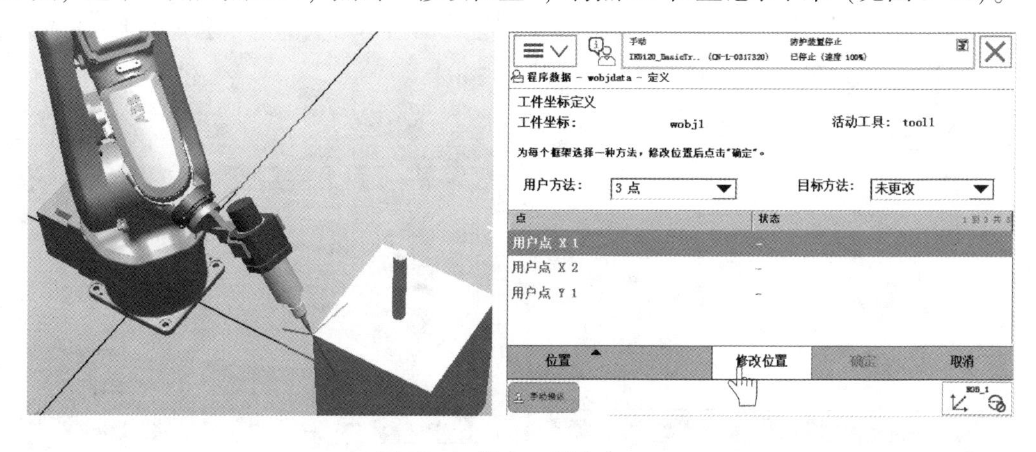


图 5-28 设定“用户点 X1”

（2）手动操作机器人的工具参考点靠近定义工件坐标系的 X2 点，选中“用户点 X2”，点击“修改位置”，将点 X2 位置记录下来；手动操作机器人的工具参考点靠近定义工件坐标系的 Y1 点，选中“用户点 Y1”，点击“修改位置”，将点 Y1 位置记录下来（见图 5-29）。

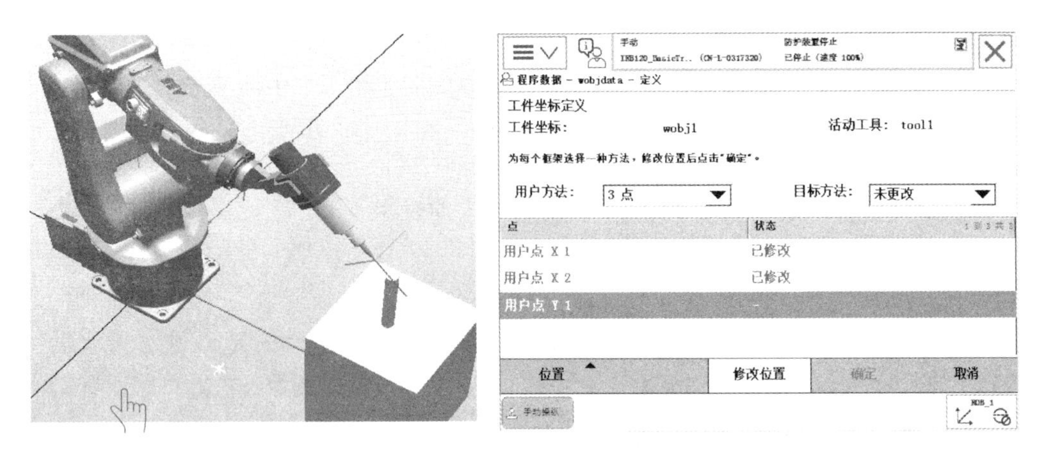


图 5-29　设定“用户点 Y1”

（3）点击“确定”，完成设定；对自动生成的工件坐标数据进行确认后，点击“确定”；选中“wobj1”，点击“确定”，动作模式选定为“线性”，坐标系选定为“工件坐标”，工件坐标选定为“wobj1”；设定手动操纵画面项目，使用线性动作模式，体验新建立的工件坐标。

#### 5.2.3.3　有效载荷 Loaddata 数据的设定

载荷数据 Loaddata 的创建步骤为：

单击左上角主菜单按钮，选择“手动操纵”，选择“有效载荷”；点击“新建”，根据需要设定数据的属性，一般是不用修改的，点击“初始值”，对有效载荷的数据根据实际的情况进行设定，点击“确定”；夹具夹紧，指定当前搬运对象的重量和重心 load1，夹具松开，将搬运对象清除为 load0（见图 5-30）。

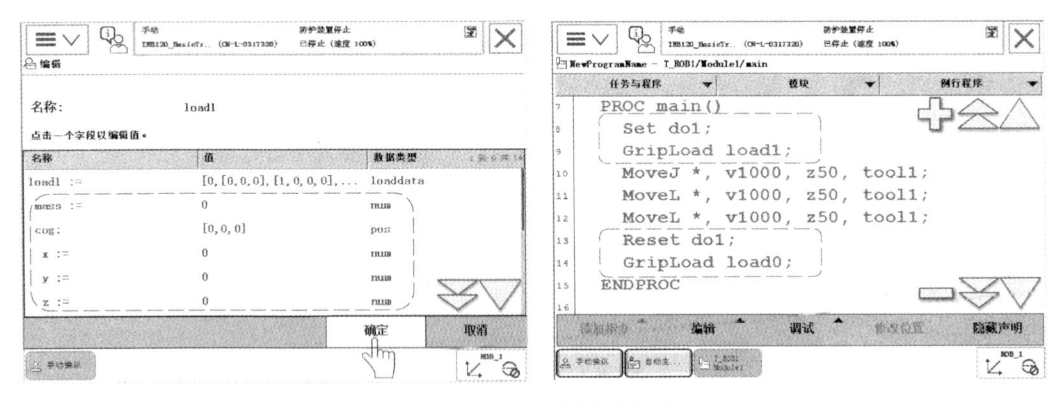


图 5-30　设定有效载荷数据

### 5.2.4　任务评价

任务评价表见表 5-4。

表 5-4 任务评价表

| 序号 | 评价内容 | 成绩占比 | 自评 | 师评 |
|---|---|---|---|---|
| 1 | 工业机器人的三个关键程序的含义 | 15 分 | | |
| 2 | 工业机器人的前两个关键程序的设定方法 | 15 分 | | |
| 3 | 工业机器人的前两个关键程序数据的实操 | 25 分 | | |
| 4 | 工业机器人的有效载荷数据的设定实操 | 25 分 | | |
| 5 | 操作工业机器人的安全注意事项 | 20 分 | | |

（1）工具动作的作用点（Tool Center Point，TCP）称为工具中心点，是工具坐标系的原点。

（2）工具中心点可设定在工具上，也可以设定在工具外。

（3）工业机器人本体默认的工具中心点位于工具安装法兰盘外表面的中心。

（4）工件坐标系对应工件，它定义工件相对于大地坐标系（或者其他坐标系）的位置。

（5）对工业机器人进行编程就是在工件坐标系中创建目标和路径。

（6）工业机器人在进行作业任务前，一定正确设定夹具的质量、重心 Tooldata 以及工作任务对象的质量和重心数据 Loaddata。

（7）工业机器人载荷数据常常定义工业机器人的有效负载或抓取对象的负载（通过指令 GripLoad 或 MechUnitLoad 来设置），即工业机人夹具所夹持的负载。

### 5.2.5 任务拓展

（1）请在示教器中找出机器人原有的工具坐标和工件坐标。

（2）请在示教器中建立一个名称为“toolrobot”的工具数据。

（3）请在示教器中建立一个名称为“wobjrobot”的工件数据。

# 项目 6　ABB 工业机器人的编程实战

本项目的学习目标是了解 ABB 工业机器人编程语言 RAPID，了解编程中任务、模块和例行程序之间的关系，以及掌握常用的 RAPID 指令的用法，并能够创建一个可以运行的完整的 RAPID 程序。

## 任务 6.1　搭建 RAPID 的程序架构

### 6.1.1　任务引入

RAPID 语言是一种基于计算机系统的高级编程语言，它易学易用，灵活性强，具有中断、错误处理、多任务处理等高级功能。

RAPID 程序是由 RAPID 编程语言的特定词汇和语法编写而成的。其所包含的指令可以移动工业机器人、设置输出、读取输入，还能实现决策、重复其他指令、构造程序以及与系统操作员交流等功能。

工业现场中一个机器人的操作任务，通常需要几十句甚至上万句指令构成。为了方便管理和阅读程序，通常对指令进行包装成例行程序，对例行程序进行模块化归类，并在模块中存储起来。

### 6.1.2　相关知识

RAPID 程序的基本架构见表 6-1。

**表 6-1　RAPID 程序架构**

| RAPID 程序（任务） | | | |
|---|---|---|---|
| 程序模块 1 | 程序模块 2 | 程序模块 3 | 系统模块 |
| 程序数据 | 程序数据 | …… | 程序数据 |
| 主程序 main | 例行程序 | …… | 例行程序 |
| 例行程序 | 中断程序 | …… | 中断程序 |
| 中断程序 | 功能 | …… | 功能 |
| 功能 | — | …… | — |

RAPID 程序架构的说明包括：

（1）RAPID 程序（任务）。通常一个 RAPID 程序也称作一个任务，这个任务是由一系列的模块组成，并由程序模块与系统模块组成。通常情况下，通过新建程序模块来构建机器人的程序，而系统模块多用于系统方面的控制之用。

（2）程序模块。可以根据不同的用途创建多个程序模块，如专门用于主控制的程序模块，用于位置计算的程序模块，以及用于存放数据的程序模块，这样的目的在于方便归类管理不同用途的例行程序与数据。

每一个程序模块包含了程序数据、例行程序、中断程序和功能四种对象，但不一定在一个模块都有这四种对象的存在。程序模块之间的数据、例行程序、中断程序和功能是可以互相调用的。

（3）main 主程序。在 RAPID 程序中，只有一个主程序 main，可以存在于任意一个程序模块中。与其他高级编程语言一样，它是作为整个 RAPID 程序执行的起点。

### 6.1.3 任务实施

以 RAPID 程序为例，其具体操作步骤为：

（1）单击左上角主菜单按钮，选择“程序编辑器”（见图 6-1），然后点击“任务与程序”（见图 6-2）。

图 6-1　选择“程序编辑器”

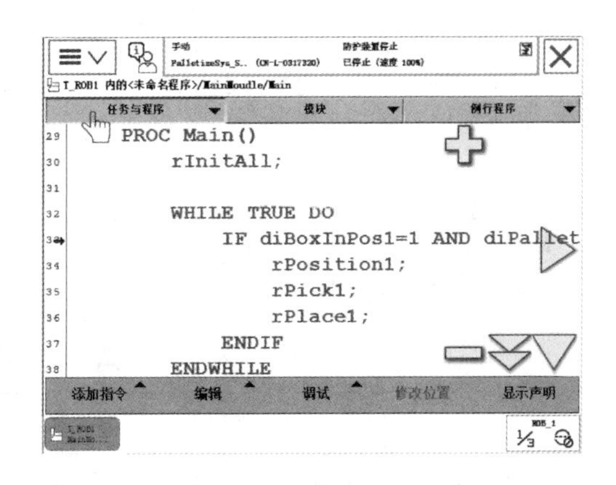


图 6-2　打开“任务与程序”

（2）可以看到一个名为 T_ROB1 任务，点击“显示模块”（见图 6-3）。该任务中有两个名为“BASE”和“user”的系统模块，一个名为“MainMoudle”的程序模块。选中“MainModule”，点击“显示模块”，就可以查看到该模块里的所有例行程序（见图 6-4）。

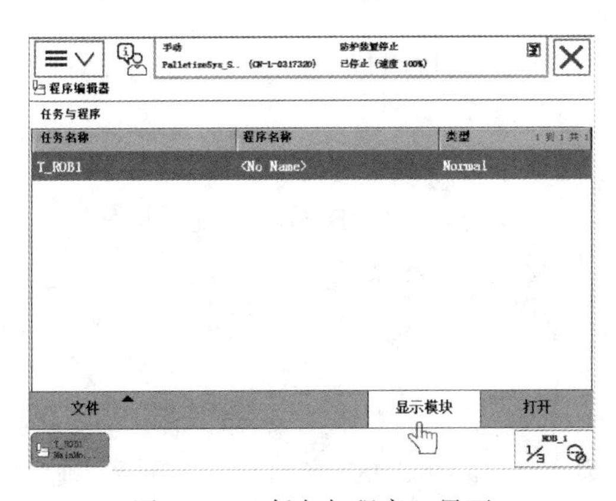


图 6-3　“任务与程序”界面

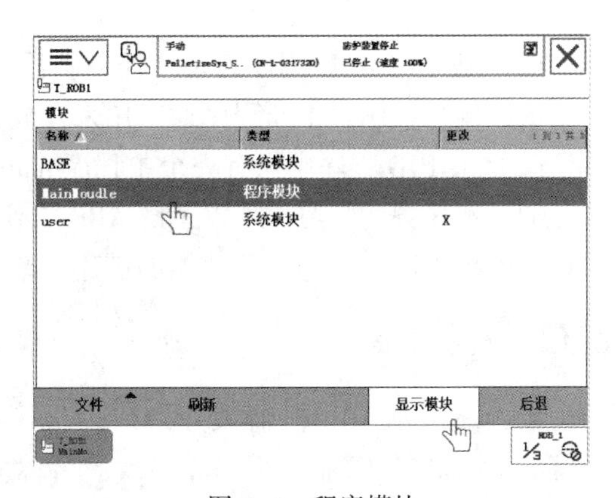


图 6-4　程序模块

（3）如图 6-5 所示，“main（）”为主程序，“rPick1（）”为例行程序，“tPallet1”为中断程序。选中某一个例行程序，点击“显示例行程序”，则可以查看其中的代码。

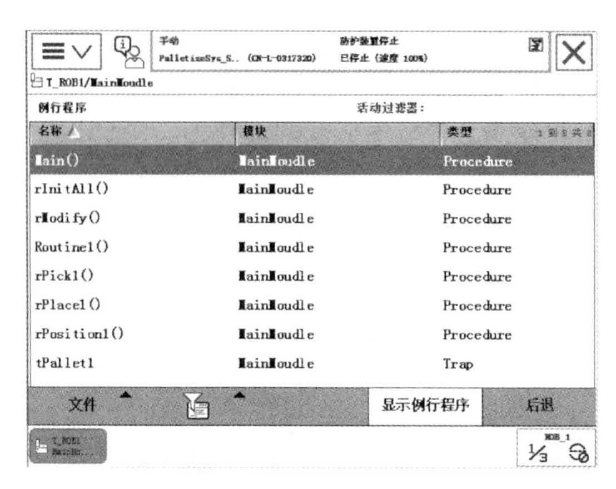

图 6-5　例行程序

了解了 RAPID 程序的架构后，可以快速地掌握整个程序所要从事的工作，为程序编写做准备。

### 6.1.4　任务评价

任务评价表见表 6-2。

**表 6-2　任务评价表**

| 序号 | 评价内容 | 成绩占比 | 自评 | 师评 |
| --- | --- | --- | --- | --- |
| 1 | 理解 RAPID 的程序组成 | 30 分 | | |
| 2 | 理解任务的定义 | 20 分 | | |
| 3 | 理解程序模块的定义 | 20 分 | | |
| 4 | 理解例行程序的定义 | 30 分 | | |

（1）一个 RAPID 程序为一个任务，任务可由若干个模块组成，模块又由若干例行程序组成，例行程序、中断程序通常用来存放机器人指令。

（2）RAPID 程序中只有一个主程序 main，它是整个程序执行的起点。

通过本任务的学习，可以了解 ABB 工业机器人编程语言 RAPID 的基本概念及其任务、模块、例行程序之间的关系，能够搭建一个完整的 RAPID 程序架构。

### 6.1.5　任务拓展

设计一个为 ABB 机器人分别数控机床和加工中心上下料的程序架构，任务名称为“Load_Unload”，添加三个程序模块，名称分别为：“MainModule”，“Lathe”和“CNC”。“MainModule”中含有“main”例行程序，“Lathe”模块中含有“Lathe_load”和“Lathe_unload”两个例行程序，“CNC”模块中含有“CNC_load”和“CNC_unload”两个例行程序。

# 任务 6.2　学习常用的 RAPID 编程指令

## 6.2.1　任务引入

在搭建完 RAPID 架构以后，就需要填充实际意义的东西（添加具体的指令），这样整个 RAPID 程序才算完整，本任务就是学习工程中常用的 RAPID 指令。

ABB 工业机器人的 RAPID 程序提供了丰富的指令，用于各种简单及复杂的应用，可以实现对工业机器人的控制操作。

## 6.2.2　相关知识

ABB 工业机器人如果要完成一个简单或者复杂的任务，就会用到各种指令。常见的 Rapid 指令包括赋值指令、机器人运动指令、I/O 控制指令、条件逻辑判断指令、等待指令等。

## 6.2.3　任务实施

下面从最常用的指令开始学习 RAPID 指令。

### 6.2.3.1　赋值指令

赋值指令（ :=）是用于对程序数据进行赋值，赋值可以是一个常量或数学表达式。以下就以添加一个常量赋值“reg1 :=5;”为例进行说明，其具体操作步骤为：

（1）在指令列表中选择“ :=”（见图 6-6）；在接下来出现的“<VAR>=<EXP>”中，点击“<VAR>”，然后选择“更改数据类型…”，选择“num”数字型数据；在“num”数据中，选择“reg1”（若“num”数据中没有“reg1”，可以选择点击“新建”来创建“reg1”）。

（2）选中“<EXP>”让其高亮显示，然后打开“编辑”菜单，选择“仅限选定内容”；通过软键盘输入〈5〉，然后点击“确定”，这样就添加好了指令（见图 6-7）。

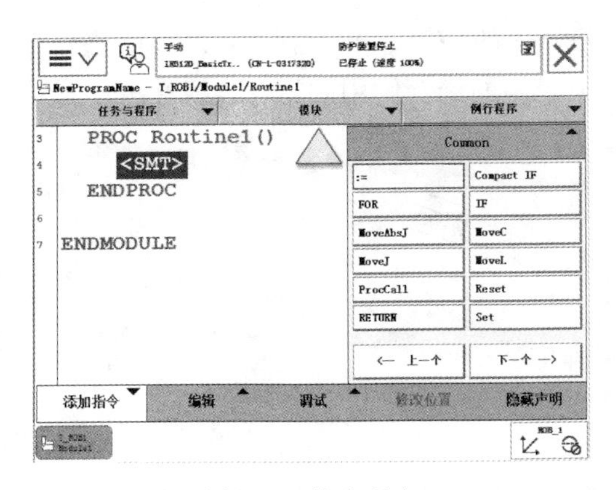


图 6-6　指令列表

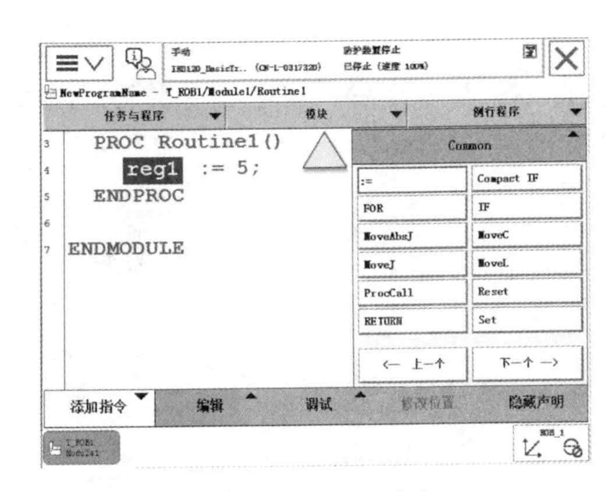


图 6-7　已添加指令

接下来以添加“reg2 :=reg1+4;”为例进行说明，其操作步骤为：

（1）在指令列表中选择“ :=”；选中“ :=”左侧“<VAR>”，选取 num 数据中的“reg2”（若 num 数据中没有“reg2”，可以选择点击“新建”来创建“reg2”）。

（2）选中右侧“<EXP>”，选取“reg1”，并点选刚添加的“reg1”让其处于选中状态，然后点击“+”按钮；在“+”的右侧，会出现一个新的“<EXP>”（见图 6-8）。

（3）选中“<EXP>”（蓝色高亮显示），随后打开“编辑”菜单，选择“仅限选定内容”；通过软键盘输入数字〈4〉，然后点击“确定”。

（4）系统会提示将添加的指令是放置在光标的上方还是下方（一般选择下方）。

添加完成的界面如图 6-9 所示。添加指令后，可点击“添加指令”，将指令列表收起来。

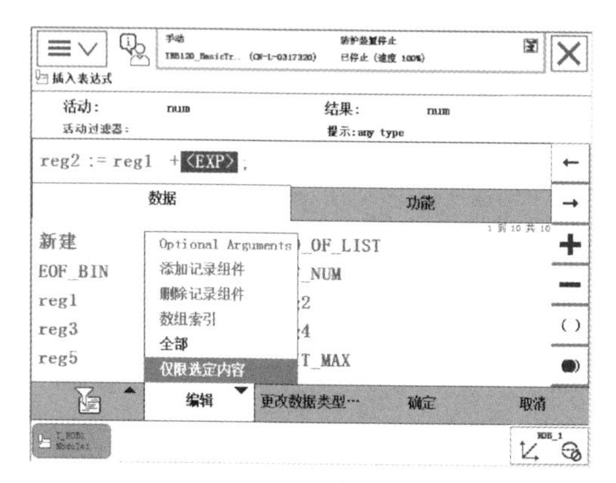


图 6-8 添加“<EXP>”

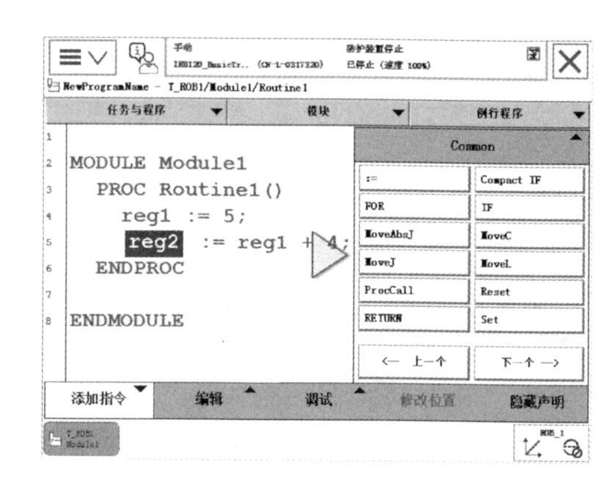


图 6-9 已添加指令

#### 6.2.3.2 工业机器人运动指令

机器人在空间中进行运动主要分为三种方式，包括关节运动（MoveJ）、线性运动（MoveL）和圆弧运动（MoveC）。

注意：在添加或修改机器人的运动指令之前一定要在手动操纵界面确认所使用的工具坐标与工件坐标。

**线性运动指令（MoveL）**

首先学习线性运动指令 MoveL，线性运动是机器人的 TCP 从起点到终点之间的路径始终保持为直线，一般如焊接、涂胶等应用对路径要求高的场合使用此指令。线性运动示意图如图 6-10 所示。

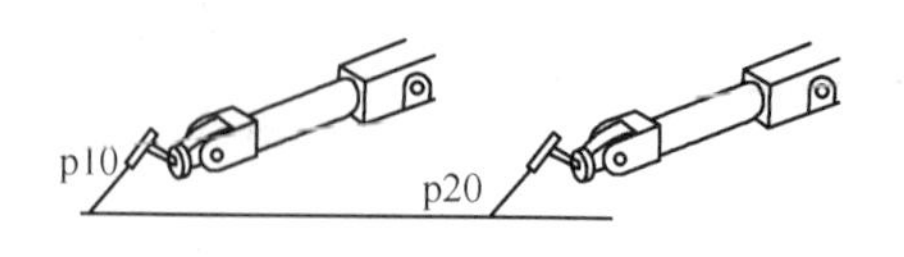


图 6-10 直线运动示意图

添加指令 MoveL 的操作步骤为：

（1）选中“<SMT>”为添加指令的位置，在指令列表中选择“MoveL”；选中“＊”

（蓝色高亮显示），再单击“ * ”，将“ * ”用目标点位置数据代替（见图 6-11）。

（2）点击“新建”，添加一个位置数据；对目标点位置数据属性进行设定后，点击“确定”。这时“ * ”已经被名为 p10 的目标点代替。

（3）点击“添加指令”将指令列表收起来；点击“减号”缩小编辑窗口，此时可以看到整条运动指令（见图 6-12）；选中“p10”，点击“修改位置”，系统将存储基于工具“tool1”和工件坐标系“wobj1”的位置信息到“p10”。

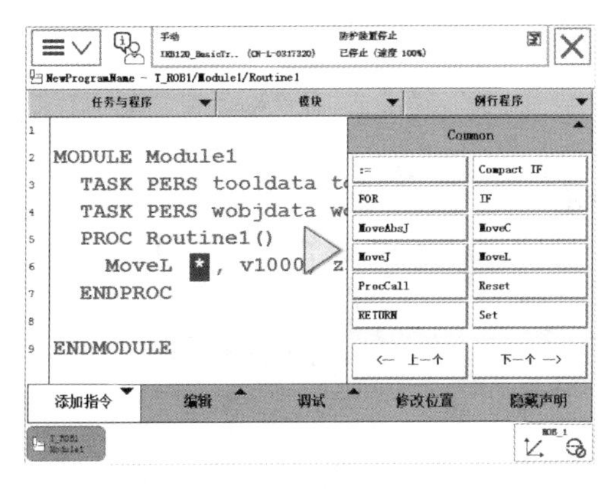


图 6-11　双击“ * ”号

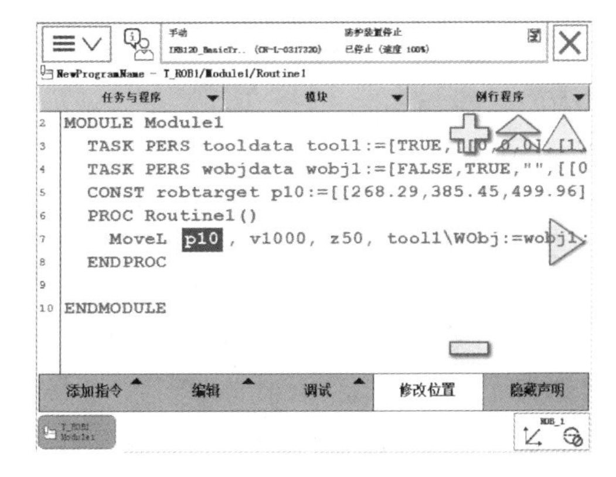


图 6-12　显示整条指令

指令示例：

```
MoveL p10,v1000,z50,tool1\Wobj:=wobj1;
```

指令说明：机器人的运动模式为直线运动，目标点为 p10，运动速度数据为 1000mm/s，转弯区域数据为 50mm，当前的工具为 tool1，当前的工件坐标数据为 wobj1。

**关节运动指令（MoveJ）**

关节运动指令是在对路径精度要求不高的情况，机器人的工具中心点 TCP 从一个位置移动到另一个位置，两个位置之间的路径不一定是直线。关节运动示意图如图 6-13 所示。

关节运动指令适合机器人大范围运动时使用，不容易在运动过程中出现关节轴进入机械死点的问题。

直线运动和关节运动典型示例如图 6-14 所示。

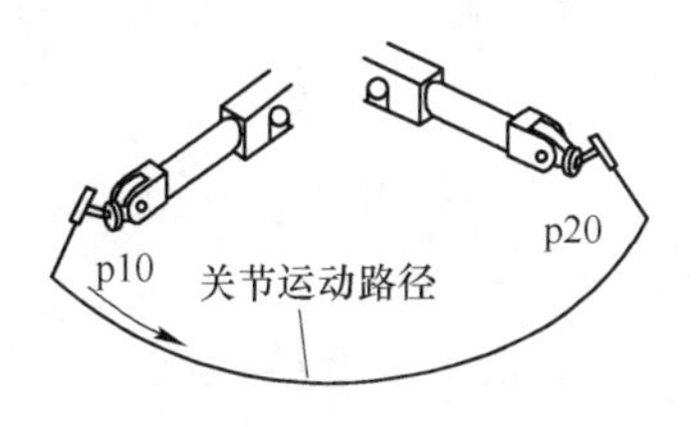


图 6-13　关节运动示意图

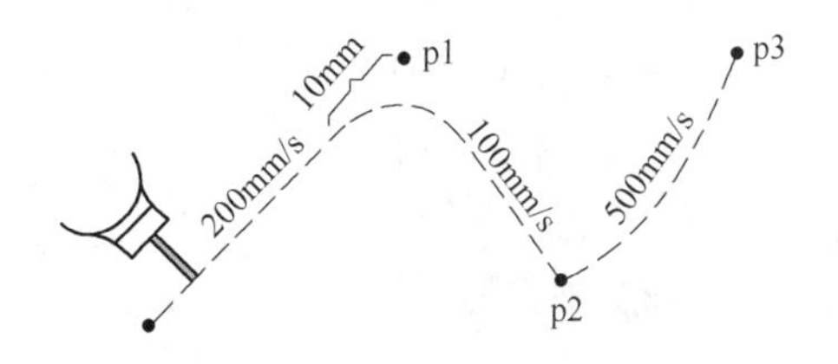


图 6-14　典型案例

该运动轨迹可由三条运动指令表示如下：

（1）指令 1：

```
MoveL p1,v200,z10,tool1\Wobj:=wobj1;
```

指令说明：机器人的 TCP 从当前位置向 p1 点以线性运动方式前进。其速度是 200mm/s，转弯区数据是 10mm，距离 p1 点还有 10mm 的时候开始转弯。使用的工具数据是 tool1，工件坐标数据是 wobj1。

（2）指令 2：

```
MoveL p2,v100,fine,tool1\Wobj:=wobj1;
```

指令说明：机器人的 TCP 从 p1 向 p2 点以线性运动方式前进。其速度是 100mm/s，转弯区数据是 fine，机器人在 p2 点稍作停顿。

（3）指令 3：

```
MoveJ p3,v500,fine,tool1\Wobj:=wobj1;
```

指令说明：添加指令的步骤与 MoveL 相同。机器人的 TCP 从 p2 向 p3 点以关节运动方式前进。其速度是 100mm/s，转弯区数据是 fine，机器人在 p3 点停止。

关于转弯区，其中“fine”指机器人 TCP 达到目标点，在目标点速度降为零，机器人动作有所停顿，然后再向下一运动。一般情况下，路径的最后一个点一定要为“fine”。转弯区数值越大，机器人的动作路径就越圆滑与流畅。

**圆弧运动指令（MoveC）**

圆弧路径是在机器人可到达的空间范围内定义三个位置点。第一个点是圆弧的起点，第二个点用于圆弧的曲率，第三个点是圆弧的终点。圆弧运动示意图如图 6-15 所示。

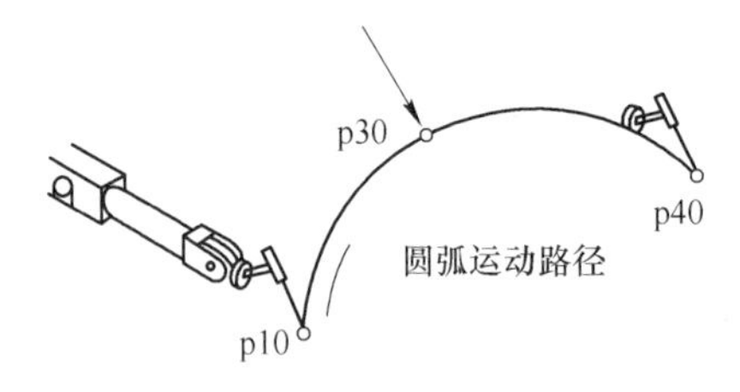


图 6-15 圆弧运动示意图

指令示例：

```
MoveL p1,v200,z10,tool1\Wobj:=wobj1;
MoveC p30,p40,v1000,z1,tool1\Wobj:=wobj1;
```

指令说明：p10 为圆弧的起点；p30 为圆弧的第二个点，根据此点决定轨迹半径；p40 为圆弧的终点。

注意：在后续的调试中，调试 MoveC 指令时，不能以此指令为开始调试，因为 MoveC 指令需要第一点。MoveC 指令前一定要有 MoveL 或 MoveJ 指令来确定第一点，否则系统会以机器人启动调试时的位置为第一点。

### 6.2.3.3 I/O 控制指令

I/O 控制指令用于控制 I/O 信号，以达到与机器人周边设备进行通讯的目的。下面介绍基本的 I/O 控制指令。

**Set 数字信号置位指令**

Set 数字信号置位指令用于将数字输出信号置位为“1”。

指令示例：

```
Set do1;
```

指令说明：设置输出信号 do1 置位。

**Reset 数字信号复位指令**

Reset 数字信号复位指令用于将数字输出信号置位为“0”。

指令示例：

```
Reset do1;
```

注意：如果在 Set、Reset 指令前有运动指令 MoveJ，MoveL，或 MoveC，这些运动指令的转变区数据必须使用“fine”才可以准确到达目标点，然后再输出 I/O 信号状态的变化。

**WaitDI 数字输入信号判断指令**

WaitDI 数字输入信号判断指令用于判断数字输入信号的值是否与目标的一致。

指令示例：

```
WaitDI di1,1;
MoveL p1,v200,z10,tool1\Wobj:=wobj1;
```

指令说明：程序执行时，等待 di1 的值为 1。如果 di1 为 1，则程序继续往下执行，直线运动到 p1 点；如果到达最大等待时间 300s（此时间可根据实际进行设定）以后，di1 的值还不为 1，则机器人会报警或进入出错处理程序。

### 6.2.3.4 逻辑判断指令

条件逻辑判断指令是用于对若干条件进行判断后，执行相应的操作。该指令是 RAPID 中重要的组成。

**紧凑型条件判断指令（Compact IF）**

当一个条件满足了以后，该指令就执行一句指令。

指令示例：

```
IF flag1=TRUE   Set do1;
```

指令说明：如果 flag1 的状态为 TRUE，则 do1 被置位为 1。

**IF 条件判断指令**

IF 条件判断指令就是根据不同的条件去执行不同的指令。

注意：条件判定的条件数量可以根据实际情况进行增加与减少。

指令示例：

```
IF num1=1 THEN
        flag1:=TRUE;
ELSE
        flag1:=FALSE;
        Set do1;
```

```
ENDIF
```

指令说明：如果 num1 为 1，则 flag1 会赋值为 TRUE，否则 flag1 会赋值为 FALSE，且执行数字输出 do1 置位为 1。

**FOR 重复执行判断指令**

FOR 重复执行判断指令是用于一个或多个指令需要重复执行确定数次的情况。此指令以 FOR 开始，以 ENDFOR 结束。

语句的用法为：

```
FOR i FROM n TO m DO
        <若干指令或若干例行程序>
ENDFOR
```

循环次数为（m−n）+1 次。

指令示例：

```
FOR i FROM 1 TO 10 DO
        Routine1;
ENDFOR
```

指令说明：例行程序 Routine1，重复执行了 10 次。

**WHILE 条件判断指令**

WHILE 条件判断指令用于在给定的条件满足的情况下，一直重复执行对应的指令。此指令以 WHILE 开始，以 ENDWHILE 结束。

语句的用法为：

```
WHILE  <某条件> DO
         <若干指令或若干例行程序>
ENDWHILE
```

指令示例：

```
WHILE  num1 > num2  DO
         num1:=num1-1;
ENDWHILE
```

指令说明：当“num1>num2”的条件满足的情况下，就一直执行“num1：=num1−1”的操作，直到条件不满足为止。

6.2.3.5 等待指令

时间等待指令（WaitTime）用于程序在等待一个指定的时间以后，再继续向下执行。此指令的用法为：WaitTime 时间（单位 s）。

指令示例：

```
WaitTime 4;
Reset do1;
```

指令说明：等待 4s 以后，程序向下执行 Reset do1 指令。

### 6.2.3.6 其他常用指令

**ProcCall 调用例行程序指令**

通过使用此指令在指定的位置调用其他的例行程序。其具体步骤为:

(1) 选中“<SMT>”为要调用例行程序的位置,在指令列表中选择“ProcCall”指令(见图 6-16)。

(2) 选中要调用的例行程序“Routine1”,然后单击“确定”(图 6-17)。

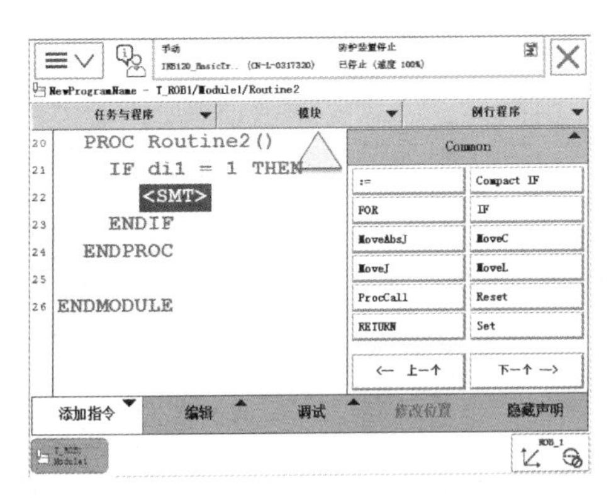


图 6-16 选择“ProcCall”指令

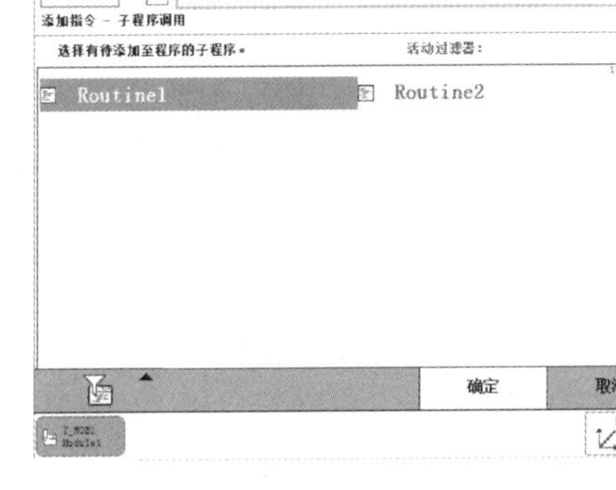


图 6-17 例行程序

调用例行程序指令执行的结果如图 6-18 所示。

指令说明:如果输入信号 di1 为 1,则执行 Routine1 的例行程序,否则程序跳过此语句,顺序往下执行其他的指令。

**RETURN 返回例行程序指令**

当此指令被执行时,则马上结束本例行程序的执行,返回程序指针到调用此例行程序的位置。指令示例如图 6-19 所示。

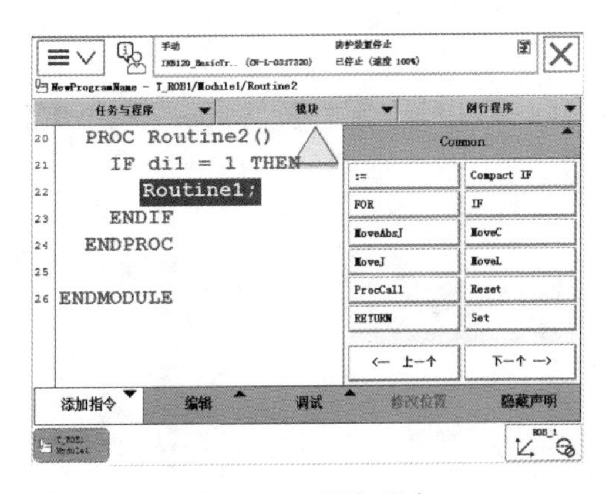


图 6-18 已添加指令

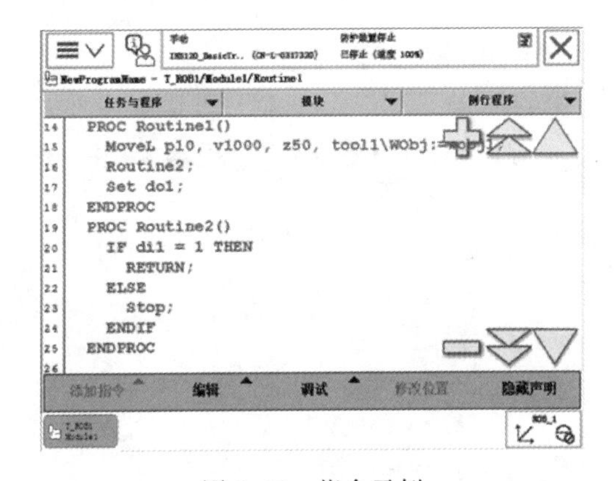


图 6-19 指令示例

指令说明:当 di1=1 时,执行 RETURN 指令,程序指针返回到调用 Routine2 例行程序的位置,并继续向下执行 Set do1 这个指令。

### 6.2.4 任务评价

任务评价表见表 6-3。

**表 6-3 任务评价表**

| 序号 | 评价内容 | 成绩占比 | 自评 | 师评 |
|---|---|---|---|---|
| 1 | 赋值指令 | 5 分 | | |
| 2 | 运动指令 | 35 分 | | |
| 3 | I/O 控制指令 | 20 分 | | |
| 4 | 条件逻辑判断指令 | 20 分 | | |
| 5 | 等待指令 | 5 分 | | |
| 6 | 其他常用指令 | 15 分 | | |

（1）RAPID 编程系统常用的运动指令有关节运动指令 MoveJ、直线运动指令 MoveL 和圆弧运动指令 MoveC。

（2）RAPID 编程系统常用 I/O 控制指令有数字信号置位指令 Set、数字信号复位指令 Reset 和数字输入信号判断指令 WaitDI。

（3）条件逻辑判断是会"思考"的核心，是程序的重要组成部分。常见的信号有紧凑型条件判断指令 Compact IF、条件判断指令 IF、重复执行指令 FOR 和条件循环指令 WHILE。

（4）调用例行程序指令 ProcCall 和返回例行程序指令 RETURN，在复杂的程序中经常用到。

### 6.2.5 任务拓展

（1）在名称为"MainModule"模块中添加一个名称为"rTrain"的例行程序。

（2）在"rTrain"例行程序中，新建"num"类型数据"count"，编辑"count :=count+1"。

（3）在"rTrain"例行程序中，分别使用 MoveL 和 MoveJ 指令，走一个正方形轨迹，四个点分别为 p1，p2，p3，p4。

（4）在"MainModule"模块中再添加一个"Main"主程序，并将"rTrain"程序调用到主程序中，调试运行整个机器人程序。

# 参考文献


［1］吕景泉，黄旭锋．工程实践创新项目教程［M］．北京：中国铁道出版社，2012.
［2］叶晖．工业机器人实操与应用技巧［M］．北京：机械工业出版社，2017.
［3］史维玉．机械创新思维的训练方法［M］．武汉：华中科技大学出版社，2013.
［4］张建忠，李自鹏．机电一体化技术应用［M］．北京：北京邮电大学出版社，2014.
［5］王静霞．单片机基础与应用［M］．北京：高等教育出版社，2016.
［6］张建民．机电一体化系统设计［M］．北京：北京理工大学出版社，2010.
［7］周德卿．机电一体化技术与系统［M］．北京：机械工业出版社，2014.
［8］贾海瀛．传感器技术与应用［M］．北京：高等教育出版社，2015.
［9］汤晓华．工业机器人应用技术［M］．北京：高等教育出版社，2015.
［10］佘明洪．工业机器人操作与编程［M］．北京：机械工业出版社，2018.